Edited by
Mikel Duke, Dongyuan Zhao,
and Raphael Semiat

Functional Nanostructured Materials and Membranes for Water Treatment

Related Titles

Lior, N.
Advances in Water Desalination

ISBN: 978-0-470-05459-8

García-Martínez, J., Serrano-Torregrosa, E. (eds.)
The Chemical Element
Chemistry's Contribution to Our Global Future

2011
ISBN: 978-3-527-32880-2

Peinemann, K.-V., Pereira Nunes, S. (eds.)
Membrane Technology
Volume 4: Membranes for Water Treatment

2010
ISBN: 978-3-527-31483-6

Li, N. N (ed.)
Advanced Membrane Technology and Applications

2008
ISBN: 978-0-471-73167-2

Kumar, C. S. S. R. (ed.)
Nanomaterials - Toxicity, Health and Environmental Issues

2006
ISBN: 978-3-527-31385-3

Edited by Mikel Duke, Dongyuan Zhao, and Raphael Semiat

Functional Nanostructured Materials and Membranes for Water Treatment

WILEY-VCH

WILEY-VCH Verlag GmbH & Co. KGaA

The Volume Editors

Prof. Dr. Mikel Duke
Victoria Univ., Inst. of
Sustainability+ Innovat., Blg. 4
Room 4.107, Hoppers Lane
Werribee, Victoria 3030
Australia

Prof. Dr. Dongyuan Zhao
Fudan University
Dept. of Chemistry
Handan Road 220
Shanghai 200233
China

Prof. Raphael Semiat
Technion - Israel Institute of
Technology, The Wolfson Chem.
Engineering Dept.
32000 Technion City, Haifa
Israel

All books published by **Wiley-VCH** are carefully produced. Nevertheless, authors, editors, and publisher do not warrant the information contained in these books, including this book, to be free of errors. Readers are advised to keep in mind that statements, data, illustrations, procedural details or other items may inadvertently be inaccurate.

Library of Congress Card No.: applied for

British Library Cataloguing-in-Publication Data
A catalogue record for this book is available from the British Library.

Bibliographic information published by the Deutsche Nationalbibliothek
The Deutsche Nationalbibliothek lists this publication in the Deutsche Nationalbibliografie; detailed bibliographic data are available on the Internet at <http://dnb.d-nb.de>.

© 2013 Wiley-VCH Verlag GmbH & Co. KGaA, Boschstr. 12, 69469 Weinheim, Germany

All rights reserved (including those of translation into other languages). No part of this book may be reproduced in any form – by photoprinting, microfilm, or any other means – nor transmitted or translated into a machine language without written permission from the publishers. Registered names, trademarks, etc. used in this book, even when not specifically marked as such, are not to be considered unprotected by law.

Print ISBN: 978-3-527-32987-8
ePDF ISBN: 978-3-527-66849-6
ePub ISBN: 978-3-527-66848-9
mobi ISBN: 978-3-527-66847-2
oBook ISBN: 978-3-527-66850-2

Typesetting Laserwords Private Limited, Chennai, India
Printing and Binding Markono Print Media Pte Ltd, Singapore
Cover Design Simone Benjamin, McLeese Lake, Canada

Printed in Singapore
Printed on acid-free paper

Editorial Board

Members of the Advisory Board of the "Materials for Sustainable Energy and Development" Series

Professor Huiming Cheng
Professor Calum Drummond
Professor Morinobu Endo
Professor Michael Grätzel
Professor Kevin Kendall
Professor Katsumi Kaneko
Professor Can Li
Professor Arthur Nozik
Professor Detlev Stöver
Professor Ferdi Schüth
Professor Ralph Yang

Contents

Foreword *XIII*

Series Editor Preface *XV*

Acknowledgments *XVII*

About the Series Editor *XIX*

About the Volume Editors *XXI*

List of Contributors *XXIII*

1 Target Areas for Nanotechnology Development for Water Treatment and Desalination *1*
Mikel Duke, Raphael Semiat, and Dongyuan Zhao
1.1 The Future of Water Treatment: Where Should We Target Our Efforts? *1*
1.2 Practical Considerations for Nanotechnology Developers *2*
1.3 The Water Treatment Market for New Nanotechnology *3*
1.4 Purpose of This Book *4*
1.5 Concluding Remarks *5*
References *6*

2 Destruction of Organics in Water via Iron Nanoparticles *7*
Hilla Shemer and Raphael Semiat
2.1 Introduction *7*
2.2 Nanoparticles as Catalysts *8*
2.2.1 Colloidal Nanoparticles *9*
2.2.2 Supported Nanoparticles *9*
2.3 Advanced Oxidation Processes *10*
2.3.1 Fenton-Like Reactions *12*
2.3.1.1 Iron Oxide as Heterogeneous Nanocatalyst *12*

2.3.2	Photo-Fenton Reactions	16
2.3.3	Nanocatalytic Wet Oxidation	17
2.4	Nano Zero-Valent Iron (nZVI)	18
2.4.1	Synthesizing Methods	20
2.4.1.1	Emulsified Zero-Valent Iron	21
2.4.2	Degradation Mechanism	22
2.4.3	Field Application of nZVI	25
2.5	Bimetallic nZVI Nanoparticles	27
2.6	Summary	29
	References	30

3	**Photocatalysis at Nanostructured Titania for Sensing Applications**	**33**
	Shanqing Zhang and Huijun Zhao	
3.1	Background	33
3.1.1	Photocatalysis at TiO_2 Nanomaterials	33
3.1.2	Photoelectrocatalysis at TiO_2 Nanomaterials	36
3.2	Fabrication of TiO_2 Photoanodes	37
3.2.1	Common Fabrication Techniques and Substrates for Photoanodes	37
3.2.2	TiO_2/BDD Photoanode	38
3.2.3	TiO_2 Mixed-Phase Photoanode	39
3.2.4	CNTs/TiO_2 Composite Photoanode	40
3.3	The Sensing Application of TiO_2 Photocatalysis	41
3.3.1	Photocatalytic Determination of TOC	42
3.3.2	Photocatalytic Determination of COD	43
3.4	The Sensing Application of TiO_2 Photoelectrocatalysis	46
3.4.1	Probe-Type TiO_2 Photoanode for Determination of COD	46
3.4.2	Exhaustive Degradation Mode for Determination of COD	50
3.4.3	Partial Oxidation Mode for Determination of COD	53
3.4.4	UV-LED for Miniature Photoelectrochemical Detectors	55
3.4.5	Photoelectrochemical Universal Detector for Organic Compounds	55
3.5	Photocatalytic Gas Sensing	56
3.5.1	The Photoelectrocatalytic Generation of Analytical Signal	57
3.5.2	Photocatalytic Surface Self-Cleaning for Enhancement of Analytical Signal	58
3.6	Conclusions	59
	References	59

4	**Mesoporous Materials for Water Treatment**	**67**
	Yonghui Deng and Dongyuan Zhao	
4.1	Adsorption of Heavy Metal Ions	68
4.2	Adsorption of Anions	73
4.3	Adsorption of Organic Pollutants	74
4.4	Multifunctional Modification of Sorbents	77
4.5	Photocatalytic Degradation of Organic Pollutants	79
4.6	Conclusions and Outlook	82

Acknowledgments 83
References 83

5 Membrane Surface Nanostructuring with Terminally Anchored Polymer Chains 85
Yoram Cohen, Nancy Lin, Kari J. Varin, Diana Chien, and Robert F. Hicks

5.1 Introduction 85
5.2 Membrane Fouling 86
5.3 Strategies for Mitigation of Membrane Fouling and Scaling 89
5.4 Membrane Surface Structuring via Graft Polymerization 91
5.4.1 Overview 91
5.4.2 Reaction Schemes for Graft Polymerization 92
5.4.3 Surface Activation with Vinyl Monomers 94
5.4.4 Surface Activation with Chemical Initiators 95
5.4.5 Irradiation-Induced Graft Polymerization 97
5.4.5.1 Gamma-Induced Graft Polymerization 97
5.4.5.2 UV-Induced Graft Polymerization 99
5.4.6 Plasma-Initiated Graft Polymerization 101
5.5 Summary 104
References 107

6 Recent Advances in Ion Exchange Membranes for Desalination Applications 125
Chalida Klaysom, Bradley P. Ladewig, G.Q. Max Lu, and Lianzhou Wang

6.1 Introduction 125
6.2 Fundamentals of IEMs and Their Transport Phenomena 125
6.2.1 Ion Transport through IEMs 127
6.2.2 Concentration Polarization and Limiting Current Density 128
6.2.2.1 The Overlimiting Current Density 128
6.2.2.2 Water Dissociation 129
6.2.2.3 Gravitational Convection 130
6.2.2.4 Electroconvection 130
6.2.3 Structure and Surface Heterogeneity of IEMs 130
6.3 Material Development 135
6.3.1 The Development of Polymer-Based IEMs 135
6.3.1.1 Direct Modification of Polymer Backbone 135
6.3.1.2 Direct Polymerization from Monomer Units 139
6.3.1.3 Charge Induced on the Film Membranes 142
6.3.2 Composite Ion Exchange Membranes 143
6.3.3 Membranes with Specific Properties 147
6.3.3.1 Improving Antifouling Property 149
6.4 Future Perspectives of IEMs 150
6.4.1 Hybrid System 150
6.4.2 Small-Scale Seawater Desalination 152

6.5	Conclusions 152
	References 154

7	**Thin Film Nanocomposite Membranes for Water Desalination** 163
	Dan Li and Huanting Wang
7.1	Introduction 163
7.2	Fabrication and Characterization of Inorganic Fillers 168
7.3	Fabrication and Characterization of TFC/TFN Membranes 172
7.3.1	Interfacial Polymerization 172
7.3.2	Interfacial Polymerization with Inorganic Fillers 175
7.3.3	Characterization of TFN or TFC Membranes 177
7.4	Membrane Properties Tailored by the Addition of Fillers 179
7.4.1	Water Permeability and Salt Rejection 179
7.4.2	Fouling Resistance, Chlorine Stability, and Other Properties 184
7.5	Commercialization and Future Developments of TFN Membranes 185
7.6	Summary 187
	References 188

8	**Application of Ceramic Membranes in the Treatment of Water** 195
	Weihong Xing, Yiqun Fan, and Wanqin Jin
8.1	Introduction 195
8.2	Membrane Preparation 196
8.2.1	Extrusion 196
8.2.2	Sol–Gel Process 196
8.3	Clarification of Surface Water and Seawater Using Ceramic Membranes 198
8.3.1	Ceramic Membrane Microfiltration of Surface Water 199
8.3.1.1	Pretreatment with Flocculation/Coagulation 199
8.3.1.2	Effect of Transmembrane Pressure (TMP) and Cross-Flow Velocity (CFV) 199
8.3.1.3	Ultrasound Cleaning 200
8.3.1.4	Hybrid Ozonation–Ceramic Ultrafiltration 201
8.3.1.5	Ceramic Membrane Applications for Industrial-Scale Waterworks 201
8.3.2	Pretreatment of Seawater RO Using Ceramic Membranes 201
8.3.2.1	Effect of Operational Parameters 201
8.3.2.2	Ceramic Membrane Application for the Industrial-Scale SWRO Plant 202
8.4	Ceramic Membrane Application in the Microfiltration and Ultrafiltration of Wastewater 202
8.4.1	Microstructure of the Membranes 204
8.4.2	Surface Properties of Ceramic Membranes 205
8.4.2.1	Wettability 205
8.4.2.2	Surface Charge Properties 206
8.4.2.3	Technical Process 208
8.4.2.4	Cost 212

8.5	Conclusions and Prospects *213*	
	References *213*	
9	**Functional Zeolitic Framework Membranes for Water Treatment and Desalination** *217*	
	Bo Zhu, Bin Li, Linda Zou, Anita J. Hill, Dongyuan Zhao, Jerry Y. S. Lin, and Mikel Duke	
9.1	Introduction *217*	
9.2	Preparation of Zeolite Membranes *219*	
9.2.1	Direct *In situ* Crystallization *220*	
9.2.2	Seeded Secondary Growth *221*	
9.2.3	Microwave Synthesis *223*	
9.2.4	Postsynthetic Treatment *228*	
9.3	Zeolite Membranes for Water Treatment *229*	
9.3.1	Zeolite Membranes for Desalination *229*	
9.3.2	Zeolite Membranes for Wastewater Treatment *235*	
9.3.3	Zeolite Membrane-Based Reactors for Wastewater Treatment *238*	
9.4	Conclusions and Future Perspectives *241*	
	Acknowledgments *241*	
	References *242*	
10	**Molecular Scale Modeling of Membrane Water Treatment Processes** *249*	
	Harry F. Ridgway, Julian D. Gale, Zak E. Hughes, Matthew B. Stewart, John D. Orbell, and Stephen R. Gray	
10.1	Introduction *249*	
10.2	Molecular Simulations of Polymeric Membrane Materials for Water Treatment Applications *249*	
10.2.1	RO Membranes: Synthesis, Structure, and Properties *250*	
10.2.2	Strategies for Modeling Polymer Membranes *255*	
10.2.3	Simulation of Water and Solute Transport Behaviors *262*	
10.2.4	Concluding Remarks *266*	
10.3	Molecular Simulation of Inorganic Desalination Membranes *267*	
10.3.1	Modeling of Zeolites *268*	
10.3.2	Behavior of Water within Zeolites *270*	
10.3.3	Zeolites and Salt Ions *276*	
10.3.4	Concluding Remarks *278*	
10.4	Molecular Simulation of Membrane Fouling *279*	
10.4.1	Molecular Modeling of Potential Organic Foulants *280*	
10.4.2	Modeling of Membrane Fouling *286*	
10.4.3	Future Directions *291*	
10.4.4	Concluding Remarks *291*	
	References *292*	

11		**Conclusions: Some Potential Future Nanotechnologies for Water Treatment** *301*
		Mikel Duke
11.1		Nanotubes *301*
11.1.1		Fast Molecular Flow *302*
11.1.2		CNTs as High Strength Fibers *302*
11.1.3		High Aspect Ratio *303*
11.1.4		Electrical Conductivity *304*
11.2		Graphene *305*
11.2.1		Graphene Barrier Material *305*
11.2.2		Desalination and Heavy Metal Adsorption *306*
11.2.3		Catalytic Assistance *306*
11.3		Aquaporins *306*
11.4		Metal–Organic, Zeolitic Imidazolate, and Polymer Organic Frameworks *307*
11.5		Conclusions *309*
		References *309*

Index *313*

Foreword

It is with great pleasure to present to you this book on the "Functional Nanostructured Materials and Membranes for Water Treatment." Nanomaterials are an emerging area in science that demonstrated achievements occurring in the last two decades. We can date its beginnings from the name, 'nanotechnology', which was originally coined by Norio Taniguchi, a professor from Tokyo University of Science, in 1974. The emergence of nanotechnology research appears to have come later with the advent of the scanning tunnelling microscope in 1981 and the discovery of fullerenes in 1985. So from that time, besides the concept, we acquired the capacity to identify and characterise nanostructured materials, and this marks the start of the field of science dedicated to the nanoscale.

Nanotechnology is basically the field of science involving the manipulation of matter or theory at the atomic or molecular scale. In doing this, we have unearthed in exciting new properties that can influence the wavelength of light, increase the efficiency of catalysts, selectively diffuse small molecules, and even allow particles to penetrate living cell membranes. It is interesting to note that the natural world has been based on nanostructured materials throughout evolution, and thus, mankind has just started to recognise the need for tools to explore properties of materials in 1 to 100 nm range. It is no wonder that governments around the world have invested billions of dollars specifically in nanotechnology. While industries can harness the different benefits of such materials for energy, foods, mining and, electronics, this book focuses on the advances for water treatment.

Water will always be an essential factor to our lives, and thus, there is an ongoing cause for researchers undertaking efforts to secure sustainable water sources and reduce water pollution. Water treatment is an area where I have spent much of my career, principally in the field of membrane science. During my work I have explored membrane transport properties and fouling mechanisms, which has contributed towards the underpinning science of the widespread membrane technology in the water industry. This is quite remarkable for a technology that was little heard of nearly half a century ago, and is recently one of the most widely adopted water treatment technologies. Over the years, membrane technology has involved working with nanostructured materials, although until now it has not been considered as a branch of nanotechnology. Now researchers are turning to nanotechnology to address the challenges to a sustainable water future, and in many cases this involves the marriage between nanotechnology and membrane technology. Many of the contributors to this book are membrane scientists and engineers who have the vision that membranes are an invaluable technology in water treatment and it can be enhanced by nanotechnology. Conventional membrane technology already includes a process of separating nano-dimensional molecules known as Nanofiltration. This technique is commercially applied, for example, to remove organic materials (e.g. natural organic matter or sugars) from salts. However, a recently successful development that deliberately combines membrane technology and nanotechnology is the inclusion of nanoparticles in desalination membranes. This work, coming out of the birthplace of reverse osmosis membranes, the University of California at Los Angeles in the USA, has now entered the market as a commercial desalination membrane. The inclusion of nanoparticles within the polymer structure of membranes has performance and practical benefits, that has also been explored in other forms of membranes, including ultra and microfiltration. While we see this activity rising rapidly into commercialisation, this book also presents work in catalysis, sensing, adsorption, membrane modification, ion exchange, inorganic membranes, and nanoscale modelling of membrane diffusion and interactions. Therefore, this book presents a comprehensive overview of the progress in nanotechnology to enhance membranes and other processes in water treatment. Whether you are an academic or working in industry, scientist or engineer, student or professional, this book will have relevance in your practice.

Professor Tony Fane
UNESCO Centre for Membrane Science and Technology
University of New South Wales, Sydney, Australia
and
Singapore Membrane Technology Centre
Nanyang Technological University, Singapore

Series Editor Preface

The Wiley Series on New Materials for Sustainable Energy and Development

Sustainable energy and development is attracting increasing attention from the scientific research communities and industries alike, with an international race to develop technologies for clean fossil energy, hydrogen and renewable energy as well as water reuse and recycling. According to the REN21 (Renewables Global Status Report 2012 p. 17) total investment in renewable energy reached $257 billion in 2011, up from $211 billion in 2010. The top countries for investment in 2011 were China, Germany, the United States, Italy, and Brazil. In addressing the challenging issues of energy security, oil price rise, and climate change, innovative materials are essential enablers.

In this context, there is a need for an authoritative source of information, presented in a systematic manner, on the latest scientific breakthroughs and knowledge advancement in materials science and engineering as they pertain to energy and the environment. The aim of the *Wiley Series on New Materials for Sustainable Energy and Development* is to serve the community in this respect. This has been an ambitious publication project on materials science for energy applications. Each volume of the series will include high-quality contributions from top international researchers, and is expected to become the standard reference for many years to come.

This book series covers advances in materials science and innovation for renewable energy, clean use of fossil energy, and greenhouse gas mitigation and associated environmental technologies. Current volumes in the series are:

Supercapacitors. Materials, Systems, and Applications
Functional Nanostructured Materials and Membranes for Water Treatment
Materials for High-Temperature Fuel Cells
Materials for Low-Temperature Fuel Cells
Advanced Thermoelectric Materials. Fundamentals and Applications
Advanced Lithium-Ion Batteries. Recent Trends and Perspectives
Photocatalysis and Water Purification. From Fundamentals to Recent Applications

In presenting this volume on Functional Nanostructured Materials and Membranes for Water Treatment, I would like to thank the authors and editors of this important book, for their tremendous effort and hard work in completing the manuscript in a timely manner. The quality of the chapters reflects well the caliber of the contributing authors to this book, and will no doubt be recognized and valued by readers.

Finally, I would like to thank the editorial board members. I am grateful to their excellent advice and help in terms of examining coverage of topics and suggesting authors, and evaluating book proposals.

I would also like to thank the editors from the publisher Wiley-VCH with whom I have worked since 2008, Dr Esther Levy, Dr Gudrun Walter, and Dr Bente Flier for their professional assistance and strong support during this project.

I hope you will find this book interesting, informative and valuable as a reference in your work. We will endeavour to bring to you further volumes in this series or update you on the future book plans in this growing field.

Brisbane, Australia   *Gao Qing Max Lu*
31 July 2012

Acknowledgments

Mikel Duke, Dongyuan Zhao and Raphael Semiat would like to thank all the authors for their hard work and commitment in producing their original contributions for this book. Also, we appreciate their responsiveness to our requirements in their ongoing comments and revision requests during the production phase of the chapters.

All contributions were peer reviewed, so we also extend a warm thanks to the reviewers for their time and effort in providing detailed and quality comments to the authors.

We would finally like to thank Esther Levy, Martin Graf and Claudia Nussbeck at the Wiley Editorial Office for their assistance, rapid response to questions, and enduring patience that enabled this book to be completed.

Mikel Duke wishes to thank the team at his group, the Institute for Sustainability and Innovation, Victoria University, for their patience and support in developing this book. He also acknowledges the funding agencies that provided research support related to the theme of this book including the Australian Research Council Linkage and Discovery Project schemes, the Victorian Smart Water Fund, the Australian Endeavour Awards, The Ian Potter Foundation, the National Centre of Excellence in Desalination Australia, the Australian Water Recycling Centre of Excellence, and the numerous industry and government partners. Max Lu's invitation to develop this book is greatly appreciated as well as the partnership of Dongyuan Zhao and Raphael Semiat in co-editing this book. Finally, Mikel wishes to thank Alicia and daughters, Eva and Lila, for their personal support.

Dongyuan Zhao would like to thank the NSF of China, the Fudan University, the China Ministry of Education, and the China Ministry of Science and Technology, who provided their support for research during the writing of the book.

Raphael Semiat would like to thank the team of the Rabin Desalination Laboratory at the Technion IIT for their help during this work.

About the Series Editor

Professor Max Lu
Editor, New Materials for Sustainable Energy and Development Series

Professor Lu's research expertise is in the areas of materials chemistry and nanotechnology. He is known for his work on nanoparticles and nanoporous materials for clean energy and environmental technologies. With over 500 journal publications in high-impact journals, including Nature, *Journal of the American Chemical Society, Angewandte Chemie, and Advanced Materials*, he is also coinventor of 20 international patents. Professor Lu is an Institute for Scientific Information (ISI) Highly Cited Author in Materials Science with over 17 500 citations (h-index of 63). He has received numerous prestigious awards nationally and internationally, including the Chinese Academy of Sciences International Cooperation Award (2011), the Orica Award, the RK Murphy Medal, the Le Fevre Prize, the ExxonMobil Award, the Chemeca Medal, the Top 100 Most Influential Engineers in Australia (2004, 2010, and 2012), and the Top 50 Most Influential Chinese in the World (2006). He won the Australian Research Council Federation Fellowship twice (2003 and 2008). He is an elected Fellow of the Australian Academy of Technological Sciences and Engineering (ATSE) and Fellow of Institution of Chemical Engineers (IChemE). He is editor and editorial board member of 12 major international journals including *Journal of Colloid and Interface Science and Carbon*.

Max Lu has been Deputy Vice-Chancellor and Vice-President (Research) since 2009. He previously held positions of acting Senior Deputy Vice-Chancellor (2012), acting Deputy Vice-Chancellor (Research), and Pro-Vice-Chancellor (Research Linkages) from October 2008 to June 2009. He was also the Foundation Director of the ARC Centre of Excellence for Functional Nanomaterials from 2003 to 2009.

Professor Lu had formerly served on many government committees and advisory groups including the Prime Minister's Science, Engineering and Innovation Council (2004, 2005, and 2009) and the ARC College of Experts (2002–2004). He is the past Chairman of the IChemE Australia Board and former Director of the Board of ATSE. His other previous board memberships include Uniseed Pty Ltd., ARC Nanotechnology Network, and Queensland China Council. He is currently Board member of the Australian Synchrotron, National eResearch Collaboration Tools and Resources, and Research Data Storage Infrastructure. He also holds a ministerial appointment as member of the National Emerging Technologies Forum.

About the Volume Editors

Professor **Mikel Duke** is the Principal Research Fellow of Membrane Science and Deputy Director of the Institute for Sustainability and Innovation at Victoria University, Australia. He has worked in membrane research for over 12 years and has 92 peer-reviewed publications in this field. His focus is on development of ceramic and polymeric membranes and their processes, specializing in molecular scale diffusion and optimizing functional material parameters. He is the recipient of an Australian Research Council Linkage International Fellowship and an Endeavour Executive Award and the founding chair of the Membrane Society of Australasia.

Professor **Dongyuan Zhao** is Cheung Kong Professor of the China Education Ministry, Vice Director of the Advanced Materials Laboratory at Fudan University and Visiting Professor at Monash University (Australia). He is an academician of the Chinese Academy of Sciences. With over 350 peer-reviewed papers earning >20 000 citations, he is the 65th Most-Cited Scientist in Chemistry (according to ISI). His research interests are in the synthesis of porous materials and their application in catalysis, separation, photonics, sorption, environmental decontamination, sensors, and so on.

Professor **Raphael Semiat** is the Yitzhak Rabin Memorial Chair in Science, Engineering and Management of Water Resources at Technion, Israel Institute of Technology. He has wide industrial experience in the research and development of chemical processes. His current interests and activities are centered on water technologies, including desalination, and chemical-environmental processes and use of nano particles for removal of organic matter and heavy metals from water. He has published more than 140 papers in scientific journals.

List of Contributors

Diana Chien
University of California
Chemical and Biomolecular
Engineering Department
Water Technology Research
Center
5531 Boelter Hall
Los Angeles, CA 90095-1592
USA

Yoram Cohen
University of California
Chemical and Biomolecular
Engineering Department
Water Technology Research
Center
5531 Boelter Hall
Los Angeles, CA 90095-1592
USA

Yonghui Deng
Fudan University
Department of Chemistry
Shanghai 200433
P.R. China

Mikel Duke
Victoria University
Institute for Sustainability and
Innovation
Hoppers Lane
PO Box 14428
Melbourne
Victoria 8001
Australia

and

Victoria University
School of Engineering and
Science
Ballarat Road
PO Box 14428
Melbourne
Victoria 8001
Australia

Yiqun Fan
Nanjing University of Technology
State Key Laboratory of
Materials-Oriented Chemical
Engineering
Membrane Science and
Technology Research Center
5 Xinmofan Road Nanjing 210009
P.R. China

Julian D. Gale
Curtin University
Department of Chemistry
Nanochemistry Research Institute
Kent Street, Bentley
PO Box U1987
Perth
WA 6845
Australia

Stephen R. Gray
Victoria University
Institute for Sustainability and Innovation
Hoppers Lane
Werribee
3030
Victoria
Australia

Robert F. Hicks
University of California
Chemical and Biomolecular Engineering Department
Water Technology Research Center
5531 Boelter Hall
Los Angeles, CA 90095-1592
USA

Anita J. Hill
The Commonwealth Scientific and Industrial Research Organization (CSIRO)
Materials Science and Engineering
Clayton South
Victoria 3169
Australia

Zak E. Hughes
Curtin University
Department of Chemistry
Nanochemistry Research Institute
Kent Street, Bentley
PO Box U1987
Perth
WA 6845
Australia

Wanqin Jin
Nanjing University of Technology
State Key Laboratory of Materials-Oriented Chemical Engineering
Membrane Science and Technology Research Center
5 Xinmofan Road Nanjing 210009
P.R. China

Chalida Klaysom
The University of Queensland
School of Chemical Engineering and Australian Institute of Bioengineering and Nanotechnology
ARC Centre of Excellence for Functional Nanomaterials
Brisbane
Queensland 4072
Australia

Bradley P. Ladewig
Monash University
Department of Chemical Engineering
Wellington Road
Monash
Victoria 3800
Australia

Bin Li
Fudan University
Department of Chemistry
Laboratory of Advanced Materials
2205 Songhu Road
Shanghai 200438
China

Dan Li
Monash University
Department of Chemical
Engineering
Wellington Road
Clayton
Victoria 3800
Australia

Jerry Y. S. Lin
Arizona State University
School for Engineering of Matter
Transport and Energy
University Drive and Mill Avenue
Tempe, AZ 85287
USA

Nancy Lin
University of California
Chemical and Biomolecular
Engineering Department
Water Technology Research
Center
5531 Boelter Hall
Los Angeles, CA 90095-1592
USA

Gao Qing Max Lu
The University of Queensland
School of Chemical Engineering
and Australian Institute of
Bioengineering and
Nanotechnology
ARC Centre of Excellence for
Functional Nanomaterials
Brisbane
4072 Queensland
Australia

John D. Orbell
Victoria University
Institute for Sustainability and
Innovation
Hoppers Lane
Werribee
3030
Victoria
Australia

and

Victoria University
School of Engineering and
Science
Faculty of Health, Engineering
and Science
Ballarat Rd
Footscray
3011
Victoria
Australia

Harry F. Ridgway
Stanford University
Department of Civil
Environmental Engineering
450 Serra Mall
Stanford, CA
94305
USA

and

AquaMem Scientific Consultants
Rodeo
New Mexico 88056
USA

Raphael Semiat
Department of Chemical Engineering
GWRI Rabin Desalination Laboratory
Technion-Israel Institute of Technology
Technion City
Haifa 32000
Israel

Hilla Shemer
Technion-Israel Institute of Technology
Department of Chemical Engineering
GWRI Rabin Desalination Laboratory
Haifa 32000
Israel

Matthew B. Stewart
Victoria University
Institute for Sustainability and Innovation
Hoppers Lane
Werribee
3030
Victoria
Australia

and

Victoria University
School of Engineering and Science
Faculty of Health, Engineering and Science
Ballarat Rd, Footscray
3011
Victoria
Australia

Kari J. Varin
University of California
Chemical and Biomolecular Engineering Department
Water Technology Research Center
5531 Boelter Hall
Los Angeles, CA 90095-1592
USA

Huanting Wang
Monash University
Department of Chemical Engineering
Wellington Road
Clayton
Victoria 3800
Australia

Lianzhou Wang
The University of Queensland
School of Chemical Engineering and Australian Institute of Bioengineering and Nanotechnology
ARC Centre of Excellence for Functional Nanomaterials
Brisbane
Queensland 4072
Australia

Weihong Xing
Nanjing University of Technology
State Key Laboratory of Materials-Oriented Chemical Engineering
Membrane Science and Technology Research Center
5 Xinmofan Road, Nanjing
210009
P.R. China

Shanqing Zhang
Centre for Clean Environment
and Energy
Environmental Futures Centre
Griffith School of Environment
Gold Coast Campus
Griffith University
Parklands Drive
QLD, 4222 Australia

Dongyuan Zhao
Fudan University
Laboratory of Advanced Materials
Department of Chemistry
2205 Songhu Road
Shanghai 200433
China

Huijun Zhao
Centre for Clean Environment
and Energy
Environmental Futures Centre
Griffith School of Environment
Gold Coast Campus
Griffith University
Parklands Drive
QLD, 4222 Australia

Bo Zhu
Victoria University
Institute for Sustainability and
Innovation
Hoppers Lane
Melbourne
Victoria 8001
Australia

Linda Zou
University of South Australia
Centre for Water Management
and Reuse
Adelaide
South Australia 5095
Australia

1
Target Areas for Nanotechnology Development for Water Treatment and Desalination

Mikel Duke, Raphael Semiat, and Dongyuan Zhao

It is of no surprise to many around the world that water is a priority research area: water is one of the most fundamental elements of our existence. We use water for drinking, cleaning, cooking, removing waste, recreation, manufacturing, cooling, and so on. These uses have criteria related to the quality for its intended purpose (e.g., drinking) and quality to minimize its harm to the environment when it is disposed after use. To meet these criteria, there is nearly always a treatment process that requires energy and chemicals. Our health and our environment are significant priorities. We have conflicting issues regarding energy usage and chemicals that lead to pollution and harmful by-products associated with their production, delivery, and disposal; hence, we need to minimize the usage of those resources while improving water quality and counter the risk of human illness or the damage to our ecosystems. This is therefore, the motivation for innovative technologies for water treatment: reduced energy and chemicals use. Of course this must be achieved at low cost.

1.1
The Future of Water Treatment: Where Should We Target Our Efforts?

Many research endeavors ranging from fundamental to applied, typically address specific issues and aim to make improvements based on relative measures. For example, we can report improvements to new types of materials to remove microcystin, or we can, for the first time, apply a commercial ceramic membrane to filter an industrial waste. But what in common drives such efforts, and where is all this heading in terms of the greater needs of the society for low energy, safe, and reliable water?

Technologies that are used for water treatment include adsorption, coagulation, reaction, heating, and filtration. These have the effect of removing or deactivating/converting unwanted elements, such as salt, organics, odors, microbes, suspended solids, and toxins. The choice of technologies and the required removal varies considerably around the world, driven by specific quality needs or regulatory requirements. Water treatment systems must

Functional Nanostructured Materials and Membranes for Water Treatment, First Edition.
Edited by Mikel Duke, Dongyuan Zhao, and Raphael Semiat.
© 2013 Wiley-VCH Verlag GmbH & Co. KGaA. Published 2013 by Wiley-VCH Verlag GmbH & Co. KGaA.

- reliably provide water fit for its intended purpose (e.g., drinking water) and
- collect contaminated water (by humans and/or industry) and remove harmful components before its release to the environment or reuse.

The connection between these points (i.e., closing the loop) increases as technologies become more efficient, reliable, and available. Ideally, we might like to take any water source, regardless of its contamination, and convert it directly to drinking quality water. This is known as *direct potable reuse* and has many incentives for a future with sustainable water [1]. This pathway to adopting such treated wastewaters for applications that include human contact (irrigation, drinking, washing, etc.) will involve not only efficient technologies but also evidence of their reliability and is thus another measure from an innovative solution.

The argument for direct potable reuse is that it is far less energy and resource intensive than indirect potable reuse (i.e., holding the water in a large "diluting" body such as a reservoir before reusing). But under increasing economic, energy, and environmental pressures, we foresee that recent developments based on nanotechnology, which might take at least five years to come to market, are likely to become part of a future headed toward direct potable reuse systems. This might give some thought to how researchers might like to steer their work. For example, energy saving systems that produce high-quality water safely and reliably will be likely successes. The alternative to this is treating water that is "fit for purpose," – it meets the minimum requirements for another purpose such as direct (nonpotable) reuse – for example, desalination of saline waste for reuse in an industrial boiler. Such opportunities require less public and government acceptance to engage (and are indeed already underway), but there must be a convenient user of this water to make fit-for-purpose treatment viable. So it appears that for technologies emerging in the next decade, we should aim to provide solutions that support the fit-for-purpose agenda, when the market for the water is known. Therefore, efforts that bring the costs and environmental impacts down to reliably deliver water for our direct consumption are most worthwhile.

In this book, we present nine chapters focused entirely on technology approaches to improve water quality. All of them essentially and ultimately aim to achieve fit-for-purpose, or even direct potable reuse, aligning with the future demands of the water industry. Specifically, the contributions have routes in nanotechnology respecting that the chemistry, materials, and thinking at this scale offers new opportunities for future water treatment. The technologies considered harness functions such as catalysis, sensing, diffusion, and adsorption.

1.2
Practical Considerations for Nanotechnology Developers

Any new nanotechnology that demonstrates virtue for water treatment must undergo a rigorous process to validate its full commercial and environmental potential. We have listed the following considerations/questions that should be determined/answered at the earliest phases of development to facilitate its success as a water treatment solution:

- Nanoparticles cannot reach people, animals, or the environment.
- No new hazardous by-products are inadvertently created.
- Unwanted materials and by-products (if created) are completely removed or mineralized.
- Is reaggregation of nanoparticles going to occur? Is this a problem?
- Is the process that must be installed to harness the nanotechnology simple? Can untrained people use it? Is it expensive?

On top of these specific considerations, the work must also consider the broader implications:

- Does the new treatment solve the problem or generate a new one?
- What is the fate of the contaminants – can we completely destroy them? Recover? Or maybe they are returned to the environment? What is the cost?

We propose that the aforementioned points be considered in any future research and in turn publications to be considered when weighing up if their nanotechnology is on the right path to becoming a practical water treatment process. As you will see in the following contributions from the authors' areas of expertise, there has been a focus on the new technologies in working toward real needs in the water industry.

1.3
The Water Treatment Market for New Nanotechnology

Clearly, with the recent scientific developments in the last decade in the field of nanotechnology, these aspire to commercial use. But how big is this market expected to be? A report published in early 2011 by BCC Research [2] looked at the current market status of nanotechnology applied to water treatment and forecasts its growth. In 2010, they estimated the market (in US dollars) for nanostructured products used in water treatment to be $1.4 billion. By 2015, they expected this to grow to $2.2 billion. Interestingly, this was mostly confined to established products including membrane technology, which is categorized as nanobased (reverse osmosis, nanofiltration, and ultrafiltration). Of the nine original contributions to this book, six are based on membrane technology, aligning with the significance of membranes picked up by the market report. For emerging nanomaterials such as nanofiber fillers, carbon nanotubes, and nanoparticles alone the market estimate was $45 million in 2010, but is expected to grow rapidly to $112 million by 2015.

We have pointed out that researchers should target their efforts at achieving potable water quality, or at best a significant and defined fit-for-purpose application. At the same time, this should be done with demonstrated opportunities for cost reduction, energy savings, and chemical reduction while no other consequences emerge as a result. With these taken into account, successful nanotechnologies are expected by economists to experience growth in the sales in the next five years. Therefore, it seems safe to say that it is an exciting time for scientists and engineers to develop new technologies and theories as it is certain that the market will take them up in future. We are currently at the phase where science is demonstrating the concepts proposed, but there is still a lot of work ahead in terms of measuring the performance and economic potential in real application.

1.4
Purpose of This Book

There are a multitude of agendas for improving desalination and water recycling deployment, for example, demand management, public perception, "simple" solutions, and of course new technologies. None of these will solve our water issues alone, but in this book, we focus on new technologies and thinking that have been borne out by exploring at the nanoscale, which has emerged only in the last decade from fundamental level research. The chapters presented in this book cover the major areas where nanotechnology has shown promise in addressing the issues in water treatment that are understood by industry.

In Table 1.1, we broadly divide water treatment into the three categories: pollutant removal, detection/monitoring, and desalination. Many water treatment efforts can be defined under these categories for relatively simple purposes as shown. Some of the chapters spread across these categories (e.g., desalination and pollutant removal). It is interesting to point out that despite these broad and highly differing purposes, the concepts behind the technology share a lot in common, such as catalysis, adsorption, materials engineering, colloidal chemistry, and molecular diffusion. These are well-known scientific pillars of nanotechnology. Therefore, the purpose of this book is not only to demonstrate working nanotechnological solutions for major water treatments but also to highlight the common sciences and achievements that bring about such solutions.

In the chapters, we show how nanotechnology leads us to develop new materials, improving existing technologies (e.g., membranes), and to enhance our understanding of complex processes (e.g., molecular simulations). Developing new materials from the bottom up offers new and exciting opportunities for efficiency improvements yet unseen by industry. Such materials are included in Chapters 2–4. Improving existing technologies is part of Chapters 5–9. For example, membrane technology has been successfully deployed at full scale for water filtration and desalination, but limitations are being realized through the ongoing issues related to fouling. Also, as current technologies such as membranes move into more challenging water treatment areas, these issues will become cost prohibitive. So the priority in research is to explore ways to enhance membrane life and performance by way of improved fouling tolerance and durability without compromising on the essential flux and selectivity features. We also broaden the thinking of nanotechnology beyond materials developing in Chapter 10, where nanodimension modeling gives fresh insight into molecular diffusion, interaction

Table 1.1 The purposes of each water treatment category aligned to the chapters presented in this book and the nanotechnology concepts applied.

Category	Purpose	Chapters in this book	Nanotechnology concepts applied
Pollutant removal	Prevent micropollutants released to environment. Improve water quality for reuse.	Chapter 2 Chapter 4 Chapter 5 Chapter 8 Chapter 9 Chapter 10	Catalysis, adsorption, materials design, surface chemistry, colloidal chemistry, molecular diffusion, and molecular dynamics.
Detection/ monitoring	Rapid and specific water quality monitoring. Improved performance of water treatment systems.	Chapter 3	Catalysis, adsorption, materials design, surface chemistry, colloidal chemistry, electromaterials, and optical physics.
Desalination	Access abundant ocean water resource. Improve water quality for reuse.	Chapter 5 Chapter 6 Chapter 7 Chapter 9 Chapter 10	Adsorption, materials design, surface chemistry, colloidal chemistry, electromaterials, molecular diffusion, and molecular dynamics.

chemistries, and fouling mechanisms. This not only contributes new science but also offers new approaches to manage plant operation to gain better performance of our current membrane technologies.

1.5 Concluding Remarks

We have identified that fit-for-purpose, and ultimately direct potable reuse, should be on the minds of nanotechnology researchers when developing their technologies for likely uptake in no less than five years. The chapters in this book cover most of the global efforts underway to bring about these water treatment agendas. The authors of this book were identified when the book was conceived to provide expert contributions from their field, but we duly acknowledge that more nanotechnology research is being carried out beyond what has been published here particularly as the field is in a state of rapid growth with creative minds continually emerging with new ideas.

Finally, we would like to make a mention of the current state of the world in which this book was written, which has given priority to water treatment research. At the time of writing, the world was undergoing major economic issues, specifically the Global Financial Crisis. This had a direct consequence to funding research that addresses our need for improving environmental and economic sustainability. Despite the uncertainty in global economies, climate change was recently accepted by politicians while society begins to witness never before weather activities such as severe and prolonged water scarcity and pollution. This is compounded by the rapid intensification of mining, resource extraction, and manufacturing that presents new water treatment challenges. So with the achievements in the fundamental science giving rise to nanotechnology in the last decade, now is an exciting time to drive emerging nanotechnology solutions to practically solve our most critical issues.

Whether you are an engineer or a scientist, a student, or working in industry or research, we hope this book serves your needs and gives you a comprehensive picture of the emerging nanotechnology and associated sciences now being applied to water treatment.

References

1. Khan, S.J. (2011) The case for direct potable reuse in Australia. *Water – J. Aust. Water Assoc.*, **38** (4), 92–96.

2. BCC Research (2011) Nanotechnology in Water Treatment, BCC Research.

2
Destruction of Organics in Water via Iron Nanoparticles

Hilla Shemer and Raphael Semiat

2.1
Introduction

Nanotechnology is expected to become a key component of future technologies. The following chapter focuses on degradation of organic pollutants by advanced oxidation processes (AOPs) catalyzed by nanoiron particles as well as nanoscale zero-valent iron (nZVI) and nZVI-based bimetallic nanoparticles (BNPs). Nanoparticles (NPs) have a large surface-to-volume ratio compared to other bulk materials, making them more reactive compared to their microsized counterparts. This characteristic also makes them excellent catalysts. Although particle size is an important consideration, many other factors such as geometry, composition, oxidation state, and chemical/physical environment can play a role in determining the nanoparticles' reactivity. Besides increased reactivity, nanoscale particles have the additional advantage of being relatively easily incorporated into support structures, potentially without significantly altering their reactivity. This broadens their potential applications.

Metallic nanoparticles are kinetically stable, because thermodynamics favors the formation of structures with a high surface/volume ratio, and consequently the bulk metal represents the lowest energy. In order to increase their stability, reactivity and/or mobility stabilizers are required to prevent the agglomeration of the nanoclusters by providing a steric and/or electrostatic barrier between particles. In addition, the stabilizers play a crucial role in controlling both the size and shape of the nanoparticles. Other possibilities to increase nanoparticles' stability, reactivity, and/or mobility include combination with other metals (such as bimetallic iron-based nanoparticles), support materials (such as activated carbon and zeolites), or embedding of the particles in organic membranes (i.e., emulsified nZVI).

Advances in nanoscience provide opportunities for developing nanoparticles with high activities for energetically challenging reactions, high selectivity to valuable products, and extended life times. The control of nanoparticle size and morphology represents a crucial goal to achieve in order to tune the physical properties of nanomaterials, which will enable full-scale implementation of environmental remediation.

Functional Nanostructured Materials and Membranes for Water Treatment, First Edition.
Edited by Mikel Duke, Dongyuan Zhao, and Raphael Semiat.
© 2013 Wiley-VCH Verlag GmbH & Co. KGaA. Published 2013 by Wiley-VCH Verlag GmbH & Co. KGaA.

2.2
Nanoparticles as Catalysts

A catalyst is a substance that increases the rate of a chemical reaction by reducing the required activation energy, yet is left unchanged by the reaction. Nanocatalysts represent the convergence of catalysts with nanotechnology, which is concerned with the synthesis and functions of materials at the nanoscale range (<100 nm) [1]. In a sense, all catalysis is in the nanoscale, as it involves chemical reactions at the nanoscale. However, in this chapter, a *nanocatalyst* is defined as a substance with catalytic properties that has nanoscale dimensions. An important feature of nanomaterials is that their surface properties can be very different from those shown by their macroscopic or bulk counterparts [2]. NPs have a large surface-to-volume ratio compared to other bulk materials. This characteristic makes them excellent catalysts. Although particle size is an important consideration, many other factors such as geometry, composition, oxidation state, and chemical/physical environment can play a role in determining NP reactivity. However, the exact relationship between these parameters and NP catalytic performance may be system dependent and is yet to be laid out for many nanoscale catalysts. Nevertheless, it is recognized that the enhanced catalytic activity and selectivity observed for small NPs is a combination of several factors acting in parallel. For instance, with decreasing NP size, the surface-to-volume ratio increases, resulting in a larger number of low-coordinated atoms available for interaction with chemical adsorbates. The distinct electronic properties of such sites are also expected to play a role in chemical reactivity, for example, by facilitating the dissociation of reactants or by stabilizing intermediate reaction species. Another parameter that plays a vital role in catalysis is the roughness of surface. Enhanced chemical activity has been observed for stepped surfaces as compared to smooth surfaces [3].

From a catalytic point of view, at the nanoscale, the reactivity of the active sites is influenced by the surrounding environment at a supramolecular level. This environment drives the local adsorption/desorption of reactants and products; for instance, local pH around the active site can be different from the bulk pH depending on the characteristics of the catalytic cavities. For this reason, adequate tuning of catalysts at the nanoscale level would allow enhancement of the catalytic properties. The nanosized catalytic materials are assembled at the microscale with the aim of obtaining optimal spatial distribution and catalyst composition in a final macrostructured catalyst (e.g., pellets, membranes, monoliths, and foams) [4]. A heterogeneous catalyst (i.e., the catalyst is in a different phase from the reactants) is characterized mainly by the relative amounts of different components (active species, physical and/or chemical promoters, and supports), shape and size, pore volume and distribution, and surface area. The nature of the active species is always the most important factor. Heterogeneous nanocatalysts are found both in colloidal and supported forms.

2.2.1
Colloidal Nanoparticles

Colloidal nanoparticles are used as catalyst in which the nanoparticles are finely dispersed in the aqueous solution. Colloidal nanoparticles tend to aggregate. The specific surface area of highly aggregated nanoparticles is likely to be very different from the specific surface area of dispersed nanoparticles. This is of importance mainly with respect to the determination of the reactive surface area and reactive sites on the particle surface. Aggregation of nanoparticles is difficult to avoid under environmental conditions. Therefore, in order to prevent aggregation, colloidal nanoparticle solutions are often stabilized. A good stabilizer is one that protects the nanoparticles during the catalytic process but does not passivate the surface of nanoparticles resulting in loss of nanoparticle catalytic activity (passivation is the formation of a nonreactive surface film). Accordingly, the preference of stabilizers should meet two challenges: (i) the development of methods for stabilizing nanoparticles by eliminating aggregation without blocking most of the active sites on the nanoparticle surfaces or otherwise reducing catalytic efficiency and (ii) controlling nanoparticle size, shape, and size distribution [5]. Some of the common stabilizers include polymers, block copolymers, dendrimers, surfactants, and other ligands. Transition-metal colloids are the most used nanocatalysts as a large number of atoms are present in the nanoparticle surface.

When compared with microparticles, nanoscale-based particles have higher reactivity because of their high specific area and more reactive surface sites. Their ability to remain in suspension enables injection of the nanoparticles into contaminated soils, sediments, and aquifers. Yet, owing to their tendency to form aggregates, it is difficult to maintain them in a suspension. Support of the nanoparticles may inhibit aggregation as well as increase their transport [6].

2.2.2
Supported Nanoparticles

According to the most common preparation procedures, the supported catalysts can be classified as follows [4]:

- Supports where the catalyst is generated as a new solid phase (e.g., precipitation, sol–gel). The formation and precipitation of a crystalline solid occurs by supersaturation with physical (e.g., variations in temperature) or chemical (e.g., addition of bases or acids) perturbations of a solution, formation of stable small particles by nucleation and growth by agglomeration of those particles. The sol–gel method consists of the transformation of a solution into a hydrated solid precursor (hydrogel). This method has gained importance because of a better control of different properties such as texture and homogeneity.
- Impregnated catalysts where the active phase is introduced or fixed on a preexisting solid (e.g., impregnation, ion exchange, adsorption, deposition–precipitation, and chemical- or physical-vapor deposition). Impregnation is based on the contact between the solid support and a certain volume of solution, which contains

the precursor of the active phase. The method is called *wet impregnation* when an excess of solution is used and incipient wetness impregnation if the volume is equal or slightly less than the pore volume of the support. The ion exchange method is based on the replacement of ions on the surface of a support by electrostatic interactions. Adsorption consists in the controlled attraction of a precursor contained in an aqueous solution by charged sites on the support. In the deposition-precipitation method, slurries are formed and the precipitation occurs in interaction with a support surface by addition of an alkali solution. The most common method of preparing heterogeneous transition-metal nanocatalysts is by adsorption onto the support.

The prepared solids are separated from the mother liquor by filtration, decantation, and/or centrifugation, usually washed with distilled water (or specific organic solvents) for complete removal of impurities, and submitted to thermal treatments (e.g., drying and calcination-heating). After calcination, modification of the nature and/or structure of the phases stabilization of mechanical properties occurs. The substrates used as support for the nanocatalyst include silica, alumina, titanium dioxide, polymeric support, and carbon. The carbon support of choice is activated carbon, followed by carbon black and graphite [7]. The support materials stabilize metal catalysts against sintering at high reaction temperatures. Some support materials, especially reducible oxides, can also promote the activity and selectivity of active metal catalysts. Many innovations have been made in designing nanostructured support materials. One example is the use of carbon nanotube–inorganic oxide hybrid nanoparticles as a support for phase-transfer reactions (reactants and products are in different phases) to simplify the separation and purification processes. These hybrid nanoparticles are amphiphilic and stabilize water–oil emulsions. The metal catalysts, immobilized on the hybrid support, preferentially stay at the water–oil interface, where catalytic phase-transfer reactions happen. The emulsions are extremely stable and can be easily separated from the biphasic liquid by a simple filtration. The recycled hybrid nanoparticles can be reused without any special treatment [8].

2.3
Advanced Oxidation Processes

NPs can be coupled with AOPs to carry out degradation of organics. AOPs refer specifically to processes in which oxidation of organic contaminants occurs primarily through reactions with hydroxyl radicals. AOPs involve two stages of oxidation.

1) Formation of strong oxidants (e.g., hydroxyl radicals (·OH)) and
2) Reaction of these oxidants with organic contaminants in water.

The reactive radicals are capable of decomposing a wide range of organic compounds. Depending on the structure of the organic compound in question, different reactions may occur including hydrogen atom abstraction, electrophilic addition, electronic transfer, and radical–radical interactions [9].

AOPs can be applied to fully or partially oxidize pollutants, usually using a combination of oxidants. In brief, Fenton process consists of hydrogen peroxide (H_2O_2) and iron as catalyst ($H_2O_2/Fe^{2+}(Fe^{3+})$); ozonation is based on the strong oxidant properties of ozone (O_3). Ozone is mostly used in conjunction with hydrogen peroxide and/or ultraviolet (UV) irradiation. Photocatalysis uses the photonic activation of the catalyst via light irradiation, producing reactive electron–holes (UV/TiO_2). UV light combined with hydrogen peroxide (UV/H_2O_2) is based on photolysis of the peroxidic bond of hydrogen peroxide by absorption of the UV irradiation. The most efficient hydroxyl radical yields are obtained when shortwave UV wavelengths (200–280 nm) are used. In catalytic wet oxidation (CWO), the pollutant molecules are oxidized with pure oxygen or air at elevated temperatures (130–250 °C) and pressures (5–50 bar). Other AOPs include sonolysis, which consists of the ultrasonically induced acoustic cavitation and electrochemical oxidation, which produces large amounts of hydroxyl radicals directly in water. Each technology requires a different catalytic material as well as the optimization of their catalytic properties.

The efficiency of the various AOPs depends both on the rate of generation of the free radicals and on the extent of contact between the radicals and the organic compound. It is assumed that a combination of oxidants results in better degradation rates and efficiencies as compared to single-oxidant processes. This assumption relies on [10]

- the similarity between the mechanisms of destruction (radical oxidation) of the different processes;
- the enhancements in the rate of generation of free radicals using combined methods; and
- the synergy among the methods to minimize the drawbacks of individual methods.

Catalyst loading, solution pH, and substrate concentration/nature are variables that should be considered among all the different catalytic-AOPs. Other parameters are specific for each process. Of importance are the oxygen partial pressure and the temperature in CWO, diffusion efficiency of gaseous O_3 into the treated water in ozonation, the light intensity in photocatalysis, and the ratio between Fe and H_2O_2 in the Fenton process. An appropriate catalyst of AOPs is the one that shows the highest level of pollutant mineralization (to carbon dioxide and water) or pollutant degradation to less toxic/refractory compounds. It is important to note that the environmental risk posed by a contaminant is not necessarily decreased by its degradation. The oxidation by-products may be more toxic or lengthen/enhance the harmful effect of a contaminant. Catalyst deactivation must be taken into consideration as well, being normally due to sintering, poisoning of active sites, dissolution of active components to the liquid phase (leaching), or fouling of the catalyst surface as a result of deposition of reaction intermediates [4]. The fate of the residual oxidants (if exist) and by-products as well as the cost including energy consumption should also be considered. For example, the heating of 1 m^3 of water to 10 °C consumes about 1 kg of fossil fuel. Augmentation of the pressure utilizes additional energy. Although energy recycle is often implemented, in processes such as CWO in which temperature and pressure are increased up to 250 °C and 50 bar, respectively, the issue of energy consumption should be properly considered.

2.3.1
Fenton-Like Reactions

Fenton-like reactions are defined as combination of hydrogen peroxide with ferric ions or other transition metals. The optimum pH for the homogeneous Fenton and Fenton-like processes ranges between 2 and 4. Yet, heterogeneous solid catalysts can mediate Fenton-like reactions over a wide range of pH values because Fe(III) species are immobilized within the structure and in the pore/interlayer space of the catalyst. As a result, the catalyst can maintain its ability to generate hydroxyl radicals from hydrogen peroxide, and iron hydroxide precipitation is prevented [11–13].

Solid nanocatalysts profile has been proposed to include [1]

- high activity in terms of pollutant removal;
- marginal leaching of active cations;
- stability over a wide range of pH and temperature;
- high hydrogen peroxide conversion with minimum decomposition; and
- reasonable cost for practical applications.

In comparison with their microsized counterparts, nanoparticles show higher catalytic activity because of their large specific surface where catalytically active sites are exposed [13]. For example, Kwon *et al.* [14] evaluated two iron-oxide catalysts for the oxidation of carbon monoxide and methane at low temperatures. One of the materials (NANOCAT®) had an average particle size of 3 nm and a specific surface area of $250 \, m^2 \, g^{-1}$, whereas the other material (Fe_2O_3PVS) had an average particle size of 300 nm and a surface area of $4 \, m^2 \, g^{-1}$. Although both catalysts were effective, the nanocatalyst showed superior activity. Valdés-Solís *et al.* [15] have developed a new catalyst for the Fenton-like reaction using nanosized particles with a high surface area. These solid nanocatalysts showed high catalytic activity over a wide range of pH values (6–13) and H_2O_2 concentrations (0.005–3 M). Their high reactivity was attributed to the active sites located on the surface. As such, they have a low diffusional resistance, and are easily accessible, to the substrate molecules.

2.3.1.1 Iron Oxide as Heterogeneous Nanocatalyst
Nanoscale iron oxides were reported to be effective catalysts of the Fenton reaction. Zelmanov and Semiat [16] showed that the rate of degradation of ethylene glycol and phenol degradation by iron(III) oxide nanoparticles was up to 35 times higher than that of homogeneous photo-Fenton reaction.

Lim *et al.* [17] reported a highly active heterogeneous Fenton catalyst using iron-oxide nanoparticles immobilized to alumina coated mesoporous silica.

Fenton reactions can be catalyzed by dissolved iron from the iron oxide nanoparticles or by the oxide itself, which acts as a heterogeneous catalyst. Iron oxides in general are compounds with low to very low solubility. The driving force for iron oxide dissolution is the extent of undersaturation with respect to the oxide. Undersaturation is thus a requirement for dissolution as is supersaturation for precipitation. The factors that influence the rate of dissolution of iron oxides are the properties of the reaction system (temperature and UV light), the composition of the solution phase (pH, redox potential, concentration of acids, reductants, and complexing agents), and the properties of the oxide (specific surface area, stoichiometry, crystal chemistry, crystal habit, and the presence of defects or guest ions). Models that take all of these factors into account are not available. In general, only the specific surface area, the composition of the solution, and in some cases the tendency of ions in solution to form surface complexes are considered. Figure 2.1 shows schematically, the activation energies (E_a) of the various steps of iron oxide dissolution by protonation; as the detachment step is rate limiting, it is assigned the highest E_a [18].

A mechanism of hydrogen peroxide decomposition on iron oxides surface ($\equiv$Fe(III); such as goethite, hematite, and ferrihydrite) was described by Kwan and Voelker [19], according to Equations 2.1–2.3, assuming that an Fe(III) site identical to that present in Equation 2.1 is regenerated by Equation 2.3.

$$\equiv Fe(III) + H_2O_2 \rightarrow \equiv Fe(III)H_2O_2 \tag{2.1}$$

$$\equiv Fe(III)H_2O_2 \rightarrow \equiv Fe(II) + HO_2 + H^+ \tag{2.2}$$

$$\equiv Fe(II) + H_2O_2 \rightarrow \equiv Fe(III) + HO\cdot + OH^- \tag{2.3}$$

The generation rate of hydroxyl radicals was found to be proportional to the product of [H_2O_2] and iron oxide surface area. This is consistent with a mechanism whose rate-limiting step involves H_2O_2 sorbed on the iron oxide surface. Hence, if Equation 2.2 is the rate-limiting step, then the rate of HO· production will be proportional to the concentration of $\equiv$Fe(III)H_2O_2.

Lin and Gurol [20] proposed a more detailed mechanism with its simplified form presented by Equations 2.4–2.8 (it should be noted that Equations 2.4 and 2.5 are unbalanced).

$$\equiv Fe(III) - OH + H_2O_2 \leftrightarrow (H_2O_2)_s \tag{2.4}$$

$$(H_2O_2)_s \rightarrow \equiv Fe(II) + HO_2^\cdot + H_2O \tag{2.5}$$

$$\equiv Fe(II) + H_2O_2 \rightarrow \equiv Fe(III) - OH + HO\cdot \tag{2.6}$$

$$HO_2^\cdot \leftrightarrow O_2^{\cdot-} + H^+ \tag{2.7}$$

$$\equiv Fe(III) - OH + HO_2^\cdot/O_2^{\cdot-} \rightarrow \equiv Fe(II) + H_2O/HO^- + O_2 \tag{2.8}$$

According to this mechanism, the reactions are initiated by the formation of a precursor surface complex of H_2O_2 with the oxide surface. $(H_2O_2)_s$ in Equation 2.4 represents the surface species of hydrogen peroxide, which might hold an inner- or an outer-sphere surface coordination. However, the progression of the reaction is expected to be through the inner-sphere formation directly with the surface

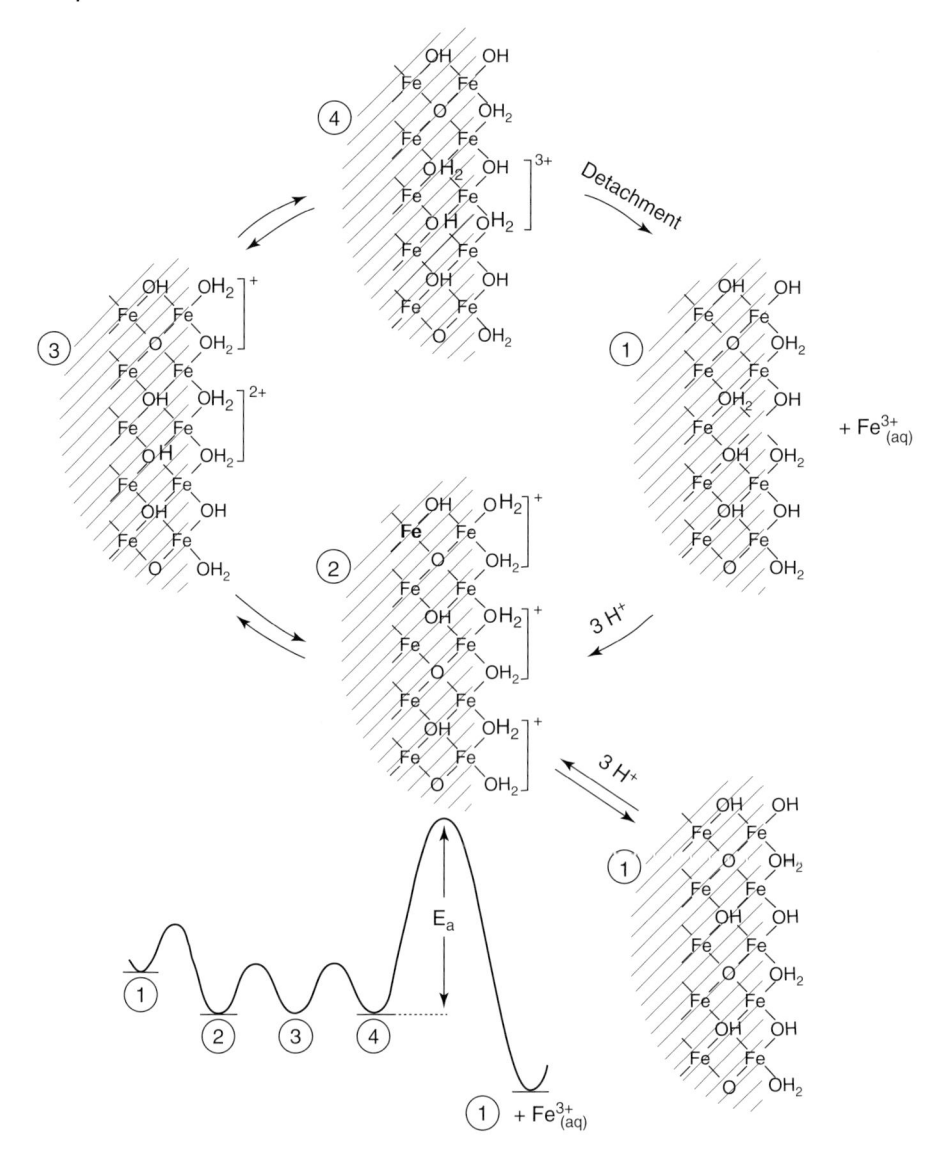

Figure 2.1 Schematic representation of the consecutive steps of dissolution by protonation of iron oxides [18].

metal centers. The surface complex may undergo a reversible electron transfer, which can be described as a ground-state electron transfer from ligand to metal within a surface complex (Equation 2.5). The electronically excited state can be deactivated through the dissociation of the peroxide radical (dissociation of the successor complex), as shown by Equation 2.6. The peroxide radical is a very active radical that can immediately react with other compounds. The reduced iron, being a reductant, can react with either H_2O_2 or oxygen (Equation 2.7). The peroxide and hydroxyl radicals produced during the reaction may react with Fe(III) and Fe(II) sites on the surface according to Equation 2.8.

In terms of the overall reaction, it was concluded that both heterogeneous and homogeneous oxidation of organic compounds occur simultaneously. While the production of active radicals, initiated by hydrogen peroxide, involves the reductive and nonreductive dissolution of iron oxide in heterogeneous processes, the iron ions (Fe^{2+}, Fe^{3+}, and complex iron species) react with hydrogen peroxide in the solution (i.e., homogeneous processes). It was further suggested that the intermediates derived from degradation of organic compounds (such as catechol and oxalic acid) promote the dissolution of iron by reductive and nonreductive pathways. Therefore, the reactivity of iron oxides in catalyzing organics degradation by hydrogen peroxide relates also to the tendency of iron to be dissolved by oxidation intermediates [21]. A schematic description of organic compound degradation by iron oxides immobilized onto a solid support is presented in Figure 2.2. Initially, hydrogen peroxide reacts with the iron oxide ($\equiv Fe^{3+}$) to generate $\equiv Fe^{2+}$, hydroperoxyl, and superoxide anion radicals. Degradation of the organic compound is initiated by its reaction with hydroxyl radicals formed by the decomposition of hydrogen peroxide. The $\equiv Fe^{2+}$ is rapidly oxidized by excess hydrogen peroxide (Fenton reaction), and the iron ion in the solution may be dissolved or in complexes with organic intermediates [21].

An important question regarding the H_2O_2/nFe oxides system is whether the oxidation/reduction reactions on the oxide surface can transform the iron particles. This transformation may lead to substantial changes in the surface characteristics

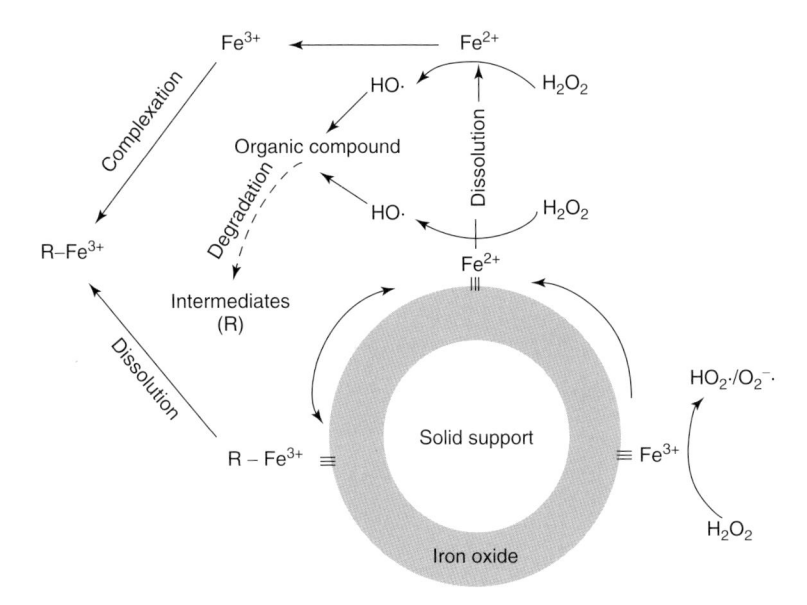

Figure 2.2 Schematic description of organic compound degradation by iron oxides immobilized onto solid support [21].

of the mineral, resulting in a different kinetic behavior and decomposition rate for H_2O_2 (i.e., each iron oxide type have different surface reactivity). In addition, iron oxide dissolution may reduce the total concentration of iron, changing the mechanism of the reaction from being surface to solution based. This matter was studied by Lin and Gurol [20] who found that the reactivity of iron oxide did not vary over several weeks of operation. These results suggest that H_2O_2 did not affect the surface reactivity of the iron oxide as well as its structure. This was further confirmed by images of the iron oxide surface taken with an electron scanning microscope prior and following exposure to H_2O_2, revealing no significant change in the iron oxide surface structure.

2.3.2
Photo-Fenton Reactions

The photo-Fenton reaction is rather similar to the Fenton one with the addition of UV irradiation. Its effectiveness is attributed to the photolysis of Fe^{3+} in acidic media yielding Fe^{2+} in conjunction with the reaction between Fe^{2+} and H_2O_2 to yield hydroxyl radicals (Fenton's reaction). Different iron-containing catalysts can be used for this process. First, bulk catalysts containing iron, such as hematite, goethite, or magnetite, may be considered. A different approach is the incorporation of iron into different supports, being zeolites, polymers, or carbon. The use of heterogeneous solid supports catalysts is considered particularly beneficial as very often complete mineralization of the organic pollutant can be reached, along with easy separation of the catalysts from the treated wastewater, not causing secondary metal ion pollution [22]. Besides iron, other transition metals, such as copper, can also catalyze the photo-Fenton reaction.

Of significant importance in determining the performance of the photo-Fenton reaction is the type of UV lamp used and its power. It is established that an increase in the UV radiation intensity results in an enhanced catalytic activity. When the UV intensity increases, a faster photoreduction of Fe^{3+} to Fe^{2+} is obtained resulting in a higher regeneration rate of Fe^{2+}. Accordingly, more hydroxyl radicals are formed resulting in higher mineralization [22].

Feng et al. [23] suggested a simple mechanism of immobilized nanoiron particles catalysis of photo-Fenton reaction in which R is the organic compound and R^* are the reaction intermediates

$$\equiv Fe(III) + h\nu \leftrightarrow \equiv Fe(II) \tag{2.9}$$

$$\equiv Fe(II) + H_2O_2 \rightarrow \equiv Fe(III) + OH\cdot + OH^- \tag{2.10}$$

$$\equiv R + OH\cdot \rightarrow \equiv R^* + OH\cdot \rightarrow CO_2 + H_2O \tag{2.11}$$

The chain of reactions is initiated by the photoreduction of Fe(III) to Fe(II) under UV irradiation. Then, the Fe(II) accelerates the decomposition of H_2O_2 in solution, generating highly oxidative hydroxyl radicals, whereas it is oxidized by H_2O_2 into Fe(III) (Fenton's reaction). The generated hydroxyl radicals attack the adsorbed organic molecules, giving rise to reaction intermediates. Finally, the reaction intermediates are mineralized into CO_2 and H_2O. The role of the adsorption of the organic molecules onto the nanoparticles and/or solid support surface is still not clear, although some researchers claim that strong adsorption of the organic reactant seems to be a necessary condition for the reaction to occur. Thus, extremely poor degradation is expected for organic compounds exhibiting no significant chemisorption [22].

2.3.3
Nanocatalytic Wet Oxidation

CWO is a reaction involving an organic compound in water and oxygen over a catalyst, either homogeneous or heterogeneous, at elevated temperatures and pressures. As in the Fenton reaction, heterogeneous catalysis is considered more efficient owing to the stability of solid catalysts compared to homogeneous or colloidal ones. In addition, a homogeneous catalyst requires a subsequent separation step, which is not required when a heterogeneous catalyst is used.

Platinum and ruthenium metals and cerium, titanium, manganese, and iron oxides deposited on zeolites, alumosilicates, ceria, alumina, and different types of carbon have been employed as catalysts in CWO reactions. Yet, the catalysts based on the platinum group metals demonstrated the higher activities. The major drawback of all these catalysts is their unstable character due to metal leaching [24].

Heterogeneous wet oxidation nanocatalysts should display the following characteristics:

- high oxidation rates;
- marginal leaching of active metal;
- stability at acidic pH and high temperatures;
- nonselectivity.

In CWO, an increase in temperature leads to a consequent increase in the oxidation rate. Possible effects of oxygen partial pressure and catalyst loading in a CWO reaction is presented in Figure 2.3, as a hypothetical behavior of the total organic carbon (TOC) conversion in a generalized reaction system at constant temperature. In the absence of a catalyst and increased oxygen partial pressure, the TOC conversion is low for nondegradable organics. Without oxygen and in the presence of a catalyst, some conversion can be achieved depending on the nature of the catalytic material. Increasing both oxygen and catalyst loading, a plateau of maximum efficiency can be observed in a region where changes in the corresponding values do not affect the global conversion. Moreover, for higher oxygen partial pressures, decrease in conversion is represented as well, in order to account for the possible occurrence of over-oxidation by the catalyst. At a constant oxygen partial pressure, decrease in conversion with higher amounts of catalyst represents the eventual acceleration of the termination step mechanism by the catalyst [4].

Nanocatalyst-based catalytic wet air oxidation (CWAO) has been studied for the removal of phenolic compounds. The nanocatalysts included suspended iron

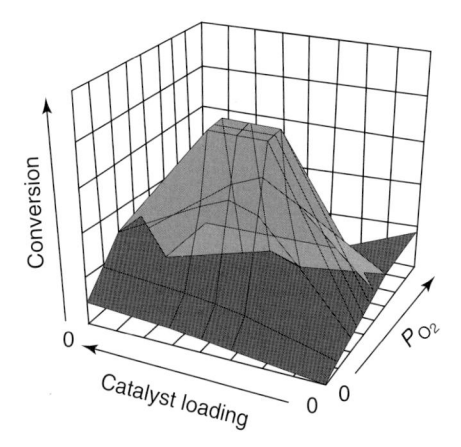

Figure 2.3 Conversion of TOC as a function of catalyst loading and oxygen partial pressure (P_{O2}), in CWO [4].

nanoparticles and colloidal nanoclusters of SnO_2 deposited on the surface of silica nanosphere. The degree of phenol removal with these catalysts did not exceed 70%, while leaching and the potential application for reuse of the recovered catalysts were not considered. But more importantly, these systems are considered unstable and tend to coagulate during the catalytic reactions, leading to catalyst deactivation [24]. Melero et al. [25] reported a nanocomposite of crystalline Fe_2O_3 and CuO particles with mesostructured SBA-15 silica as an active catalyst for wet peroxide oxidation processes. They found that the presence of copper prevents the leaching of iron species and increased TOC degradation. Recently, Botas et al. [26] prepared iron-containing catalysts on mesostructured SBA-15 silica and nonordered mesoporous silica for the oxidation of phenol aqueous solution in a catalytic fixed bed reactor in the presence of hydrogen peroxide. Sulman et al. [24] demonstrated that Pt-containing nanoparticles (size of 2.1–2.3 nm), synthesized in the pores of hyper-cross-linked polystyrene, displayed very different catalytic properties, in phenol CWAO, based on the amount of the incorporated Pt species (Pt(0), Pt(II), and Pt(IV)). With observed shielding of the catalytic sites when the Pt load was too high. Chemical oxygen demand (COD) removal ranged from 54 to 94%, depending on the Pt content, at catalyst concentration of 5.15×10^{-3} mol(Pt) l^{-1}; phenol concentration of 0.44 mol l^{-1}; temperature of 95 °C; pressure of 0.1 MPa; reaction time of 5 h; and oxygen flow rate of 0.018 $m^3 h^{-1}$.

2.4
Nano Zero-Valent Iron (nZVI)

Many different nanoscale materials have been explored for the remediation of environmental contaminants, such as nanoscale zeolites, metal oxides, carbon nanotubes and fibers, enzymes, various noble metals (mainly as BNPs), and

2.4 Nano Zero-Valent Iron (nZVI)

titanium dioxide [6]. Of these, nZVI is currently the most widely used for removal of organic and inorganic contaminants. In this chapter, the discussion concentrates only on organic compounds.

Zero-valent iron is a moderate reducing agent. In an aqueous medium, nZVI will react with dissolved oxygen, water (Equations 2.11 and 2.12), and if present, with other oxidants such as nitrate and possible organic and inorganic contaminants. Owing to the redox reactions taking place, the application of nZVI results in an increase in the pH and a decrease in the oxidation–reduction potential (ORP). The consumption of oxygen further leads to an anaerobic environment while the reduction of water yields hydrogen [6].

$$Fe^0_{(s)} + 2H_2O_{(aq)} \rightarrow Fe^{2+}(aq) + H_{2(g)} + 2OH^-_{(aq)} \quad (2.12)$$

$$2Fe^0_{(s)} + 4H^+_{(aq)} + O_{2(aq)} \rightarrow 2Fe^{2+}_{(aq)} + 2H_2O_{(l)} \quad (2.13)$$

The aforementioned equations are the classical electrochemical/corrosion reactions by which iron is oxidized by exposure to oxygen and water. The corrosion reactions can be accelerated or inhibited by manipulating the solution chemistry and/or solid (metal) composition [27].

nZVI has a greater affinity to reduce contaminants from aqueous solution by virtue of its large surface area, with 2–5 nm particles having surface area of approximately $142\,m^2\,g^{-1}$. The zero-valent iron surface area changes dramatically below 10 nm (Figure 2.4). Qualitatively similar trends apply to related properties such as the ratio of surface/bulk atoms and the fraction of particle volume composed of a surface layer of finite thickness. While trends such as these exhibit "edges" that fall around 10 nm, the curves extend continuously to larger particle sizes. It is in this region between 10 and 100 of nanometers that the nZVI is currently

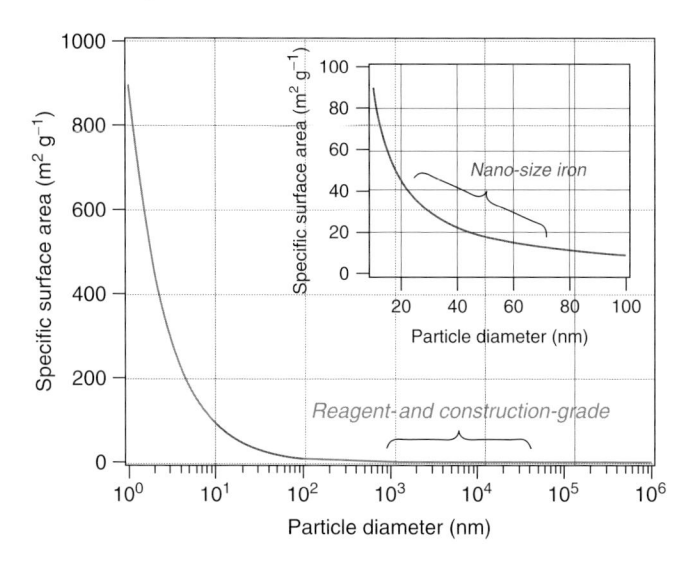

Figure 2.4 Particle surface area calculated from diameter assuming spherical geometry and density of $6.7\,g\,cm^{-3}$ (based on the average of densities for pure Fe^0 and Fe_3O_4) [28].

used in remediation applications. Even under laboratory conditions, particles of nZVI tend to aggregate, producing clusters that may approach micrometers in size. These considerations suggest that nZVI, and related materials that are used in environmental remediation applications, will not exhibit the extraordinary properties that apply to "true" nanosized particles and will behave in most respects such as environmental colloids [28].

In addition to having an increased surface area, nanoscale metal particles have high surface energies and thus are more reactive than would be expected from just the increase in surface area [29].

As mentioned earlier, the greater reactivity of nanoparticles is generally attributed to larger overall surface area, greater density of reactive sites on the particle surfaces, and/or higher intrinsic reactivity of the reactive surface sites. Together, these factors have produced three operationally distinct effects for nZVI [28]:

1) Degradation of contaminants that do not react detectably with larger particles of similar material (e.g., polychlorinated biphenyls).
2) More rapid degradation of contaminants that already react at useful rates with larger particles (e.g., chlorinated ethylenes). The reaction rates of nZVI were reported to be 25–30 times faster compared with granular iron [30] and 10–1000 times more reactive [31].
3) More favorable products from contaminants that are rapidly degraded by larger materials but yield undesirable by-products (e.g., carbon tetrachloride).

Of these three types of reactivity effects, the second is noted most frequently. Tratnyek and Johnson [28] suggested that the higher rate of organic degradation by nZVI resulted from its high surface area and not from the greater abundance of reactive sites on the surface or the greater intrinsic reactivity of surface sites. Yet, the available data is inconclusive.

Apart from iron, zinc is the most extensively studied zero-valent metal. Zinc (and tin) can degrade halogenated hydrocarbons faster and more completely than iron, because of the greater reduction potential of zinc and its high hydrogen overvoltage. Nevertheless, zinc reacts with oxygen to form passivating zinc oxides, as the reactants cannot penetrate the passivating layer which then inhibits the reactivity of the zinc. Iron oxides are nonpassivating, hence allowing the reactants to penetrate the oxide layer and react with the active iron surface [32].

2.4.1
Synthesizing Methods

nZVI can be synthesized by several methods, with the most widely used method for organics destruction purpose being the borohydrate reduction of Fe(II) or Fe(III) ions in aqueous media. In this method, sodium borohydride (NaBH$_4$) is added, drop wise to a solution of Fe(II) or Fe(III) according to the following reaction [33]

$$Fe^{2+} + 2BH^- + 6H_2O \rightarrow Fe^0 + 2B(OH)_3 + 7H_2 \uparrow \qquad (2.14)$$

The particles produced by this method are referred to as *Fe(B)*. Using this method, amorphous nZVI are formed at a size of 10–100 nm with a mean size of 50 ± 15 nm and a specific surface area of 10–50 $m^2\,g^{-1}$. The oxidative nature of Fe(B) nanoparticles was established by Liu et al. [34] in terms of total hydrogen generation. The authors demonstrated the catalytic effect of Fe(B) nanoparticles in dechlorinating trichloroethylene (TCE) by its ability to utilize the hydrogen generated in the oxidation process.

Other nZVI production methods are as follows [31]:

- heating of iron pentacarbonyl to about 200–250 °C. At this temperature, the iron pentacarbonyl dissociates into nZVI and carbon monoxide. The size of the nanoparticles formed in this reaction is ~5 nm;
- crystalline type of nanoiron particles are produced by gas-phase reduction of FeOOH. It has an average size of 70 nm with a surface area of 29 $m^2\,g^{-1}$. These nanoparticles are coated by polymaleic or polyacrylic acid;
- milling of aggregates or microscale particles.

Iron particles produced by the varying methods differ in their structural configurations, size distribution, and surface area. Thus, they also have different reactivity and aggregation properties [31].

Zero-valent iron is unstable under atmospheric conditions having the tendency to form Fe_2O_3, Fe_3O_4, and FeOOH [35]. Therefore, nZVI synthesis is mostly conducted under inert conditions or in the presence of solvents such as ethanol. Yet, it was observed that nZVI particles consist of a zero-valent core and an oxide shell (core–shell structure), while the surface characterization is not always clear. For example, the Fe(B) surface (synthesized by Equation 2.14) was reported to consist 5 nm and of FeOOH oxide shell [36], an iron–boron noncrystalline alloy [33], and magnetite/maghemite (Fe_3O_4/γ-Fe_2O_3) [37]. Liu et al. [34] showed that Fe(B) surface contained mostly borate with some boride. The boron content could affect the iron-nanoparticle properties and reactivity [30]. Aggregation into chain or floc structure shell was also observed in these Fe(B) nanoparticles.

It is assumed that the reactivity of core–shell nanoparticles is driven by oxidation of the zero-valent iron core because oxidation strongly depends on the particle size when using an ultrafine particle passivation layer of oxide, or when a stable noble metal shell is being applied to protect the iron nanoparticle from oxidation.

2.4.1.1 Emulsified Zero-Valent Iron

nZVI is efficient in the destruction of chlorinated organic compounds. Yet, in field application, it has a limited mobility and needs to be in the presence of water to promote reductive dehalogenation. The injection of nZVI into a source zone of dense nonaqueous-phase liquid (DNAPL) thus only treats the dissolved phase at the edges of the DNAPL. Emulsified zero-valent iron (EZVI) was designed for *in situ* treatment DNAPLs. EZVI is composed of an iron core (microparticles or nanoparticles) surrounded by water and packed in a droplet surrounded by food-grade surfactants and biodegradable vegetable oil, which form an oil–liquid membrane. The droplet is about 15 µm in diameter when using an nZVI-core. Laboratory studies indicated that a more stable and reactive emulsion can be produced with smaller iron particles. The hydrophobic membrane allows the DNAPL to diffuse through the membrane where it undergoes reductive dechlorination by the nZVI in the aqueous phase [31]. It is also believed that the final degradation by-products from the dechlorination reaction are driven by an increase in concentration inside the aqueous emulsion droplet to diffuse out into the nonaqueous phase (oil and TCE), and then out into the surrounding aqueous phase. While the zero-valent iron (ZVI) particles in the aqueous emulsion droplet remain reactive, the chlorinated compounds are continually degraded within the aqueous emulsion droplets, thus maintaining a concentration gradient across the oil membrane and establishing a driving force for additional TCE migration into the aqueous emulsion droplet, where additional degradation can occur. The primary application of the EZVI technology is in the treatment of DNAPL source zones, but it is also capable of treating dissolved-phase chemicals [38]. As EZVI is also a DNAPL, it will move in the subsurface in the same way as the target DNAPL, which increases the contact between the EZVI and the target DNAPL. Unlike other reductive technologies, EZVI is capable of working in sites with high dissolved oxygen or under saline conditions. The oil membrane around the iron particles protects them from corrosion [31].

In addition to the abiotic degradation associated with the ZVI, the injection of EZVI containing vegetable oil and surfactant will result in the sequestration of the chlorinated ethenes into the oil and the biodegradation of dissolved chlorinated ethenes. Chlorinated solvents will preferentially dissolve into the oil component of the EZVI, thereby reducing the aqueous phase concentrations. The chlorinated solvents may then be degraded by the ZVI in the EZVI. The vegetable oil and surfactant can also act as electron donors to promote anaerobic biodegradation of the chlorinated solvents. For example, if the amount of ZVI is not sufficient to completely degrade all of the TCE in the source area to ethane, then the vegetable oil and surfactant can act as a slow-release electron donor for biodegradation processes [38].

2.4.2
Degradation Mechanism

Nanoiron particles can reduce almost all halogenated hydrocarbons to benign compounds such as hydrocarbons, chloride, and water. A list of organic contaminants that are transformed by nZVI is presented in Table 2.1.

There are two pathways for the degradation of chlorinated hydrocarbons by nZVI. The first is hydrogenolysis (C–Cl cleavage and C–H formation) through sequential dehalogenation such as tetrachloroethene (PCE) to TCE to dichloroethylene (DCE) to vinyl chloride (VC) to ethene. The second is β-elimination in which a chlorinated ethene is converted into an ethyne group through the removal of Cl_2 (TCE to chloroacetylene to acetylene to ethene). The latter occurs in 70–90% of the reactions with nZVI – primarily when the contaminant comes in direct contact with the nZVI [6]. The organic pollutant must first be adsorbed onto the iron surface before any reaction can occur. Once sorbed, the highly electronegative

Table 2.1 Organic contaminants transformed by nanoiron particles [39].

Chlorinated methanes 　Carbon tetrachloride (CCl_4) 　Chloroform ($CHCl_3$) 　Dichloromethane (CH_2Cl_2) 　Chloromethane (CH_3Cl)	Halogenated ethenes 　Tetrachloroethene (C_2Cl_2) 　Trichloroethene (TCE; C_2HCl_3) 　Dichloroethene ($C_2H_2Cl_2$) 　Vinyl chloride (C_2H_3Cl)
Benzenes 　Hexachlorobenzene (C_6Cl_6) 　Pentachlorobenzene (C_6HCl_5) 　Tetrachlorobenzene ($C_6H_2Cl_4$) 　Trichlorobenzene ($C_6H_3Cl_3$) 　Dichlorobenzene ($C_6H_4Cl_2$) 　Chlorobenzene (C_6H_5Cl) 　Nitrobenzene ($C_6H_5NO_2$)	Hexachloroethane (C_2Cl_6) 　Trichlorofluoroethane (C_2FCl_3) 　Dibromoethane ($C_2H_2Br_2$) 　Trichlortrifluoroethane ($C_2Cl_3F_3$) 　Polychlorinated hydrocarbons 　Polychlorinated biphenyl (PCBs) 　Dioxins 　Pentachlorophenol (C_6HCl_5O)
Pesticides 　Dichloro diphenyl trichloroethane (DDT) ($C_{14}H_9Cl_5$) 　Lindane ($C_6H_6Cl_6$) 　Molinate ($C_9H_{17}NOS$) 　Chrysoidine ($C_{12}H_{13}ClN_4$) 　Tropaeolin O ($C_{12}H_9N_2NaO_5S$)	Others 　2,4,6-Trinitrotoluene (TNT) ($C_7H_5N_3O_6$) 　N-nitrosodimethylamine ($C_4H_{10}N_2O$) 　Trichloroethane ($C_2H_3Cl_3$) 　Dichloropropane ($C_3H_6Cl_2$) 　Trichloropropane ($C_3H_5Cl_3$)
Trihalomethanes 　Bromoform ($CHBr_3$) 　Dibromochloromethane ($CHBr_2Cl$) 　Dichlorobromomethane ($CHBrCl_2$)	

chlorine substituents act as the electron acceptors while Fe^0 serves as the electron donor. The reaction of nZVI with chlorinated hydrocarbons is presented in the following equation:

$$R - Cl + Fe^0 + H_2O \rightarrow R - H + Fe^{2+} + Cl^- + OH^- \quad (2.15)$$

Reduction of alkyl halides (such as the pesticide lindane) typically proceeds via a series of two discrete single-electron transfers. In the first step, an electron from Fe^0 is donated to the surface-associated alkyl halide, forming a neutral radical as one chloride ion is simultaneously ejected. Another electron is then lost from the transient Fe^+ species to the reacting carbon center of the radical, which then undergoes double bond formation and simultaneous loss of chloride from the beta carbon (Figure 2.5). A similar sequence of single-electron transfers is expected to occur during the generation of the subsequent degradation products. It appears that the initial reduction step is the rate controlling one. The two subsequent dihaloelimination steps are assumed to occur more rapidly [40].

Because nanoscale particles are so small, Brownian movement or random motion, rather than wall effects, dominates their physical movement or transport in water. The movement of micrometer-scale particles, especially microscale metal particles, is largely controlled by gravity-induced sedimentation because of their size and high density. Under flow mixing and in the absence of significant surface

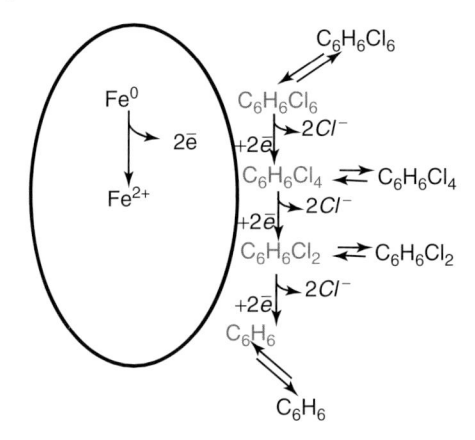

Figure 2.5 Schematic model of degradation by nanoscale zero-valent iron. In black, organics in the bulk solution and in dark gray, adsorbed compound (based on [40]).

electrostatic forces, nanosized particles can be easily suspended in water during the design and manufacturing stages, thus providing a versatile remediation tool that allows direct injection as a liquid into the subsurface where contaminants are present [6]. Materials used to support nZVI (such as zeolites, silica, and activated carbon) may have a strong effect on the reaction rates despite being inert to redox reactions. This is due to the evenness of the dispersion on the support surface and the surface roughness, which affects specific loading. The use of high-surface-area supports was found to reduce soluble metal ions much faster than unsupported zero-valent iron [41].

Joo et al. [42] proposed an oxidative degradation pathway of molinate ($C_9H_{17}NOS$) by nZVI under atmospheric conditions (i.e., in the presence of air). He was the first to study the oxidative mechanism by nZVI rather than the reductive one. This pathway is much different than that of chlorinated hydrocarbons, as shown in Equations 2.16–2.27 assuming pH > 4.8 such that the superoxide radical is deprotonated and ">" represents surface species.

$$Fe^0 + O_2 \rightarrow > O_2 \tag{2.16}$$

$$Fe^0 + > O_2 \rightarrow > Fe^{2+} + 2 > e^- \tag{2.17}$$

$$> O_2 + > e^- \rightarrow > O_2^{\cdot -} \tag{2.18}$$

$$> Fe^{2+} \rightarrow Fe^{2+} \tag{2.19}$$

$$> O_2^{\cdot -} \rightarrow O_2^{\cdot -} \tag{2.20}$$

$$O_2^{\cdot -} + O_2^{\cdot -} + 2H^+ \rightarrow H_2O_2 + O_2 \tag{2.21}$$

$$Fe^{2+} + O_2 + H_2O \rightarrow FeOH^{2+} + O_2^{\cdot -} + H^+ \tag{2.22}$$

$$Fe^{2+} + H_2O_2 \rightarrow FeOH^{2+} + OH \tag{2.23}$$

$$Fe^{2+} + OH \rightarrow FeOH^{2+} \tag{2.24}$$

$$FeOH^{2+} + OH^- \rightarrow Fe(OH)_{3(s)} \tag{2.25}$$

$$\text{molinate} + \text{OH}^- \rightarrow \text{mol}\cdot + \text{H}_2\text{O} \qquad (2.26)$$

$$\text{mol}\cdot + \text{O}_2 \rightarrow \text{keto-molinate isomers} \qquad (2.27)$$

The identification of the reaction end product (i.e., keto-molinate isomer) indicates that nZVI (average size of 50 nm) initiated an oxidation rather than a reduction process. Indeed, the keto-molinate isomer has previously been reported in molinate degradation using photocatalysis with TiO_2 as a consequence of hydrogen abstraction resulting from hydroxyl radicals attack on the N-alkyl chain. Hydroxyl radicals may be well generated in these systems as a result of Fenton's reagent processes arising from the presence of both ferrous iron and hydrogen peroxide. It is highly likely that many of the key steps occur at the surface of the nZVI particles.

Although hydrogen peroxide was not added to the system, it was produced by the reaction of superoxide radicals and protons. Combined with the oxygen-induced oxidation of nZVI to form ferrous ions this sets up the Fenton chemistry responsible for the degradation of molinate. The major advantage of this system over an ordinary Fenton system is that while normally the reaction stops once all the added ferrous ions are used, the ZVI provides a constant source of the ions and hydrogen peroxide. However, this reaction, and in particular the production of hydrogen peroxide, is limited by the solubility of oxygen in water nZVI reduction was also studied as an enhancement to Fenton reagent oxidation.

For contaminants that are difficult to either oxidize or reduce, such as TNT (2,4,6 trinitrotoluene) and RDX (hexahydro-1,3,5-trinitro-1,3,5-triazine), a combination of nZVI reduction and Fenton oxidation was shown to augment their degradation. By combining these two processes ∼20–60% degradation enhancement was obtained for TNT and RDX, respectively, compared to Fenton oxidation. Both compounds showed an overall removal of ∼96% in about 90 min, with 92 and 79% nitrogen recovered as NH_4^+ and NO_3^- for TNT and RDX, respectively [32].

2.4.3
Field Application of nZVI

The first field application of nZVI was reported in 2000. Nanoparticles have been shown to remain reactive in soil and water for up to eight weeks and can flow with the groundwater for >20 m enabling it to treat larger areas of the affected aquifers away from the injection point [6]. The realistic degradation efficiencies of contaminants, listed in Table 2.1, in the field are between 60 and 80% within one year. But reductions of up to 99% were achieved in batch and field experiments. For example, Zhang [27] showed a 99% reduction of TCE within a few days of nZVI injection. Obviously, site characteristics play a major role in successful field application. The site geology, hydrogeology, geochemistry, and microbiology should be evaluated. The hydrogeology influences the transportability and the lifetime of the particles, whereas the geochemistry shows potential substances that nZVI could react with (other than the target compounds). Field applications have been conducted in porous aquifers, fractured rocks, and unconsolidated sediments. It is generally agreed that remediation in dense geological formations is less efficient [31].

Despite their high efficiency in the removal and/or reduction of contaminants, field application of nZVI for water remediation holds some drawbacks including the following:

- iron passivation (nontarget reactions);
- high cost associated with substantial amount of zero-valent iron nanoparticles required to treat large volume of polluted water. Depending on the amount ordered and the modifications, nZVI (in 2009) cost around €100–125 kg of pure iron [31];
- iron nanoparticles are positively charged within natural pH, and hence, tend to repel each other making them stable;
- for *in situ* applications where the nZVI plume may be directly injected into an aquifer of groundwater and it is expected to migrate to contaminated zones [28]. A high mobility of nanoparticles is essential in direct injection applications. But because of colloidal properties of nZVI, particles tend to precipitate and they cannot be driven for more than a few meters by the migration of water and probably cannot reach the contaminated zone;
- storage of iron nanoparticles for extended time is a problem as thermodynamically zero-valent iron tends to react with oxygen;
- transfer from the laboratory (excellent results) to the field: it is difficult to predict reactivity, particle distribution, hydro and geochemistry, hydrology, geology, and microbiology; and thus, the design of the injection technology, monitoring, up-scaling regarding mixing, well conditioning, pretreatment of aquifer with additives, and so on to facilitate nZVI distribution;
- changes by environmental factors such as light, oxidants, and microorganisms, which may result in chemical or biological modifications or degradation of the NP-functionalized surface or coating of the surface with natural compounds, have not been evaluated thoroughly [6].

On the other hand, nZVI application holds several advantages, which are summarized as follows:

- fast reaction;
- short treatment time;
- complete reduction pathway to nontoxic end products is possible;
- *in situ* treatment.

In addition to self-aggregation, nanoparticles could associate with suspended solids or sediment, where they could bioaccumulate and enter the food chain or drinking water sources during or after completion of the desired reaction. These fate processes depend on both the characteristics of the particle and the characteristics of the environmental system. The use of nanoparticles in environmental remediation will inevitably lead to the release of nanoparticles into the environment and subsequent ecosystems. To understand and quantify the potential risks, the mobility, bioavailability, toxicity, and persistence of manufactured nanoparticles need to be studied. Nevertheless, to pose a risk, nanoparticles must be hazardous and have a route for exposure [6].

2.5
Bimetallic nZVI Nanoparticles

Generally, bimetallic systems consisting of a base metal and a second dopant metal as the catalytic agent. Typically, palladium, platinum, silver, copper, or nickel are added as a catalyst to nZVI, of which palladium has been found to be the most active catalyst. The second metal creates a catalytic synergy between itself and Fe and also aids in the nanoparticles' distribution and mobility. The second metal is usually less reactive and is believed to promote Fe oxidation or electron transfer [6]. Specifically, when nZVI is attached to a noble metal, galvanic cells are formed in the coupled bimetallic particles. Iron serves as the electron donor and preferably becomes reacted with the organic pollutants, whereas the noble metal (cathode) is protected [30].

As discussed previously, nZVI have a core–shell structure with an Fe^0 core and an oxide or noble metal shell. Decline in iron reactivity over time is often attributed to the formation of surface oxide layers. Noble metals, such as Pd, Ni, or Ag, formed on the iron particle surface oxidize less rapidly than Fe^0 thus preserving the Fe^0 core for organic pollutants degradation. Therefore, coating nZVI with a thin layer of noble metal greatly increases its reactivity [30]. BNPs were shown to have higher reaction rates, by orders of magnitude, compared with the corresponding monometallic nanoparticles. It was hypothesized that nanostructured particles with catalytic dopant are able to alter dechlorination pathways [30, 43]. Apart from catalyzing dechlorination of organic pollutants, bimetallic particles have been found to reduce the amount of toxic chlorinated by-products, indicating that the dechlorination reactions continue further toward completion.

Physical mixing of two metals does not increase the rate of a reaction; the catalysts must be doped onto the surface. For example, doping palladium on the surface of nZVI sets up a galvanic couple, which increases the rate of corrosion of the iron, thus increasing the rate of dechlorination [44]. Some noble metals, particularly palladium, catalyze dechlorination and hydrogenation [6]. Bimetallic iron nanoparticles were reported to transform various organic pollutants, as listed in Table 2.2.

A schematic depiction of the bimetallic nanoparticle-mediated reduction of TCE in water is presented in Figure 2.6 [27].

Palladized Fe particles are prepared by soaking freshly prepared nanoscale iron particles with an ethanol solution containing 1% (wt) of palladium acetate ($[Pd(C_2H_3O_2)_2]_3$). This causes the reduction and subsequent deposition of Pd on the Fe surface [29]

$$Pd^{2+} + Fe^0 \rightarrow Pd^0 \downarrow + Fe^{2+} \tag{2.28}$$

Similar methods were used to prepare Fe/Pt, Fe/Ag, Fe/Ni, Fe/Co, Fe/Cu bimetallic particles [45]. Measurement of over 150 Fe/Pd particles yielded a mean value of particle diameter at 66.6 ± 12.6 nm. The median size was 60.2 nm. Most particles (>80%) had diameters less than 100 nm, with 30% less than 50 nm. The average specific BET surface area of the nanoscale Pd/Fe particles was about $35 \pm 2.7 \, m^2 \, g^{-1}$ [27].

Table 2.2 Organic contaminants transformed by bimetallic iron-NP.

Compound	BNP	
Chloromethane	Fe/Pd	[30]
Dichloromethane	Fe/Pd	[30]
Tetrachloromethane	Fe/Pd	[30]
Tetrachloroethene	Fe/Pd	[44, 45]
Trichloroethene	Fe/Pd; Fe/Ni	[29–31, 42, 43]
Dichloroethene	Fe/Pd	[44, 45]
Vinyl chloride	Fe/Pd	[30, 44, 45]
Hexachlorobenzene	Fe/Ag	[46]
Tetrachlorobenzene	Fe/Ag	[46]
Trichlorobenzene	Fe/Ag	[46]
Pentachlorobenzene	Fe/Ag	[46]
Carbon tetra chloride	Fe/Pd	[45]
Chloroform	Fe/Pd	[45]
Pentachlorophenol	Fe/Pd; Fe/Ni; Fe/Pt	[47]
DDT (1,1,1-trichloro-2,2-bis(p-chlorophenyl)ethane)	Fe/Ni	[48]
Trichloroethylene	Fe/Ni	[49]
Polychlorinated biphenyls (PCBs)	Fe/Pd	[30]
Orange G (monoazo dye)	Fe/Ni	[50]

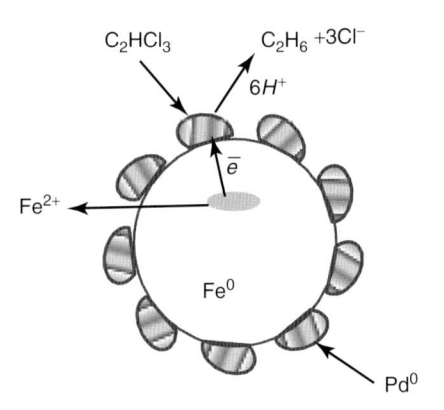

Figure 2.6 Schematic depiction of the bimetallic iron-nanoparticle-mediated reduction of trichloroethene in water (based on [27]).

Doping of nZVI with other metals does not always increase the rate of degradation. Inhibitive effects have been observed on pentachlorophenol degradation when ZVI was modified with Pd, Pt, Ni, and Cu [46].

The synergistic characteristics of bimetallic systems involved the oxidation of iron as the active reactant in generating the electrons and hydrogen necessary for the degradation process (Equation 2.14). The electrons and hydrogen generated are then utilized by a second dopant metal to catalyze the reaction through the formation of active surface metal hydride as a powerful reductant, represented as follows [43]:

$$2M + H_2 \rightarrow 2M-H \tag{2.29}$$

where M may be nickel or palladium.

pH is a crucial parameter affecting the degradation rate of chlorinated compounds by BNP. At higher pH, less atomic hydrogen or hydride on the catalyst surface attacks chlorinated molecules to replace the chlorine atom and to form dechlorinated molecule and chlorine ion. Meanwhile, the ferrous and hydroxyl ions form ferrous hydroxide and precipitate. The surface passivating layer of the ferrous hydroxide and precipitate could hinder the transport of the chlorinated molecules and block the reactive sites on the BNP and hence depress the overall reaction rate [47].

The kinetics of dechlorination of organic pollutants by bimetallic iron nanoparticles follow pseudo first-order, as presented by the following equations [48]:

$$-\frac{dC}{dt} = k_{SA}\alpha_s\rho_m C \tag{2.30}$$

where C is the concentration of chlorinated pollutant, k_{SA} is the specific reaction rate constant ($l\,h^{-1}\,m^{-2}$), α_s is the specific surface area of the bimetallic iron nanoparticle ($m^2\,g^{-1}$), and ρ_m is the mass concentration of the bimetallic iron nanoparticle ($g\,l^{-1}$). Variations in the k_{SA} value are attributed to impurities within the metal matrix, which can affect the distribution of active sites [47].

Schrick et al. [51] reported that the enhanced degradation kinetics of Ni/Fe nanoparticles was accompanied by hydrogen generated by anaerobic iron oxidation of the bimetallic system. On the contrary, the nano and bulk iron, which had reaction rates that were 2 orders of magnitude lower than Ni/Fe, had significantly lower hydrogen generation.

Tee et al. [43] found that a maximum rate of dechlorination is achieved with a Ni content of 25 wt%, similar to the Ni/Fe ratio designed by Schrick et al. [51]. The decrease in reaction rates as the Ni content increases beyond this level was attributed to high levels of Ni with respect to Fe at the particle edges, which hinder the ability of Fe to transfer electrons to the reaction sites.

2.6
Summary

The reactive iron-based nanoparticles discussed in this chapter may be applied to transform and detoxify organic pollutants, such as in the case of nZVI and BNPs, or catalyze degradation/mineralization by AOP, such as in the case of nanoscale iron oxides (immobilized or suspended). These nanomaterials have properties that enable both chemical reduction and catalysis to mitigate organic pollutants of concern.

The ongoing development of nanoscale particles is focused on morphology, reactivity, and mobility. Yet, the current understanding of the basic processes involved in this technology is still evolving and incomplete. The use of NP-based processes is expected to grow significantly within the next years, in particular the application of nZVI *in situ* remediation of contaminated groundwater. Nevertheless, the existing knowledge makes it difficult to move forward with the engineering of full-scale implementations, especially since it is very difficult to assess the risks that this technology might have to human and/or ecological health, while maintaining low-cost operation.

References

1. Garrido-ramírez, E.G., Theng, B.K., and Mora, M.L. (2010) Clays and oxide minerals as catalysts and nanocatalysts in Fenton-like reactions – A review. *Appl. Clay Sci.*, **47**, 182–192.
2. Theng, B.K.G. and Yuan, G. (2008) Nanoparticles in the soil environment. *Elements*, **4**, 395–399.
3. Cuenya, B.R. (2010) Critical review: synthesis and catalytic properties of metal nanoparticles: size, shape, support, composition, and oxidation state effects. *Thin Solid Films*, **518**, 3127–3150.
4. Manuel, A. and da Silva, T. (2009) Environmental catalysis from nano- to microscale. *Mater. Technol.*, **43**, 113–121.
5. Crooks, R.M., Zhao, M., Sun, L., Chechik, V., and Yeung, L.K. (2001) Dendrimer-encapsulated metal nanoparticles: synthesis, characterization, and applications to catalysis. *Acc. Chem. Res.*, **34**, 81–190.
6. Karn, B., Kuiken, T., and Otto, M. (2009) Nanotechnology and in situ remediation: A review of the benefits and potential risks. *Environ. Heal. Perspect.*, **117**, 1823–1831.
7. Stuüber, F., Fonta, J., Fortunyb, A., Bengoaa, C., Eftaxiasa, A., and Fabregat, A. (2005) Carbon materials and catalytic wet air oxidation of organic pollutants in wastewater. *Top. Catal.*, **33**, 3–50.
8. Li, Y.M. and Somorjai, G.A. (2010) Nanoscale advances in catalysis and energy applications. *Nano Lett.*, **10**, 2289–2295.
9. Nogueira, R.F.P., Trovó, A.G., da-Silva, M.R.A., Villa, R.D., and de Oliveira, M.C. (2007) Fundaments and environmental applications of Fenton and photo-Fenton processes. *Quím. Nova*, **30**, 400–408.
10. Gogate, P.R. and Pandit, B.A. (2004) A review of imperative technologies for wastewater treatment. I. Oxidation technologies at ambient conditions. *Adv. Environ. Res.*, **8**, 501–551.
11. Chen, J. and Zhu, L. (2006) Catalytic degradation of Orange II by UV-Fenton with hydroxyl- Fe-pillared bentonite in water. *Chemosphere*, **65**, 1249–1255.
12. Caudo, S., Centi, G., Genovese, C., and Perathoner, S. (2007) Copper- and iron-pillared clay catalysts for the WHPCO of model and real wastewater streams from olive oil milling production. *Appl. Catal., B*, **70**, 437–446.
13. Nurmi, J., Tratnyek, P.G., Sarathy, V., Baer, D.R., Amonette, J.E., Pecher, K., Wang, C., Linehan, J., Watson, D., Penn, R., and Driessen, M. (2005) Characterization and properties of metallic iron nanoparticles: spectroscopy, electrochemistry and kinetics. *Environ. Sci. Technol.*, **39**, 1221–1230.
14. Kwon, S.C., Fan, M., Wheelock, T.D., and Saha, B. (2007) Nano- and micro-iron oxide catalysts for controlling the emission of carbon monoxide and methane. *Sep. Purif. Technol.*, **58**, 40–48.
15. Valdés-Solís, T.P., Valle-Vigón, P., Álvarez, S., Marbán, G., and Fuertes, A.B. (2007) Manganese ferrite nanoparticles synthesized through a nanocasting route as a highly active Fenton catalyst. *Catal. Commun.*, **8**, 2037–2042.
16. Zelmanov, G. and Semiat, R. (2008) Iron(3) oxide-based nanoparticles as catalysts in advanced organic aqueous oxidation. *Water Res.*, **42**, 492–498.
17. Lim, H., Lee, J., Jin, S., Kim, J., Yoon, J., and Hyeon, T. (2006) Highly active heterogeneous Fenton catalyst using iron oxide nanoparticles immobilized

in alumina coated mesoporous silica. *Chem. Commun.*, **4**, 463–465.
18. Cornell, R.M. and Schwertmann, U. (2003) *The Iron Oxides: Structure, Properties, Reactions, Occurrences and Uses*, Wiley-VCH Verlag GmbH, Weinheim.
19. Kwan, W.P. and Voelker, B.M. (2003) Rates of hydroxyl radical generation and organic compound oxidation in mineral-catalyzed Fenton-like systems. *Environ. Sci. Technol.*, **37**, 1150–1158.
20. Lin, S.S. and Gurol, M.D. (1998) Catalytic decomposition of hydrogen peroxide on iron oxide: kinetics, mechanisms, and implication. *Environ. Sci. Technol.*, **32**, 1417–1423.
21. Huang, C. and Huang, Y. (2008) Comparison of catalytic decomposition of hydrogen peroxide and catalytic degradation of phenol by immobilized iron oxides. *Appl. Catal. Gen.*, **346**, 140–148.
22. Henry-Ramirez, J., Vicente, M.A., and Madeira, L.M. (2010) Heterogeneous photo-Fenton oxidation with pillared clay-based catalysts for wastewater treatment: A review. *Appl. Catal., B*, **98**, 10–26.
23. Feng, J., Hu, X., Yue, P.L., Zhu, H.Y., and Lu, G.Q. (2003) A novel laponite clay-based Fe nanocomposite and its photo-catalytic activity in photo-assisted degradation of Orange II. *Chem. Eng. Sci.*, **58**, 679–685.
24. Sulman, E.M., Matveeva, V.G., Doluda, V.Y., Sidorov, A.I., Lakina, N.V., Bykov, A.V., Sulman, M.G., Valetsky, P.M., Kustov, L.M., Tkachenko, B.D., and Bronstein, L.M. (2010) Environmental Efficient polymer-based nanocatalysts with enhanced catalytic performance in wet air oxidation of phenol. *Appl. Catal., B*, **94**, 200–221.
25. Melero, J.A., Calleja, G., Martinez, F., and Molina, R. (2006) Nanocomposite of crystalline Fe_2O_3 and CuO particles and mesostructured SBA-15 silica as an active catalyst for wet peroxide oxidation processes. *Catal. Commun.*, **7**, 478–483.
26. Botas, J.A., Melero, J.A., Martinez, F., and Pariente, M.I. (2010) Assessment of Fe_2O_3/SiO_2 catalysts for the continuous treatment of phenol aqueous solutions in a fixed bed reactor. *Catal. Today*, **149**, 334–340.
27. Zhang, W.X. (2003) Nanoscale iron particles for environmental remediation: An overview. *J. Nanopart. Res.*, **5**, 323–332.
28. Tratnyek, P.G. and Johnson, R.L. (2006) Nanotechnologies for environmental cleanup. *Nanotoday*, **1**, 44–48.
29. Wang, C. and Zhang, W. (1997) Nanoscale metal particles for dechlorination of PCE and PCBs. *Environ. Sci. Technol.*, **31**, 2154–2156.
30. Li, L., Fan, M., Brown, R.C., Wang, J., Wang, W., Somg, Y., and Zhang, P. (2006) Synthesis, properties and environmental application of nanoscale iron-based materials: A review. *Crit. Rev. Environ. Sci. Technol.*, **36**, 405–431.
31. (2010) Observatory NANO Focus ReportNano Zero Valent Iron – THE Solution for Water and Soil Remediation? http://www.observatorynano.eu. (accessed 3 November 2010).
32. McDowall, L. (2005) Degradation of Toxic Chemicals by Zero-Valent Metal Nanoparticles – A Literature Review, Human Protection & Performance Division; DSTO Defence Science and Technology Organisation, Victoria.
33. Wang, W., Jin, Z., Li, T., Zhang, H., and Gao, S. (2006) Preparation of spherical iron nanoclusters in ethanol–water solution for nitrate removal. *Chemosphere*, **65**, 1396–1404.
34. Liu, Y.-Q., Majetich, S.A., Tilton, R.D., Sholl, D.S., and Lowry, G.V. (2005) TCE dechlorination rates, pathways, and efficiency of nanoscale iron particles with different properties. *Environ. Sci. Technol.*, **39**, 1338–1345.
35. Noubactep, C., Meinrath, G., Dietrich, P., Sauter, M., and Merkel, B.J. (2005) Testing the suitability of zero valent iron materials for reactive walls. *Environ. Chem.*, **2**, 71–76.
36. Li, X. and Zhang, W. (2007) Sequestration of metal cations with zero valent Iron nanoparticles A study with high resolution X-ray photoelectron spectroscopy (HR-XPS). *J. Phys. Chem.*, **111**, 6939–6946.
37. Lien, H., Jhuo, Y., and Chen, L. (2007) Effect of heavy metals on dechlorination of carbon tetrachloride by iron nanoparticles. *Environ. Eng. Sci.*, **24**, 21–30.

38. O'Hara, Z., Krug, T., Quinn, J., Clausen, C., and Geiger, C. (2006) Field and laboratory evaluation of the treatment of DNAPL source zones using emulsified zero-valent Iron. *Remediation*, **16**, 35–56.
39. (2010) Nanoiron http://www.nanoiron.cz/en/contaminants (accessed 3 November 2010).
40. Elliott, D.W., Lien, H.L., and Zhang, W.X. (2009) Degradation of Lindane by zero-valent iron nanoparticles. *J. Environ. Eng. ASCE*, **135**, 317–324.
41. Ponder, S.M., Ford, J.R., Darab, J.G., and Mallouk, T.E. (1999) Ferragels: a new family of materials for remediation of aqueous metal ion solutions. *Mater. Res. Soc. Symp. Proc.*, **556**, 1269–1276.
42. Joo, S., Feitz, A.J., and Waite, T.D. (2004) Oxidative degradation of the carbothioate herbicide, molinate, using nanoscale zero-valent iron. *Environ. Sci. Technol.*, **38**, 2242–2247.
43. Tee, W.H., Bachas, L., and Bhattacharyya, D. (2009) Degradation of trichloroethylene and dichlorobiphenyls by iron-based bimetallic nanoparticles. *J. Phys. Chem., C Nanomater. Interface*, **113**, 9454–9464.
44. Zhang, W.X., Wang, C.B., and Lien, H.L. (1998) Treatment of chlorinated organic contaminants with nanoscale bimetallic particles. *Catal. Today*, **40**, 387–395.
45. Elliott, D.W. and Zhang, W.X. (2001) Field assessment of nanoscale bimetallic particles for groundwater treatment. *Environ. Sci. Technol.*, **35**, 4922–4927.
46. Xu, Y. and Zhang, W. (2000) Subcolloidal Fe/Ag particles for reductive dehalogenation of chlorinated benzenes. *Ind. Eng. Chem. Res.*, **39**, 2238–2244.
47. Kim, Y.H. and Carraway, E.R. (2000) Dechlorination of pentachlorophenol by zero valent iron and modified zero valent irons. *Environ. Sci. Technol.*, **34**, 2014–2017.
48. Tian, H., Li, J., Mu, Z., Li, L., and Hao, Z. (2009) Effect of pH on DDT degradation in aqueous solution using bimetallic Ni/Fe nanoparticles. *Sep. Purif. Technol.*, **66**, 84–89.
49. Schrick, B., Blough, J.L., Jones, A.D., and Mallouk, T.E. (2002) Hydrodechlorination of trichloroethylene to hydrocarbons using bimetallic nickel-iron nanoparticles. *Chem. Mater.*, **14**, 5140–5148.
50. Bokare, A.D., Chikate, R.C., Rode, C.V., and Paknikar, K.M. (2008) Iron-nickel bimetallic nanoparticles for reductive degradation of azo dye Orange G in aqueous solution. *Appl. Catal., B*, **79**, 270–278.
51. Chen, J.L., Al-Abed, S.R., Ryan, J.A., and Li, Z.B. (2001) Effects of pH on dechlorination of trichloroethylene by zero-valent iron. *J. Hazard. Mater.*, **83**, 243–254.

3
Photocatalysis at Nanostructured Titania for Sensing Applications

Shanqing Zhang and Huijun Zhao

3.1
Background

Since the discovery by Fujishima and Honda in 1972 that water could be decomposed into hydrogen and oxygen over a ultraviolet (UV)-illuminated titanium dioxide semiconductor electrode [1], the exploration of new applications of this material has been carried out enthusiastically by generations of research scientists. Because of the fascinating physicochemical properties of nanometer-sized materials and their potential applications, nanomaterials synthesis has attracted enormous attention in science community [2, 3]. Many types of TiO_2 nanomaterials have been synthesized and reported in the literature [4], including the synthesis of nanoparticles, nanotubes, nanofibers, nanowires, nanoclusters, nanorods, or nanocomposites. In the past two decades, significant progress has been achieved in environmental wastewater processing using the TiO_2 nanomaterials [5], owing to the enhanced mass transport between nanostructured TiO_2 and reactants (i.e., organic and microbial pollutants). Titanium dioxide photocatalysis has been researched as a possible alternative/complementary water treatment method, owing to its excellent photocatalytic activity, superior oxidation power, cheap production cost, being chemically stable throughout a wide pH range, and robustness against photocorrosion, and most importantly, being environmental friendly.

3.1.1
Photocatalysis at TiO_2 Nanomaterials

Owing to its powerful oxidation capability (+3.2 V) of photohole that is generated under UV illumination at the TiO_2 surface, the photohole is able to mineralize all types of organic compounds [6], microbial pollutants [7], and even whole cells of microorganism [8] in both contaminated water and air [9]. The photocatalytic oxidation process is initiated through the photogeneration of electron/hole pairs under UV light (Figure 3.1). Illumination on TiO_2 with photons, whose energy is equal to or greater than the bandgap energy (i.e., $E_g = 3.2$ eV), will lead to the promotion of an electron from the valence band (VB) to the conduction band (CB)

Functional Nanostructured Materials and Membranes for Water Treatment, First Edition.
Edited by Mikel Duke, Dongyuan Zhao, and Raphael Semiat.
© 2013 Wiley-VCH Verlag GmbH & Co. KGaA. Published 2013 by Wiley-VCH Verlag GmbH & Co. KGaA.

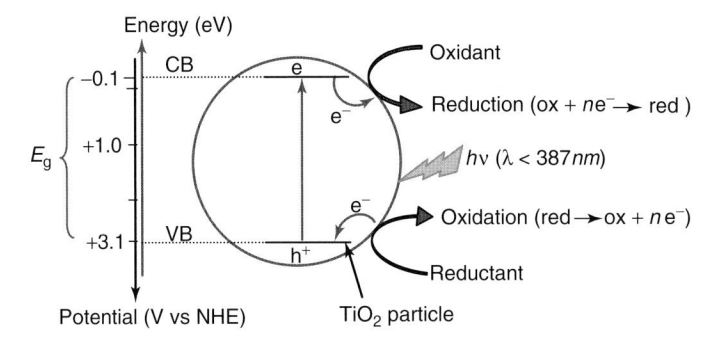

Figure 3.1 Schematic energy band diagram and the photocatalytic processes at a TiO_2 nanoparticle. CB, conduction band; VB, valence band; and NHE, normal hydrogen electrode.

(Equation 3.1), resulting in free photoelectrons in the CB (e_{cb}^-) and photoholes in the VB (h_{vb}^+)

$$TiO_2 + h\nu \longrightarrow h_{vb}^+ + e_{cb}^- \tag{3.1}$$

It is well established that the photohole is capable of capturing an electron from almost any species adsorbed to the solid semiconductor (R-H$_{(ads)}$ in Equation 3.2), or any species found in natural waters and wastewaters (R-H$_{(aq)}$ in Equation 3.3) [10] to an intermediate (R*). Subsequently, mineralization can be achieved (Equation 3.4).

$$h_{vb}^+ + \text{R-H}_{(ads)} \longrightarrow R^*_{(ads)} + H \tag{3.2}$$

$$h_{vb}^+ + \text{R-H}_{(aq)} \longrightarrow R^*_{(aq)} + H \tag{3.3}$$

$$h_{vb}^+ + \text{R-H} \longrightarrow CO_2 + H_2O \tag{3.4}$$

In this pathway, oxidation of organic compounds can be achieved directly on TiO_2 surface [11].

In the other pathway, water is also commonly oxidized during this process and generates hydroxyl radicals efficiently because of the high concentration of water (Equation 3.5).

$$h_{vb}^+ + H_2O \longrightarrow OH\cdot + H^+ \tag{3.5}$$

The hydroxyl radical is a well-known strong oxidant that can degrade a wide range of organic compounds. Overall, via these two pathways, TiO_2 photocatalysis has been found to be highly effective for mineralization of essentially all organic compounds [12, 13].

Traditionally, the photocatalytic degradation of organics is commonly carried out using a slurry system (i.e., TiO_2 colloidal particle suspension), which has the advantage of enhanced mass transport. The overall reaction can be represented as [14]

$$C_xH_yO_zX_q + \{x + (y - q - 2z)/4\} O_2 \longrightarrow xCO_2 + xH^+ + qX^- + \{(y - x)/2\} H_2O \tag{3.6}$$

where X represents a halogen element. The oxidation number (n) in the complete oxidation process is equal to $4x + y - q - 2z$.

However, there are two major drawbacks of the TiO_2 slurry photocatalytic system. The first one is the costly posttreatments that aim to separate and recycle the TiO_2 catalysts from the reaction media. This has been recognized as an inherent technical barrier that prohibits the practical application of the slurry reaction system [15]. In order to overcome this barrier, the TiO_2 has been immobilized on many kinds of substrates, such as titanium substrates [16], polymer substrates (e.g., polystyrene [17] polyethylene terephthalate [18]), stainless steel [19, 20], glazed ceramic tiles [21], and silicon substrate. In order to obtain a photocatalytic reactor for analytical application, the substrates are commonly required to have the following characteristics: (i) mechanically strong and chemically inert, (ii) UV transparency to allow UV illumination, (iii) small dead volume to achieve reasonable analytical resolution, and (iv) long lifetime and low cost.

The second drawback is the insufficient photocatalytic degradation efficiency at the TiO_2 surface, which depends on the degree of recombination of photoelectrons and holes [22, 23]. The electron–hole recombination reaction (Equation 3.7) is the back reaction of the electron–hole dissociation reaction (Equation 3.1).

$$h_{vb}^+ + e_{cb}^- \longrightarrow \text{heat} \tag{3.7}$$

The recombination reaction is a very rapid reaction. In order to achieve chemically useful photocatalysis (i.e., to facilitate the above oxidation reaction (Equation 3.6) to occur) recombination of the electron–hole pair (Equation 3.7) must be suppressed. This can be achieved by removing the photoelectron at the CB (i.e., e_{cb}^- generated from the TiO_2 surface). Dissolved oxygen in sample solution is a typical electron acceptor that reacts with the electrons, and forms hydrogen peroxide or water (Equations 3.8 and 3.9) [6].

$$2e_{cb}^- + 2H^+ + O_2 \longrightarrow H_2O_2 \tag{3.8}$$

or

$$4e_{cb}^- + O_2 + 4H^+ \longrightarrow 2H_2O \tag{3.9}$$

Unfortunately, this process is often a rate-limiting step in the overall photocatalytic process as the photoelectron in TiO_2 is a relatively weak reducing agent, and the solubility of oxygen in aqueous solution is commonly low (<10 ppm at 25 °C, 1 atm) [23]. As a result of the recombination reaction (Equation 3.7), the quantum yields for this process in aqueous suspensions of TiO_2 particles are low, typically 0.1 [24]. It is for this reason that O_2 sparging is a common method employed to accelerate the photocatalytic oxidation reaction in the slurry reaction system [6]. However, in sensing process, this will distort the analytical signal.

The immobilization of TiO_2 onto a conducting substrate and the application of potential bias on TiO_2 can be used to tackle the aforementioned problems of the traditional photocatalysis systems.

3.1.2
Photoelectrocatalysis at TiO$_2$ Nanomaterials

Application of electric field is an effective tool to separate photoholes and photoelectrons, which improves the photocatalytic efficiency [10, 25]. This apparently requires the immobilization of TiO$_2$ nanoparticles on a conducting substrate, or the assembly of nanostructured TiO$_2$ electrodes. Traditionally, commercially available TiO$_2$ nanoparticles (e.g., P25) or TiO$_2$ sol–gel can be immobilized on supporting substrates using the dip-coating technique [22, 23, 26–29].

In a TiO$_2$ photoelectrochemical system, a TiO$_2$ film electrode is used as the working electrode, namely, the photoanode. By applying an appropriate potential bias (i.e., electric field) to the working electrode (Figure 3.2), it becomes more favorable for the photoelectron to be transferred to the external circuit, rather than to the adsorbed O$_2$ [23]. The photoelectrons are subsequently forced to pass into the external circuit and deliver to the auxiliary electrode, where the reduction reaction takes place (Figure 3.3). Surface recombination reactions (such as Equation 3.7) are less likely in this configuration, as the oxidation and reduction reactions are physically separated. The photocurrent (or charge) can be monitored and it gives a direct measure of the rate of photooxidation reaction and extent of oxidative degradation [10], which can be used as an analytical signal [10].

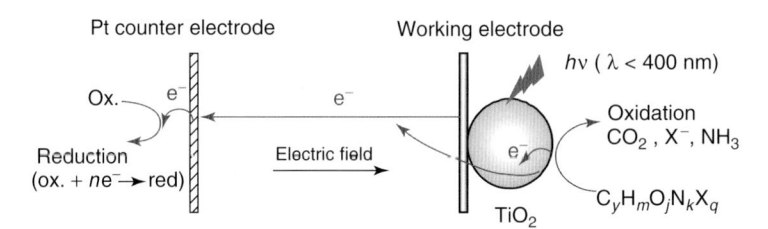

Figure 3.2 A schematic diagram of the photoelectrocatalytic processes at a TiO$_2$ electrode and simultaneous reduction at a Pt counter electrode.

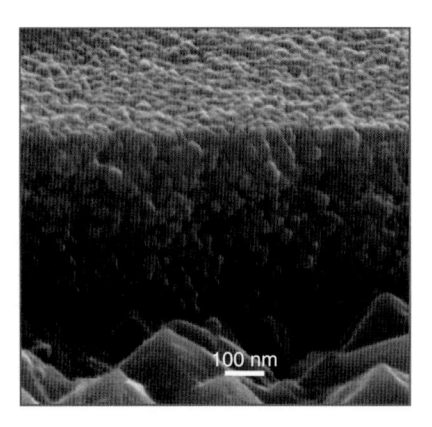

Figure 3.3 The cross-section image of TiO$_2$ film immobilized on BDD surface [56].

A further advantage of the proposed approach is that the rate of degradation is independent of O_2 concentration as the rate of reduction at the auxiliary electrode will never be the rate-limiting step of the overall degradation process [23]. This is in strong contrast with the traditional slurry system where the supply of high concentration of O_2 is essential for rapid photocatalytic degradation.

The general equation for mineralization at the TiO_2 photoanode can be summarized as follows [23]:

$$C_yH_mO_jN_kX_q + (2y-j)H_2O \longrightarrow yCO_2 + qX^- + kNH_3 \\ + (4y - 2j + m - 3k)H^+ + (4y - 2j + m - 3k - q)e^- \quad (3.10)$$

where the elements present are represented by their atomic symbols and X represents a halogen atom, respectively. The stoichiometric ratio of elements in the organic compound is represented by the coefficients y, m, j, k, and q. The oxidation number (n) in the complete oxidation process is equal to $4y - 2j + m - 3k - q$. The mineralization reaction has been extensively used to determine the chemical oxygen demand (COD) [11, 22, 23, 29–31].

3.2 Fabrication of TiO_2 Photoanodes

3.2.1 Common Fabrication Techniques and Substrates for Photoanodes

Various techniques have been employed to fabricate the TiO_2 photoanodes, including sol–gel dip coating [22], spin coating [32, 33], sputtering [34, 35], spray pyrolysis [36, 37], atomic layer deposition [38, 39], chemical vapor deposition [40, 41], electrodeposition [42], and anodic oxidation [30, 31, 43]. Each of these techniques has advantages and disadvantages.

The benefits derived from preparing TiO_2 by sol–gel, including synthesis of nanosize crystallized powder of high purity at a relatively low temperature, possibility of stoichiometry controlling process, preparation of composite materials, and production of homogeneous materials have driven many researchers to use the method in preparing TiO_2-based photocatalysts [44]. The sol–gel dip-coating method is probably the most common method for TiO_2 immobilization because it has the advantages of easy composition control, low equipment cost, and simplicity in processing. Moreover, TiO_2 sol–gel can be used by other coating techniques such as inkjet printing technique [45, 46] to achieve production automation and lower cost.

Sputtering deposition can produce films with high adhesion and hardness but relatively low photocatalytic activity and is more expensive [34]. The chemical vapor deposition technique is capable of creating films on flexible substrates, including the inner surface of pipes, and the resultant films possess good adhesion but relatively low photocatalytic activity [40].

As a sensor, the analytical performance of the TiO_2 photoanodes can be improved by using new generation substrates (such as boron-doped diamond (BDD)), manipulation of TiO_2 crystalline phase composition, and forming CNTs/TiO_2 nanocomposite.

3.2.2
TiO_2/BDD Photoanode

The BDD is one of the most promising advanced electrode materials in the field of electroanalysis with special characteristics such as a very low background current, wide working potential window, resilient mechanical strength, robust resistance against corrosion (even being anodic polarized in acidic solutions), and long-term durability and stability [47–50]. The boron doping makes this p-type semiconductor electrically as conductive as common conductors at room temperature [51]. In conjunction with n-type semiconductors, such as TiO_2, BDD can be a highly desired electrode substrate for the fabrication of a new generation of sensing device [52]. In addition, TiO_2 nanoparticle is an n-type semiconductor. Thus, the incorporation of the n-type TiO_2 with the p-type BDD could lead to the formation of a p–n heterojunction, which can act as an internal electrostatic potential in the space charge region to facilitate efficient separation of the photoinduced electrons and holes [53]. This is an added advantage of using BDD as the conducting substrate over other materials, such as Indium tin oxide (ITO) and metal (e.g., Pt or Ti). Recently, various pure-anatase TiO_2/BDD electrodes were fabricated and used to photoelectrocatalytically degrade organic compounds such as ethanol [53], acid orange II and 2,4-dichlorophenol [54], and reactive yellow and hexavalent chromium [52].

TiO_2 nanoparticles were immobilized onto the BDD electrodes by a dip-coating technique after the BDD was specially treated [55, 56]. Figure 3.3 shows that continuous and uniform mixed-phase (anatase and rutile) and pure-anatase TiO_2/BDD electrodes were obtained after calcination processes at 700 and 450 °C, respectively. The particle sizes of both types of TiO_2 film range from 20 to 30 nm. In comparison with TiO_2/ITO electrode, the TiO_2/BDD electrode demonstrates a higher photoelectrocatalytic activity toward the oxidation of organic compounds, such as glucose and potassium hydrogen phthalate (KHP). Among all the tested TiO_2 electrodes, the mixed-phase TiO_2/BDD electrode demonstrated the highest photoelectrocatalytic activity, which attributes to the formation of the p–n heterojunction between TiO_2 and BDD. The electrode was subsequently used to detect a wide spectrum of organic compounds in aqueous solution using a steady-state current method. Excellent linear relationship between the steady-state photocurrents and equivalent organic concentrations was attained. The steady-state oxidation photocurrents of the mixed-phase TiO_2/BDD electrode were insensitive to pH in the range of pH 2–10. Furthermore, the electrodes exhibited excellent robustness under strong acidic conditions that the TiO_2/ITO electrodes cannot stand. These characteristics bestow the mixed-phase TiO_2/BDD electrode to be versatile material for sensing of organic compounds.

3.2.3
TiO$_2$ Mixed-Phase Photoanode

Titanium dioxide occurs in nature as well-known minerals, namely *rutile, anatase,* and *brookite*. The most common form is rutile, which is also the most stable form. Both anatase and brookite can be converted to rutile on heating [57]. Brookite occurs rarely compared to anatase and rutile. Manipulation of the phase composition of TiO$_2$ (i.e., the ratio of anatase/rutile) is an effective way to facilitate the separation of photoelectron–hole pair using the synergistic effect of anatase and rutile [57].

The thermal treatment of immobilized TiO$_2$ films can be adopted to manipulate the physical properties of the films (such as the crystal structure, the porosity, and the microstructures). Calcination conditions have an important effect on the physiochemical properties, crystalline phase formation, mechanical stability, and photocatalytic activity of the films [58, 59]. The calcination process also serves other purposes, including burning away organic pore-creating agents, obtaining better electric connection among TiO$_2$ particles, and between TiO$_2$ particles and ITO conductive glass [9].

When TiO$_2$ nanostructured electrodes are calcined at 700 °C or above, various portion of the anatase phase TiO$_2$ can be converted into rutile phase, TiO$_2$, depending on the calcination time. This leads to the formation of mixed-phase TiO$_2$ and results in the synergetic effect because of the different energy bandgap (Figure 3.4). Because energy bandgap of anatase phase (3.2 eV) is higher than that of rutile phase (3.0 eV), the photoelectrons preferentially move to rutile side at a rutile and anatase TiO$_2$ junction [60]. As a result, the electrons and holes are spatially separated. This spatial separation of electrons and holes will increase the lifetime of photohole, which suppresses the recombination of photohole and photoelectron [61, 62]. Different organic compounds, despite their difference in chemical entities, can be stoichiometrically mineralized at the mixed-phase TiO$_2$ electrode under diffusion-controlled conditions [57], which was achieved at the

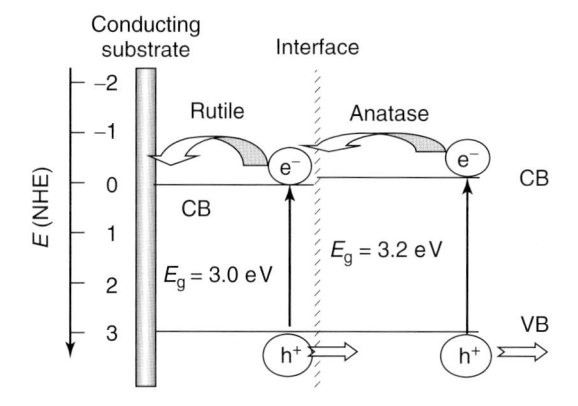

Figure 3.4 The schematic energy diagram of rutile form in contact with anatase form TiO$_2$ and the photoelectron pathway at pH 7.0 in aqueous solution.

pure anatase phase TiO_2 electrode. The exceptional ability of the mixed-phase TiO_2 electrodes for mineralization of organic compounds and their remarkable resistance to the inhibition by aromatic compounds at higher concentration was ascribed to the synergetic effect of the rutile and anatase phases. This electrode can be used as a sensing probe for organic compounds [28, 46].

3.2.4
CNTs/TiO_2 Composite Photoanode

Another advantage of the sol–gel is that they can readily be mixed with other types of materials such as carbon nanotubes (CNTs) for modifying the characteristics of the sol–gel.

CNTs have been considered as one of the most popular functional materials in recent years. Taking advantage of the large surface area, extraordinary electrical conductivity, robust mechanical strength, and thermal stability of the CNTs, the CNTs/TiO_2 composite materials have been employed for sensing devices [63–67]. Llobet et al. [64] reported that the CNTs/TiO_2 hybrid film sensor was significantly more sensitive to oxygen than to pure TiO_2 sensor. Moreover, the hybrid sensors could also be operated at a lower operating temperature (350 °C) than the pure TiO_2 sensors (500 °C). Ueda et al. [67] developed the CNTs/TiO_2 hybrid gas sensor toward NO gas, which could be operated at room temperature, and showed a good sensitivity. The CNTs/TiO_2 composite materials have also been used in the NH_3 sensing [65], H_2O_2 sensing [63], and even biosensing for cancer cells [66]. In the sensing system and the photocatalytic degradation of pollutants in water treatment, the enhancements observed in the CNTs composite systems in comparison with the nonhybrid ones can be attributed to several possible merits contributed from CNTs: (i) higher adsorption capacity to reactant species because of the high surface area nature of the CNTs [68, 69]; (ii) photo-excitation under a longer wavelength light [68, 69]; and (iii) the relatively low recombination of photogenerated electron–hole pairs facilitated by the extraordinary electrical conductivity of the CNTs [70]. CNTs/TiO_2 composite photoanodes were fabricated by a dip-coating technique, followed by subsequent calcination for sensing application [71]. The carbon nanotubes were successfully incorporated with the TiO_2 nanoparticulates without damage and that the resultant TiO_2 nanoparticles consisted of anatase and rutile. The CNTs/TiO_2 photoanodes were capable of oxidizing various types of organic compounds (e.g., glucose, KHP, and phenol) in aqueous solutions in a photoelectrochemical bulk cell. In comparison with the pure TiO_2 photoanode, the sensitivity of the photoanode for the detection of organic compounds has been improved by 64% in comparison with the pure TiO_2 photoanode, while the background current was reduced by 80% because of the introduction of the CNTs (Figure 3.5). These advantages can be ascribed to the improved adsorptivity to organic compounds, increased absorption of UV light, and enhanced electron transport at the CNTs/TiO_2 photoanode because of the introduction of the CNTs [71].

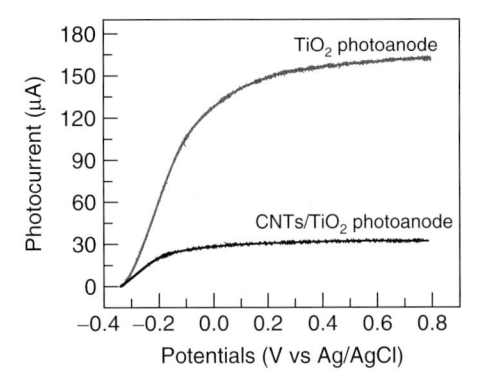

Figure 3.5 CNTs/TiO$_2$ photoanode shows a much lower background current than pure TiO$_2$ photoanode in 0.1 M NaNO$_3$ electrolyte under UV illumination. Ag/AgCl is used as a reference electrode [71].

3.3
The Sensing Application of TiO$_2$ Photocatalysis

It is practically very hard to determine and/or control the extent of the photocatalytic oxidation reaction (Equation 3.6) and the photoelectrocatalytic oxidation reaction (Equation 3.10). For example, some organic compounds such as hydroxyl organic compounds (such as alcohols and glucose) can be easily oxidized and mineralized in a short time, while some organic compounds such as aromatic compounds and large molecules are oxidized with more difficulty and demand longer reaction time, depending on the photocatalytic activity of the TiO$_2$ catalyst. Furthermore, various intermediates may form and their polymerization may take place during the degradation reaction [72]. In contrast, it is a much easier task to assure that all types of organic compounds are completely degraded by prolonging reaction time, providing sufficient UV light intensity, supplying sufficient amount of TiO$_2$ photocatalyst, increasing the volume to surface ratio of the photoelectrochemical cell (such as thin-layer cell), and injecting less amount of organic compounds into the reaction system. This process can be analogous to "wet burning." This limits the application of this process in determining the individual identities of the organic compounds independently. This process, however, has great potential in determining the amount of aggregative organic compounds, including total organic compounds (TOCs), COD, and volatile organic compounds (VOCs), and universal detection of individual organic compound in combination with other analytical separation tools, such as high-performance liquid chromatography (HPLC).

TOC and COD reflect the aggregative content of organic compounds in all types of water bodies. They have been recognized as the most important parameters for water quality for drinking water, river water, lake, and wastewaters, whereas VOCs refer to organic chemical compounds that have significant vapor pressures, which can affect the environment and human health [73].

3.3.1
Photocatalytic Determination of TOC

As aforementioned, the carbon in the organic matters can be completely oxidized and converted into CO_2 by the UV illuminated titania nanoparticulates (Equation 3.6). TiO_2 catalyst can be used as a reagent without the need of being recycled in analytical process because of low TiO_2 consumption in analytical process, cheap availability, and nontoxic nature of TiO_2. In order to make use of the photocatalytic oxidation reaction, a suspension (e.g., $1\,g\,l^{-1}$) of titanium dioxide (Degussa P25) can be used to catalyze the oxidation process for the measurement of TOC under UV light. The CO_2 gas produced in the mineralization process was detected by a CO_2 gas permeable membrane electrode (i.e., CO_2 detector) [74]. This method has been used for online measurement of 11 organic compounds. SGE International Pty. Ltd (www.anatoc.com) has made use of this principle and developed a series of TOC analyzers. In the analyses, the samples are mixed with a certain amount of TiO_2 photocatalyst and the produced CO_2 are analyzed using a dual wavelength nondispersive infrared (NDIR) detector. These methods provide reliable analysis at negligible cost per sample without the use of hazardous chemicals or clumsy, expensive, compressed gases in the traditional combustion based analytical methods.

On the basis of the principle of Equation 3.6, the TOC can be determined using a TiO_2 immobilized coil reactor coupled with a conductivity meter (Figure 3.6) to

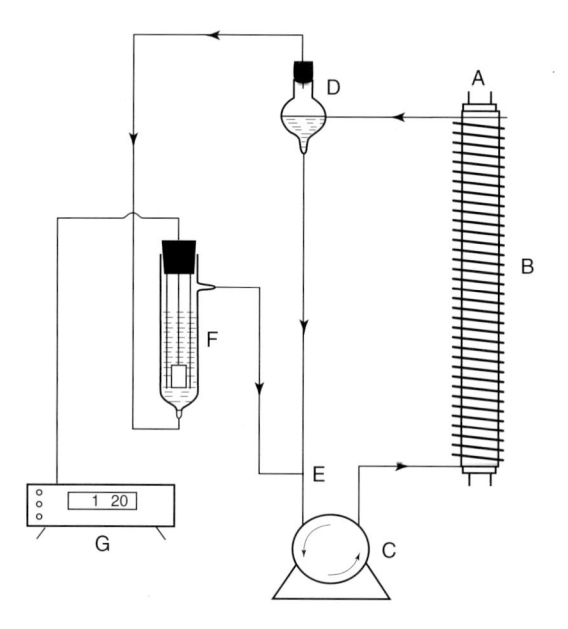

Figure 3.6 Schematic diagram of the apparatus proposed by Matthews et al. [75]. A = UV light; B = spiral photocatalytic reactor; C = peristaltic pump; D = loading port; E = Tee junction; F = CO_2 sensor (conductivity cell); and G = conductivity meter.

analyze the produced CO_2 and subsequently the concentration value of TOC [75]. This method was suitable for analysis of water containing 0.1–30 μg ml^{-1} organic carbon and sample volumes of 1–40 ml. Its main advantages were low cost, no need for TiO_2 addition, and ease of operation. Such analytical method has also been commercialized by Analyticon Instruments Corporation (www.analyticon.com).

3.3.2
Photocatalytic Determination of COD

In the photocatalytic mineralization (Equation 3.6), O_2 plays an essential role in the conversion of CO_2 through formation. Oxygen acts as an acceptor of electrons during the mineralization reaction. As a reactant, O_2 consumed in the reaction corresponds to the consumption of organic compounds, and therefore, the O_2 consumed is proportional to the COD concentrations.

Karube's group at the University of Tokyo (Japan) constructed a disposable oxygen sensor to monitor the change of oxygen concentration in TiO_2 nanoparticles suspended sample solution for COD determination [76]. Under UV irradiation, a decrease in the dissolved oxygen concentration in the sample was well correlated to COD concentration of the sample. The sensing performance was evaluated using artificial and real water samples from lakes in Japan. The response time of the sensor was ∼3 ± 4 min with a detection limit of 0.118 ppm.

Using the same analytical principle, a photocatalytic sensor was developed by combining TiO_2 beads in the photochemical column and a single oxygen electrode [77] (Figure 3.7) or dual oxygen electrodes [78] as the sensing part in flow injection analysis (FIA). These methods provide a good correlation with the standard COD values; however, they had a very narrow linear range (less than 12 ppm COD).

Karube's group, the University of Tokyo (Japan), advanced this concept by the construction of a COD probe that is made by attaching a TiO_2 fine particles

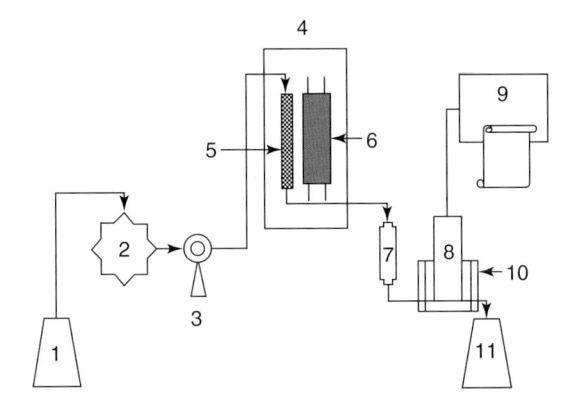

Figure 3.7 Schematic diagram of the apparatus for determination of COD [77]: (1) deionized water supply, (2) pump, (3) injector, (4) reflector, (5) photochemical column consisting of TiO_2 beads, (6) UV lamp (λmax 365 nm), (7) air damper, (8) oxygen electrode, (9) integrator, (10) thermostatic water bath, and (11) waste tank.

adsorbed Polytetrafluoroethylene (PTFE) membrane on the tip of an oxygen probe [14]. The operation characteristics of the sensor are demonstrated using artificial wastewater and real water samples from lakes in Japan. This method has a slightly increased linear range of about 20 ppm and is considered to be reliable in that the observed parameter is close to the theoretical COD value. The sensor also showed a long-term stability (relative standard deviation (RSD): 5.7%, 30 days). Simple instrumentation with no sample pretreatment and the short response time (3–5 min) is another aspect of this sensor system.

Narrow linear range is the major drawback of the aforementioned COD methods based on the photocatalytic oxidation reaction (Equation 3.6). This is mainly due to the low oxidation percentage and inadequate sensitivity of the oxygen probe. The problem of low oxidation percentage is typically stemmed from the insufficient photocatalytic activity, light intensity, and low O_2 solubility in the water solution.

The low O_2 solubility problem was addressed by replacing the electron acceptor, that is, O_2 with other more oxidative electron acceptors, such as dichromate ions ($Cr_2O_7^{2-}$) [79], Cerium ions (Ce^{4+}) [80], and permanganate (MnO_4^-) [81, 82]. Jin's group at East China Normal University (China) developed a TiO_2–Cr(VI) system in which TiO_2 was deposited on a quartz support as oxidizing photocatalyst for monitoring COD [79]. In the detection, $K_2Cr_2O_7$ could stoichiometrically accept the photoelectrons (Equations 3.11 and 3.12) at the CB and enhance the ability of photocatalytic degradation of organic compounds.

$$2Cr_2O_7^{2-} + 28H^+ + 12e^- \longrightarrow 4Cr^{3+} + 14H_2O \quad (3.11)$$

or

$$6H^+ + 2CrO_4^{2-} + 6e^- \longrightarrow 2Cr^{3+} + 8H_2O \quad (3.12)$$

Therefore, the COD value of a given sample could be assessed by tracing the change of $Cr_2O_7^{2-}$ concentration. The optimized operation conditions were studied by using glucose as a standard substance. The application range was 20–500 mg l^{-1}, and the detection limit was 20 mg l^{-1}.

Alternatively, the COD concentration can be calculated by monitoring the change in Cr^{3+} concentration produced by the photocatalytic reduction using colorimetric method [83]. Under the optimized experiment conditions, the detection range was 20–500 mg l^{-1}, and the detection limit was 20 mg l^{-1}. These proposed COD measurement methods had many advantages, such as mild operation conditions, short analysis time, and no requirement of expensive reagents.

Ce^{4+} can be the electron acceptor for the photocatalytic oxidation reaction (Equation 3.13). Nano-TiO_2–$Ce(SO_4)_2$ photocatalytic system [80] was proposed and used for determination of COD values of water samples. Accordingly, the COD measurement is based on direct determination of the concentration change of Ce^{4+} using UV spectrophotometric detector during photocatalytic oxidation of organic compounds.

$$Ce^{4+} + e^- \longrightarrow Ce^{3+} \quad (3.13)$$

Accepting the photogenerated electrons, Ce^{4+} can reduce the recombination of the photogenerated electrons and holes and thus accelerate the reaction rate of the photocatalytic degradation reaction. During the course of detection, COD value was found to be proportional to the decreased Ce^{4+} concentration. Under the optimum conditions, a good calibration curve for COD values between 1.0 and $12\,mg\,l^{-1}$ was obtained and the limit of detection achieved was as low as $0.4\,mg\,l^{-1}$. When determining the real samples, the results were in good agreement with those from the conventional methods. Alternatively, the produced Ce^{3+} can be detected by fluorescence method, using the maximum excitation and emission wavelengths at 264.8 and 362.1 nm, respectively. Subsequently, the Ce^{4+} concentration change and the COD values were determined [84]. It was found that for TiO_2 dosage, initial solution pH, initial Ce^{4+} concentration, the time of UV light irradiation, and the temperature of solution affected the degradation reaction. The fluorescence intensity change was linear with COD concentration in the range of $0–100.0\,mg\,l^{-1}$. The limit of detection was $0.9\,mg\,l^{-1}$. This method was applied in detection of COD in samples of lake water and artificial sewage.

Zhu and his group at Shanghai Jingtong University (China) proposed to use a fluorinated-TiO_2–$KMnO_4$ system (Figure 3.8) for the determination of COD [85]. In this system, MnO_4^- can capture the photoelectrons at the CB effectively (Equation 3.14). Combining with the surface fluorination of TiO_2, the system can enhance the photocatalytic process and lower the detection limit, in comparison with the TiO_2–O_2 system. In the detection, a linear correlation is observed between the amount of dissolved organic matter and the amount of MnO_4^- consumed by the coupled reduction process.

$$MnO_4^- + 8H^+ + 5e^- \longrightarrow Mn^{2+} + 4H_2O \qquad (3.14)$$

Thus, the COD determination is transformed to a simple and direct determination of the deletion of MnO_4^-. This makes the method in being rapid, environment friendly, and easy for operation. Under optimized conditions, this method can

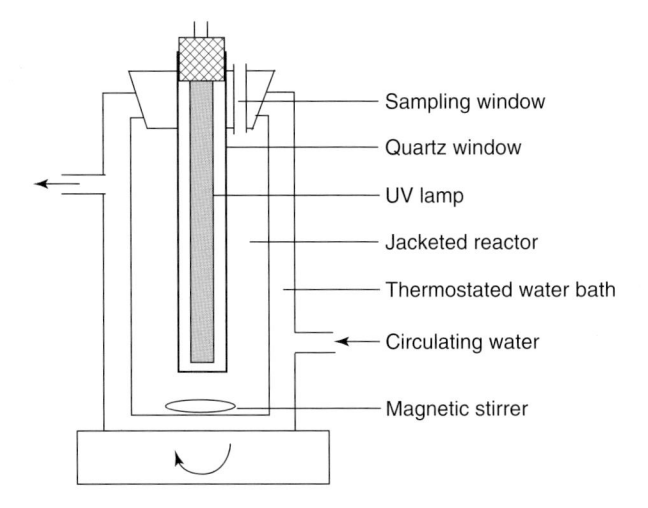

Figure 3.8 Schematic diagram of the fluorinated-TiO_2–$KMnO_4$ photocatalytic reactor [85].

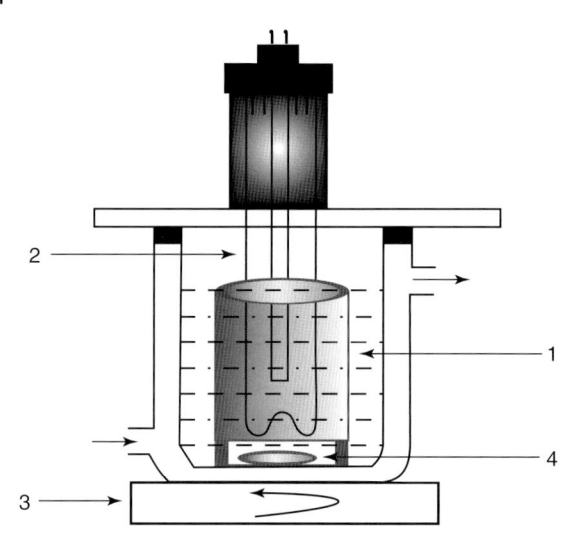

Figure 3.9 Schematic diagram of the nano-ZnO/TiO$_2$ photocatalytic reactor [82]. (1) quartz tube with nano-ZnO/TiO$_2$ film, (2) UV lamp, (3) magnetic stirrer, and (4) stirrer.

respond linearly to COD of KHP in the range of 0.1–280 mg l^{-1}, with a detection limit of 0.02 mg l^{-1} COD.

A nano-ZnO/TiO$_2$ composite film as photocatalyst was fabricated with vacuum vaporized and sol–gel methods by Jin's group at East China Normal University [82]. In the proposed system (Figure 3.9), the nano-ZnO/TiO$_2$ film could improve the separation efficiency of the charge and extend the range of light absorption spectrum; therefore, a higher photocatalytic activity was achieved, when compared with the pure nano-TiO$_2$ and nano-ZnO film. Photometric method was adopted for the determination of the Mn(VII) concentration decrease (instead of the decrease of O$_2$), which resulted from photocatalytic oxidation of organic compounds on the nano-ZnO/TiO$_2$ film. The COD values were calculated from the corresponding decrease of Mn(VII). Under the optimal operation conditions, the detection limit of 0.1 mg l^{-1}, COD values with the linear range of 0.3–10.0 mg l^{-1} were achieved. The results were in good agreement with those from the conventional COD methods.

3.4
The Sensing Application of TiO$_2$ Photoelectrocatalysis

3.4.1
Probe-Type TiO$_2$ Photoanode for Determination of COD

In order to detect the COD value of a water sample, the probe must be able to oxidize indiscriminately the whole spectrum of organic compounds in the water sample. So far, only nanostructured anatase–rutile mixed-phase photoanodes are reported to meet such a challenging requirement. With these photoanodes, the

3.4 The Sensing Application of TiO$_2$ Photoelectrocatalysis

determination of COD does not require exhaustive oxidation of all the organic compounds in the water sample.

Besides the requirement of the photoanodes, the analytical principle of the probe-type COD sensor based on the nonexhaustive photocatalytic oxidation model is developed according to the following postulates: (i) the bulk solution concentration remains essentially constant before and after the detection; (ii) all organic compounds at the photoanode surface are stoichiometrically oxidized [57, 86]; (iii) the overall photocatalytic oxidation rate is controlled by the transport of organics to the photoanode surface and can reach a steady state within a reasonable time frame (steady-state mass transfer limited process); and (iv) the applied potential bias is sufficient to draw all photoelectrons generated from the photocatalytic oxidation of organics into external circuit (100% photoelectron collection efficiency).

A photoelectrochemical probe for rapid determination of COD (i.e., PeCOD probe) was developed using a nanostructured mixed-phase TiO$_2$ photoanode (Figure 3.10). A UV-LED light source and a USB microelectrochemical station (Figure 3.11) were powered and controlled by a laptop computer, making the probe portable for on-site COD analyses. When the photoelectrocatalytic degradation is carried on at a nanoparticulate TiO$_2$ photoanode in a bulk cell, the typical photocurrent profiles (Figure 3.11) can be collected. Figure 3.11 displays a set of photocurrent–time profiles obtained in the presence and absence of organic compounds at the TiO$_2$ photoanode in the bulk cell under UV illumination. Under a constant applied potential sufficient to draw all the electrons into the external circuit, the current was around zero when the UV light was switched off. Under illumination, the current increased rapidly and then decreased to a steady value. For the blank sample without any organic compounds, the steady photocurrent (i_{blank}) was generated from the oxidation of water, whereas the total photocurrent of steady state (i_{total}) for a sample containing organics resulted from the oxidation of

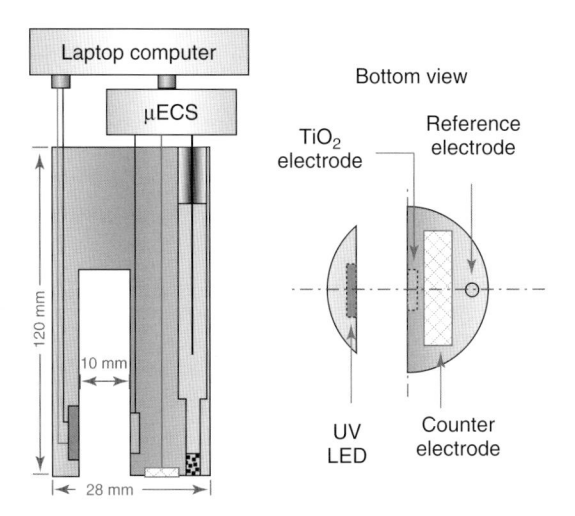

Figure 3.10 Schematic diagram of the photoelectrochemical setup and PeCOD probe.

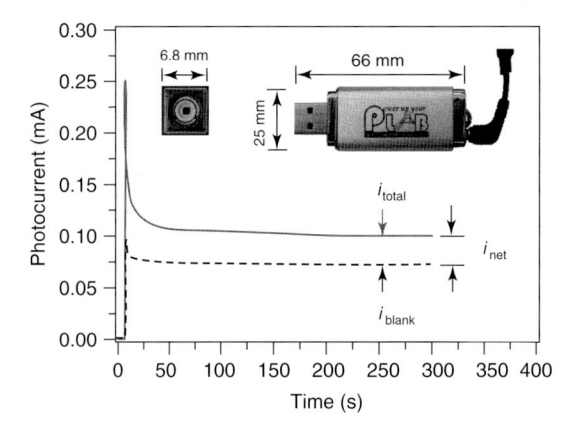

Figure 3.11 Typical photocurrent response of a 0.1 M NaNO$_3$ blank solution (i_{blank}, dashed line) and a 0.1 M NaNO$_3$ solution containing organic compounds (i_{total}, solid line). i_{net} was the difference of the two the steady-state currents. The insert shows the images and dimensions of the UV-LED (left) and the μECS (microelectrochemical system) (right) [28].

water and organic compounds. Therefore, the net steady-state photocurrent (i_{net}), generated from the oxidation of organics, can be obtained by subtracting the blank photocurrent (i_{blank}) from the total photocurrent (i_{total}) (Equation 3.15).

$$i_{net} = i_{total} - i_{blank} \tag{3.15}$$

Under diffusion-controlled conditions, the net steady-state photocurrent (i_{net}) is given by Equation 3.16 according to the semiempirical treatment of steady-state mass transfer method [57, 87].

$$i_{net} = \frac{nFAD}{\delta} C_b \tag{3.16}$$

where n is the number of electrons transferred for the complete mineralization of organic compounds, F is the Faraday constant, A is the surface area of the photoanode, and D and δ are the diffusion coefficients of the compound and the thickness of the effective diffusion layer, respectively, and C_b is the bulk concentration of the compound. For a specific organic compound, the i_{net} should increase linearly with its molar concentration (C_b) according to Equation 3.16, which has also been demonstrated in low-molar concentration ranges in the literature [57].

The linear relationship between the net steady-state photocurrent and concentration suggests that the photoelectrochemical bulk cell with TiO$_2$ photoanode has a promising potential as a sensing application for organic compounds.

Equation 3.16 defines the quantitative relationship between the net steady-state photocurrent and the molar concentration of an individual analyte. Convert the molar concentration into the equivalent COD concentration (mg l^{-1} of O$_2$), using the oxidation number n and the definition of COD [73], it can be converted into

$$i_{net} = \frac{FAD}{\delta} \times \frac{1}{8000} [COD] \tag{3.17}$$

Equation 3.17 is valid for determination of COD in a sample that contains a single organic compound. The COD of a sample contains more than one organic species which can be represented as

$$i_{net} \approx \frac{FA\overline{D}}{\overline{\delta}} \times \frac{1}{8000}[COD] \quad (3.18)$$

where $\overline{D}$ and $\overline{\delta}$ can be considered as an average diffusion coefficient and average diffusion layer thickness, respectively. $\overline{D}$ and $\overline{\delta}$ are dependent on the sample composition. If they are relatively constant, it can be concluded that i_{net} is directly proportional to COD concentration. That is

$$i_{net} \propto [COD] \quad (3.19)$$

Using the calibration method with sample with known COD concentration, the COD concentration of unknown sample can be determined without the need to find out the actual value of $\overline{D}$ and $\overline{\delta}$.

Figure 3.12 shows that the net steady-state photocurrent, i_{net}, is directly proportional to the COD value of glucose/glutamic acid (GGA) and the mixture made of six organic compounds with equal molar ratio. This demonstrates that the proposed analytical principle (Equation 3.19) can be used to determine the COD values of the individual organic compounds as well as their mixtures. The result also suggests that the effect of $\overline{D}/\overline{\delta}$ values on i_{net} is not significant.

The photoelectrochemical measurement of COD was optimized in terms of light intensity, applied bias, and pH. Under the optimized conditions, the net steady-state photocurrents originated from the oxidation of organic compounds were directly proportional to COD concentrations. A practical detection limit of 0.2 mg l^{-1} of O_2 and a linear range of 0–120 mg l^{-1} of O_2 were achieved. The analytical method using the portable PeCOD probe has the advantages of being rapid, low cost, robust, user friendly, and environmental friendly. It was successfully applied to

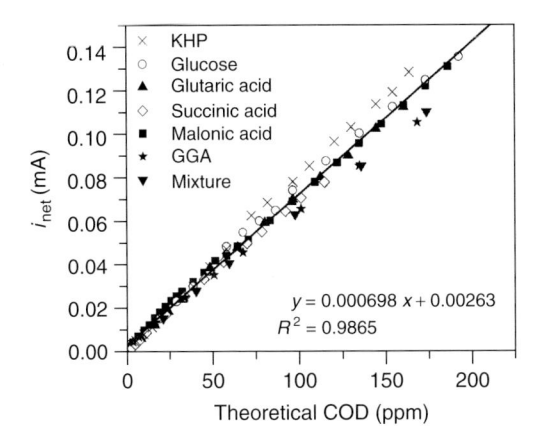

Figure 3.12 Validation of analytical principle: the plot of i_{net} against the theoretical COD values of individual organic compounds, GGA and their mixtures with equal COD concentration of each [57].

determine the COD values of the synthetic samples and real samples from various industries. Excellent agreement between the proposed method and the standard dichromate method was achieved.

3.4.2
Exhaustive Degradation Mode for Determination of COD

Zhao's research group at Griffith University (Australia) was the pioneer who designed and developed a photoelectrochemical thin-layer cell incorporating TiO_2 photoanode for photoelectrochemical determination of COD, namely, PeCOD technology [22, 23]. The thin-layer photoelectrochemical cell (Figure 3.13) can be used in two modes, stop flow for exhaustive oxidation mode and continuous flow for partial oxidation mode.

In the exhaustive model [22, 23], the photoelectrocatalytic reaction at a mixed-phase TiO_2 photoanode is carried out in a thin-layer cell following Equation 3.10. The final photoelectrocatalytic oxidation product of the nitrogen-containing in Equation 3.10 is ammonia (NH_3), which corresponds to the standard COD method [73], where the ammonia, presented either in the waste or liberated from nitrogen-containing organic matter is not oxidized by dichromate in highly acidic media. Figure 3.14 shows a set of photocurrent responses at a TiO_2 photoanode in a thin-layer cell under UV illumination. A constant potential of about +0.3 V was sufficient for 100% photoelectron collection that was applied for the purposes of improving oxidation efficiency and collecting electrons during the degradation reaction. The current was around zero without UV illumination. When the light was switched on, both the photocurrent from the blank solution (i_{blank}, solid line) and that from the sample containing the organic compound (i_{total}, dashed line) increased rapidly and then decreased to a common steady-state current. i_{blank} was originated from the photoelectrocatalytic oxidation of water, whereas i_{total} resulted from the photoelectrocatalytic oxidation of water and organics. The common steady-state current of the blank and the sample indicates that all the organics

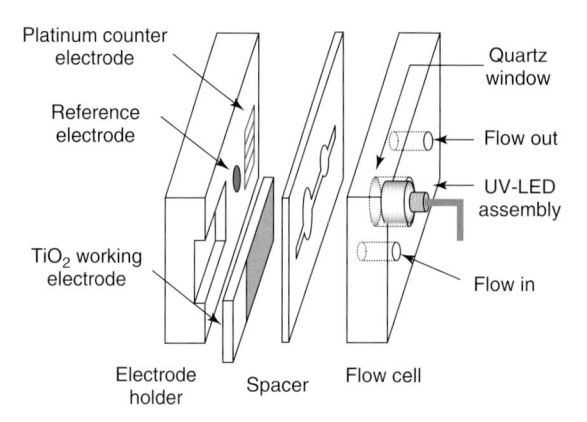

Figure 3.13 Schematic diagram of the design of a thin-layer photoelectrochemical cell [88].

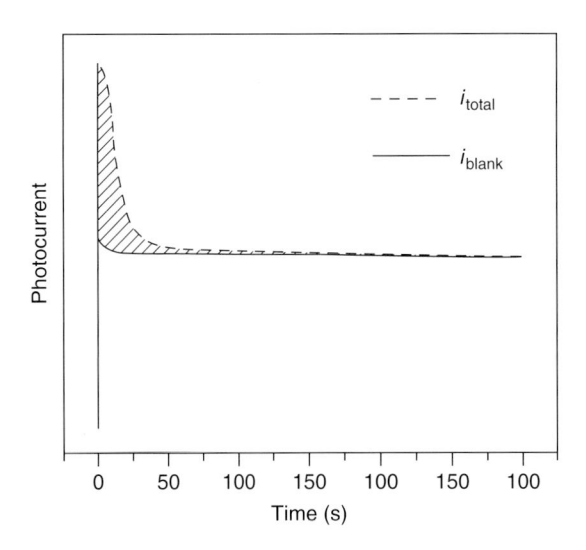

Figure 3.14 Photocurrent responses of a blank solution (solid line) and a sample containing organic compounds (dashed line) in photoelectrochemical thin-layer cell [88].

had been exhaustively oxidized, showing the endpoint of the photoelectrocatalytic degradation reaction and the determination time required in a single analysis. By integration of photocurrents i_{blank} and i_{total} with time, the blank charge Q_{blank} and the total charge Q_{total} can be obtained (Equations 3.20 and 3.21). Subsequently, the net charge Q_{net} (the shaded area in Figure 3.14), generated from the oxidation of organics, can be obtained by subtracting Q_{blank} from Q_{total} (Equation 3.22) [22].

$$Q_{blank} = \int i_{blank} dt \tag{3.20}$$

$$Q_{total} = \int i_{total} dt \tag{3.21}$$

$$Q_{net} = Q_{total} - Q_{blank} \tag{3.22}$$

When the originated charge Q is the result of photoelectrocatalytic oxidation of organics, the Faraday's law can be used to quantify the concentration by measuring the charge (Equation 3.23) [22]

$$Q = \int i dt = nFVC \tag{3.23}$$

where i is the photocurrent generated from the photoelectrocatalytic oxidation of organics; n is the number of electrons transferred in the photoelectrocatalytic oxidation; F and V are the Faraday constant and the sample volume, respectively; and C is the molar concentration of the organics. For a given TiO_2 photoanode, the net charge value should increase linearly with organics concentration according to Equation 3.23, which has been experimentally demonstrated in the literature [22]. The linear relationship of Q versus concentration enables the TiO_2 photoanode at a thin-layer cell to be a sensor for organic compounds.

Q is a direct measure of the total amount of electrons transferred that result from the complete degradation of all compounds in the sample. As one oxygen molecule is equivalent to four electrons transferred, the measured Q value can be easily converted into an equivalent O_2 concentration (or oxygen demand). The equivalent COD value can therefore be represented as

$$\text{COD}(\text{mgl}^{-1}O_2) = \frac{Q}{4FV} \times 32000 \qquad (3.24)$$

On the basis of this analytical principle, the PeCOD is a direct and absolute method that requires no calibration. The effects of important experimental conditions, such as light intensity, applied potential bias, supporting electrolyte concentration, and oxygen concentration, on analytical performance have been investigated, and optimum experimental conditions were obtained. Analytical linear range of 0–360 ppm COD with a practical detection limit of 0.2 ppm COD was achieved [88]. The results demonstrated that the measured COD values using the PeCOD and the standard methods were in an excellent agreement.

Using the exhaustive oxidation model, Zhou's group at Shanghai Jiaotong University fabricated highly organized arrays of TiO_2 nanotubes (Figure 3.15) [30, 31] and integrated the arrays into a thin-cell photoelectrocatalytic reactor for the rapid determination of COD in wastewater samples. The high photocatalytic activity

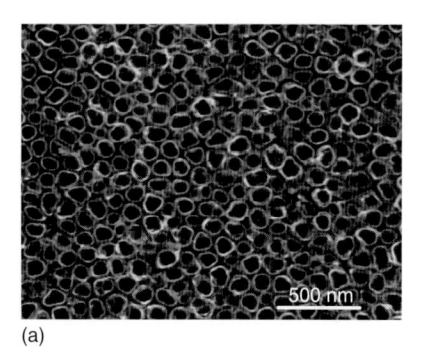

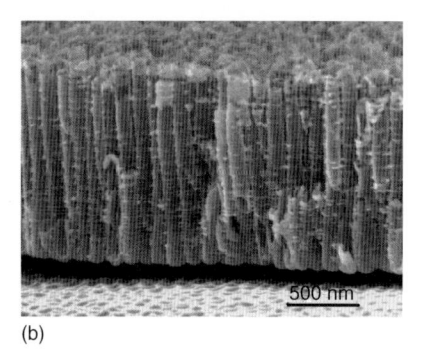

Figure 3.15 The SEM morphology (a) and cross-section (b) of a typical TiO_2 nanotubes array used for COD sensing [31].

of the nanotubes arises from the efficient separation of photogenerated electrons and holes and enables the COD determination with a wider dynamic working range, up to 700 mg l^{-1}.

3.4.3
Partial Oxidation Mode for Determination of COD

The thin-layer photoelectrochemical detector (PECD) based on the photoelectrocatalytic oxidation (Equation 3.10) is a consumption type detector as the organic compounds in the sample are oxidized at the TiO_2 working electrode. Zhang et al. [11] proposed that COD can be determined using the photoelectrochemical oxidative degradation principle in a thin-layer under a continuous flow mode. The thin-layer configuration is essential for achieving a large (electrode area)/(solution volume) ratio that ensures the rapid degradation of injected sample.

Figure 3.16 shows a typical photocurrent–time profile obtained during the degradation of organic compounds under continuous flow conditions and can be used to illustrate how Q_{net} is obtained. The flat baseline (blank) photocurrent ($i_{baseline}$) observed from the carrier solution is originated from water oxidation, whereas the peak response observed from the sample injection is the total current of two different components, one is originated from photoelectrocatalytic oxidation of organics (i_{net}) and the other is originated from water oxidation, which was the same as the blank photocurrent. The net charge, Q_{net}, originated from oxidation

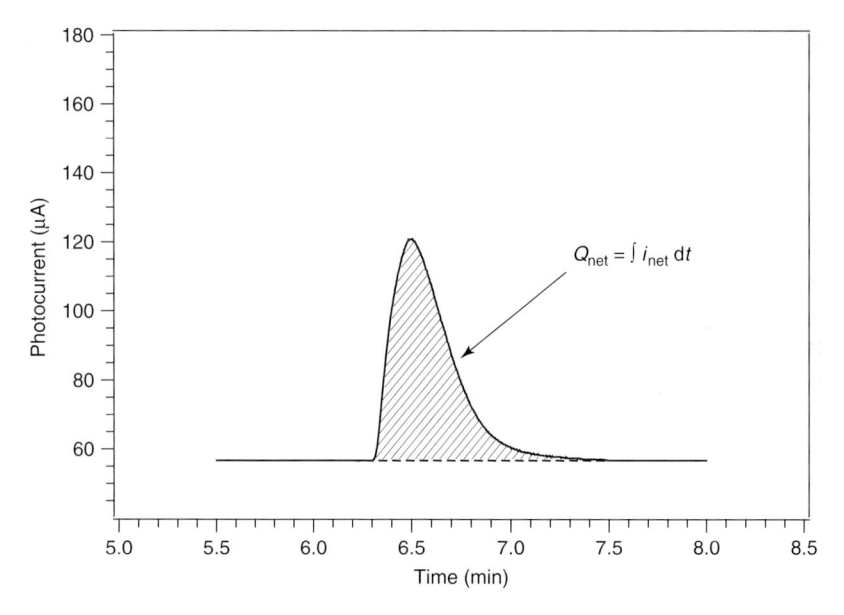

Figure 3.16 Typical photocurrent responses in continuous flow analysis. The shaded area indicates the charge (i.e., Q_{net}) originated from the oxidation of the organic compounds [11].

of organic compounds can be obtained by integration of the peak area between the solid and the dash line, that is, the shade area as indicated in Figure 3.16.

In the continuous-flow mode under controlled conditions, only a portion of organic compounds in the sample have been consumed. This consumed portion can be represented by α, the oxidation percentage, which is defined as

$$\alpha = \frac{Q_{net}}{Q_{theoretical}} \tag{3.25}$$

where Q_{net} is the amount of electrons captured during the continuous flow detection, while $Q_{theoretical}$ refers to the theoretical charge required for the mineralization of the injected sample.

If all organic compounds can be oxidized indiscriminately, it can be assumed that the oxidation percentage is a constant, which is similar to the situation in a consumption type detection under continuous flow mode [89]. The amount of electrons captured by the detector can be written as

$$Q_{net} = \alpha FV \sum_{i=1}^{m} n_i C_i \tag{3.26}$$

As each oxygen molecule equals to four transferred electrons

$$O_2 + 4H^+ + 4e^- \longrightarrow 2H_2O \tag{3.27}$$

and according to the COD definition, the Q_{net} can be readily converted into equivalent COD value [23, 22, 90].

$$COD(mgl^{-1} of O_2) = \frac{Q_{net}}{4\alpha FV} \times 32000 = kQ_{net} \tag{3.28}$$

Equation 3.28 can be used to directly quantify the COD value of a sample when Q_{net} is obtained, as k, the slope, can be obtained by calibration curve method or standard addition calibration method. The proposed method was successfully applied to determine the COD of real samples from various industrial wastewaters [11]. The COD value of real samples determined by this method agreed well the standard dichromate method. The assay time of 1–5 min per sample can be readily achieved. A practical detection limit of 1 mg l^{-1} COD with a linear range of 1–100 mg l^{-1} was achieved under the optimum conditions.

Jin and coauthors prepared Ti/TiO$_2$ photoelectrodes, using laser-assisted technique [91]. The prepared TiO$_2$ film consists of anatase TiO$_2$ nanoparticles and exhibits a superior photocatalytic activity. The electrode is employed as a sensor for FIA to determine COD using the thin-layer photoelectrochemical cell. The measuring principle is based on the photocurrent responses on the electrode, which are proportional to the COD values. The linear range is 50–1000 mg l^{-1}, and the detection limit is 15 mg l^{-1} (S/N = 3). This method has the advantages of short analysis time, simplicity, low environmental impact, and long lifetime of the sensor. In addition, the COD values obtained from the proposed and the conventional method agree well as demonstrated by the significant correlation between the two sets of COD values ($R = 0.9961$, $N = 20$).

3.4.4
UV-LED for Miniature Photoelectrochemical Detectors

Most of the photoelectrochemical systems consist of a xenon light source. For example, 150 or 500 W xenon light are typically employed as the light source [23]. The apparatus is large, heavy, and consumes a significant amount of electricity. Besides this, the application of xenon light into the sensing application introduces several practical drawbacks, such as high cost (both the Xenon lamp and the high-voltage ignition electrical power source are very expensive), short lifetime (only a few thousand hours), high running cost, and the requirement of excellent ventilation and a dry environment. High humidity in an outdoor environment may destroy the system. These characteristics are not ideal for a portable field-based monitoring device [88].

In order to convert the PeCOD technology into a truly portable and a on-site technology for COD measurement, UV-LED (NCCU033 from Nichia) was used to replace the xenon lamp [28, 88], after a careful screening of all commercially available UV-LEDs, because of its ample light intensity and suitable wavelength (250 mW cm^{-2} at 365 nm). The UV-LED PeCOD system gathers the advantages of the modern LED technology (long life, low-power consumption, and low cost), the microelectrochemical system, and merits of the original PeCOD technology, while eradicating the drawbacks of the traditional xenon light optical system and conventional photoelectrochemical station. This makes the technology portable and more user friendly. The successful incorporation of UV-LED into the miniature photoelectrochemical thin-layer cell and experimental validation [28, 88] makes the commercialization of the PeCOD technology possible. The proposed UV-LED PeCOD technology will very likely make a revolutionary improvement for the conventional COD analysis and may be widely used in water quality monitoring industry.

3.4.5
Photoelectrochemical Universal Detector for Organic Compounds

FIA and HPLC have become two of the most common analytical techniques in many fields, such as in pharmaceutical, environmental, forensic, clinical, food, flavor sciences and related industries [92–95]. The availability of a large variety of detectors supports the applications of the FIA and HPLC. The detectors can be classified into specific or nonspecific (i.e., universal) detectors [96].

Nowadays, the most common detectors, such as UV–vis detectors, photodiode array detectors, amperometric electrochemical detectors, and fluorescent detectors, are specific detectors, only giving response to the specific properties of the analytes. Mass spectroscopy (MS) detectors and reflective index detectors (RIDs) are more universal. But the MS detectors are more expensive and are not sensitive to low-molecular-weight compounds [92], while RID commonly has unsatisfactory sensitivity [97]. Low-cost and sensitive universal detectors are ideal for chromatographic application. In this regard, the flame ionization detector (FID) for gas chromatography (GC) is a good representative. The development of low-cost and sensitive universal detectors is much needed to expand the applications of FIA and HPLC.

Fox and Tien [98] have reported the development of photocatalytic oxidation based detectors for HPLC. These detectors, however, were only sensitive to the oxidizable functional groups (such as aniline and benzyl alcohol). Therefore, they were selectively responsive to compounds that possess such groups in the effluent from an HPLC column. The detectors are therefore considered as specific detectors because they are not "universal."

Zhang and coworkers developed a miniature universal PECD for organic compounds using the mixed-phase TiO_2 photoanode and the UV-LED thin-layer photoelectrochemical cell [99]. The photocurrent generated at the TiO_2 photoanode was measured at +0.3 V versus Ag/AgCl by the electrochemical workstation.

Because the sample solution flows through the thin-layer cell continuously during the determination process, it can be expected that the organic compounds in the solution can be partially oxidized. Similar to the case in the online determination of COD in FIA system [11], the oxidation percentage can be defined as α and Q is the amount of electrons captured during the detection process, and $Q_{theoretical}$ is the theoretical charge required for the complete oxidation of the injected sample.

The relationship between measured charge Q and analyte concentration C can be represented by Equation 3.29

$$Q = \alpha \times Q_{theoretical} = \alpha \times nFVC = kC \tag{3.29}$$

During the FIA and HPLC analysis, if the light intensity, applied potential, flow rate, and injection volume are well controlled, α can be considered as a constant in a certain concentration range. If these conditions are achieved, the Q will be directly proportional to C as indicated in Equation 3.28.

This analytical principle was validated by quantitative determination of a variety of organic compounds including sugars, amino acids, straight chain carboxylic acids, aliphatic alcohol, and aromatic carboxylic acid using FIA [99]. The PECD was also successfully coupled with the HPLC for determining sugars (Figure 3.17). Compared with the UV–vis detection method, PECD achieved a higher sensitivity and a larger, linear range for the determination of glucose and sucrose. With further optimization of the cell design and improvement of the TiO_2 photoanode, the performance of the PECD can be improved and it can be expected that the PECD can be commercially available in the near future.

3.5
Photocatalytic Gas Sensing

Metal oxides are well known for their capacity in gas sensing. TiO_2 nanomaterials have been used for gas sensing extensively. Titanium dioxide (TiO_2)-based sensors have been applied for measuring many gases including oxygen [100], carbon

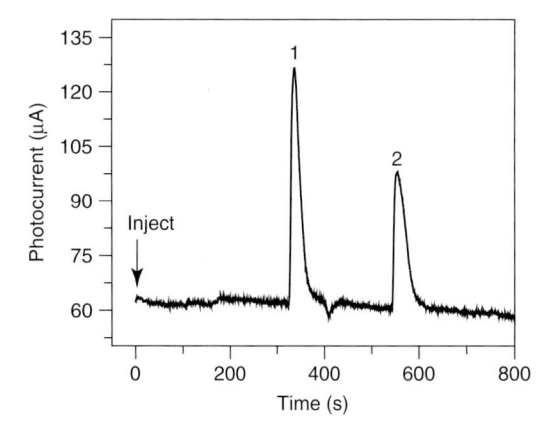

Figure 3.17 Typical responses of a sugar mixture obtained from the HPLC-PECD system. Mobile phase of the HPLC is water with a flow rate of 0.5 ml min^{-1} and an injection volume of 20 μl. Peak 1: glucose (60 mM) and Peak 2: sucrose (30 mM) [99].

monoxide [101] hydrogen [102], nitrous/nitric oxide [103], water vapor [104], and hydrocarbon gases [105].

Most of them, however, achieve the sensing purpose using the electrical property change (e.g., the change of resistivity) rather than the photocatalysis characteristic at the TiO$_2$ surface. In fact, there are scarce reports on the enhancement of gas-sensing properties by photostimulation [106].

The applications of TiO$_2$ photocatalysis in gas sensing can be classified into two types. One is to use the TiO$_2$ photocatalysis to generate analytical signal [106], whereas the other is to use surface self-cleaning properties of TiO$_2$ nanomaterial under UV light to regenerate the sensing surface and subsequently enhance sensitive and reproducible gas sensing [107–110].

3.5.1
The Photoelectrocatalytic Generation of Analytical Signal

The UV irradiation can enhance the analytical signal of the TiO$_2$-based gas sensor [106]. A porous TiO$_2$ electrode was fabricated by immobilizing the nanocrystalline TiO$_2$ on a gold electrode. Without the UV irradiation, the change in the sensitivity is about 0.03/10 ppm of gaseous CO. Under the UV illumination, the change in the sensitivity is about 5.3/10 ppm in a range of 200–300 ppm of CO. The sensitivity increment indicated that the sensitivity was enhanced more than 170 times by the UV irradiation, which is believed to be caused by the increase of the surface activity [106].

It is well established that VOCs can be mineralized when the TiO$_2$ nanomaterial is illuminated under UV light [109, 110]. Utilizing the photocatalytic reaction, the TiO$_2$-based microsensor can be used to identify specific gaseous organic compounds and allow operation at ambient temperatures. The TiO$_2$ microsensors were produced using thick-film lithographic methods and cermet materials. The

TiO_2 and platinum films were deposited on an alumina substrate. Tests were conducted in a closed quartz glass cell at ambient temperatures and pressures. As the sensor was exposed to a variety of gaseous organic constituents in the presence of UV light, electrical characteristics of the sensor were measured. Individual gases including methylene chloride, ethanol, benzene, acetone, xylene, and isopropanol produced unique signatures as they were oxidized on the sensor surface. Sensors were renewable and reusable.

3.5.2
Photocatalytic Surface Self-Cleaning for Enhancement of Analytical Signal

Self-cleaning is a well-known TiO_2 property that originated from the photocatalytic oxidation power of the TiO_2 semiconductor under UV illumination. Fung's group at the university of Hong Kong developed a piezoelectric crystal sensor for the detection of organic vapors using nanocrystalline TiO_2 films and β-cyclodextrine [108]. Owing to the rigidity of the cast material, comparatively thick coatings could be used without a significant attenuation of the acoustic wave, and the porous morphology of the nanocrystalline TiO_2 films yields a relatively fast response even for thick coatings. Piezoelectric gas sensors successfully detected organic vapors (e.g., o-xylene and methanol) with detection limits down to 0.05 ppm. In strong contrast with the porous films prepared from SiO_2, Al_2O_3, or graphite particles, under UV illumination, the nanocrystalline TiO_2 films produce holes, and hydroxyl radicals, which could oxidize and decompose organic materials. As fresh surface of porous TiO_2 could be produced by irradiating UV light onto the β-cyclodextrine/TiO_2 composite films, piezoelectric crystal sensors fabricated on porous TiO_2 substrate were found to give reproducible results. The use of such coating for fabricating piezoelectric crystal chemical and biosensors leads to significant improvements in the sensor performance and a wide scope of application.

Obee and Brown [104] found that increasing the humidity level increased the TiO_2 oxidation rate of formaldehyde up to a maximum, then decreased with increasing humidity for concentrations in the range of 10 ppm and above. For low formaldehyde concentration, 1.5 ppm and below, the oxidation rate had a weak dependence on humidity. The oxidation rate also depended on the formaldehyde concentration, increasing to a maximum at ∼100 ppm. The formaldehyde photooxidation rate decreased with increasing temperature. The presence of other VOCs in the air will lead to competitive interaction at the TiO_2 surface adsorption site. Noguchi et al. [111] found that oxidation rate of formaldehyde was higher than that of acetaldehyde because TiO_2 has higher adsorption capacity for formaldehyde.

Glancing angle deposition was used to fabricate TiO_2 thin film relative humidity sensors [112]. These sensors have been shown to have an extremely high capacitive response along with subsecond response times. However, the thin film morphology that gives rise to the strong performance also renders the film vulnerable to aging processes, which can degrade sensor performance over a period of days. The response of our sensors to UV irradiation suggested that the photocatalytic self-cleaning property of the TiO$_2$ surface not only reverses the aging of humidity sensor but also improves the overall device response [112].

3.6 Conclusions

Utilizing the extraordinary photocatalytic oxidation power at the TiO$_2$ surface under UV illumination, the nanostructured TiO$_2$-based sensors are able to determine the aggregative organic compounds in aqueous media, such as TOC, COD, and VOCs, detect individual organic compound by coupling with HPLC or FIA, and facilitate the detection of various gases. By combining modern technologies, such as microfabrication technology (e.g., lab-on-chip fabrication), microelectronic devices (e.g., UV-LED and USB powered electrochemical station), the photocatalysis, and the photoelectrocatalysis at nanostructured TiO$_2$ materials could be widely used in water quality monitoring and gas sensing.

References

1. Fujishima, A. and Honda, K. (1972) Electrochemical photolysis of water at a semiconductor electrode. *Nature (London)*, **238**, 37–38.
2. Zhu, H.Y., Gao, X.P., Song, D.Y., Bai, Y.Q., Ringer, S.P., Gao, Z. et al. (2004) Growth of boehmite nanofibers by assembling nanoparticles with surfactant micelles. *J. Phys. Chem. B*, **108**, 4245–4247.
3. Liu, G., Wang, L.Z., Yang, H.G., Cheng, H.M., and Lu, G.Q. (2010) Titania-based photocatalysts-crystal growth, doping and heterostructuring. *J. Mater. Chem.*, **20**, 831–843.
4. Centi, G. and Perathoner, S. (2011) Creating and mastering nano-objects to design advanced catalytic materials. *Coord. Chem. Rev.*, **255**, 1480–1498.
5. Low, G.K.C., McEvoy, S.R., and Matthews, R.W. (1991) Formation of nitrate and ammonium ions in titanium dioxide mediated photocatalytic degradation of organic compounds containing nitrogen atoms. *Environ. Sci. Technol.*, **25**, 460–467.
6. Hoffmann, M.R., Martin, S.T., Choi, W., and Bahnemann, D.W. (1995) Environmental applications of semiconductor photocatalysis. *Chem. Rev. (Washington, DC)*, **95**, 69–96.
7. Wolfrum, E.J., Huang, J., Blake, D.M., Maness, P.-C., Huang, Z., Fiest, J. et al. (2002) Photocatalytic oxidation of bacteria, bacterial and fungal spores, and model biofilm components to carbon dioxide on titanium dioxide-coated surfaces. *Environ. Sci. Technol.*, **36**, 3412–3419.
8. Jacoby, W.A., Maness, P.C., Wolfrum, E.J., Blake, D.M., and Fennell, J.A. (1998) Mineralization of bacterial cell mass on a photocatalytic surface in air. *Environ. Sci. Technol.*, **32**, 2650–2653.
9. Hagfeldt, A. and Graetzel, M. (1995) Light-induced redox reactions in nanocrystalline systems. *Chem. Rev. (Washington, DC)*, **95**, 49–68.
10. Jiang, D., Zhao, H., Zhang, S., and John, R. (2004) Kinetic study of photocatalytic oxidation of adsorbed carboxylic acids at TiO2 porous film by photoelectrolysis. *J. Catal.*, **223**, 212–220.
11. Zhang, S., Jiang, D., and Zhao, H. (2006) Development of chemical oxygen demand on-line monitoring system based on a photoelectrochemical degradation principle. *Environ. Sci. Technol.*, **40**, 2363–2368.
12. Lam, S.W., Chiang, K., Lim, T.M., Amal, R., and Low, G.K.C. (2005) The

role of ferric ion in the photochemical and photocatalytic oxidation of resorcinol. *J. Catal.*, **234**, 292–299.
13. Matthews, R.W. (1990) Purification of water with near-U.V. illuminated suspensions of titanium dioxide. *Water Res.*, **24**, 653–660.
14. Kim, Y.C., Lee, K.H., Sasaki, S., Hashimoto, K., Ikebukuro, K., and Karube, I. (2000) Photocatalytic sensor for chemical oxygen demand determination based on oxygen electrode. *Anal. Chem.*, **72**, 3379–3382.
15. Molinari, R., Borgese, M., Drioli, E., Palmisano, L., and Schiavello, M. (2002) Hybrid processes coupling photocatalysis and membranes for degradation of organic pollutants in water. *Catal. Today*, **75**, 77–85.
16. Oliva, F.Y., Avalle, L.B., Santos, E., and Cámara, O.R. (2002) Photoelectrochemical characterization of nanocrystalline TiO2 films on titanium substrates. *J. Photochem. Photobiol., A*, **146**, 175–188.
17. Yang, J.-H., Han, Y.-S., and Choy, J.-H. (2006) TiO2 thin-films on polymer substrates and their photocatalytic activity. *Thin Solid Films*, **495**, 266–271.
18. Suzuki, K., Christie, A.B., and Howson, R.P. (1986) Interface structure between reactively ion plated TiO2 films and PET substrate. *Vacuum*, **36**, 323–328.
19. Duminica, F.D., Maury, F., and Hausbrand, R. (2007) Growth of TiO_2 thin films by AP-MOCVD on stainless steel substrates for photocatalytic applications. *Surf. Coat. Technol.*, **201**, 9304–9308.
20. Vigil, E., Dixon, D., Hamilton, J.W.J., and Byrne, J.A. (2009) Deposition of TiO2 thin films on steel using a microwave activated chemical bath. *Surf. Coat.Technol.*, **203**, 3614–3617.
21. Marcos, P.S., Marto, J., Trindade, T., and Labrincha, J.A. (2008) Screen-printing of TiO_2 photocatalytic layers on glazed ceramic tiles. *J. Photochem. Photobiol., A*, **197**, 125–131.
22. Zhao, H., Jiang, D., Zhang, S., Catterall, K., and John, R. (2004) Development of a direct photoelectrochemical method for rapid determination of chemical oxygen demand. *Anal. Chem.*, **76**, 155–160.
23. Zhang, S., Zhao, H., Jiang, D., and John, R. (2004) Photoelectrochemical determination of chemical oxygen demand based on an exhaustive degradation model in a thin-layer cell. *Anal. Chim. Acta*, **514**, 89–97.
24. Choi, W., Termin, A., and Hoffmann, M.R. (1994) The role of metal ion dopants in quantum-sized TiO2: correlation between photoreactivity and charge carrier recombination dynamics. *J. Phys. Chem.*, **98**, 13669–13679.
25. Zhao, H. (2003) Photoelectrochemical determination of chemical oxygen demand. Australia Provisional Patent 2003901589.
26. Low, G.K.C. and Matthews, R.W. (1990) Flow-injection determination of organic contaminants in water using an ultraviolet-mediated titanium dioxide film reactor. *Anal. Chim. Acta*, **231**, 13–20.
27. Low, G.K.C. and McEvoy, S.R. (1996) Analytical monitoring systems based on photocatalytic oxidation principles. *TrAC, Trends Anal. Chem.*, **15**, 151–156.
28. Zhang, S., Li, L., and Zhao, H. (2009) A portable photoelectrochemical probe for rapid determination of chemical oxygen demand in wastewaters. *Environ. Sci. Technol.*, **43**, 7810–7815.
29. Zhao, H. and Zhang, S. (2007) Water analysis using photoelectrochemical method. PCT Int. Patent. WO 2007016740.
30. Zhang, J., Zhou, B., Zheng, Q., Li, J., Bai, J., Liu, Y. *et al.* (2009) Photoelectrocatalytic COD determination method using highly ordered TiO2 nanotube array. *Water Res.*, **43**, 1986–1992.
31. Zheng, Q., Zhou, B., Bai, J., Li, L., Jin, Z., Zhang, J. *et al.* (2008) Self-organized TiO_2 nanotube array sensor for the determination of chemical oxygen demand. *Adv. Mater. (Weinheim)*, **20**, 1044–1049.
32. Kasanen, J., Suvanto, M., and Pakkanen, T.T. (2009) Self-cleaning, titanium dioxide based, multilayer coating fabricated on polymer and glass surfaces. *J. Appl. Polym. Sci.*, **111**, 2597–2606.

33. Watanabe, T., Nakajima, A., Wang, R., Minabe, M., Koizumi, S., Fujishima, A. et al. (1999) Photocatalytic activity and photoinduced hydrophilicity of titanium dioxide coated glass. *Thin Solid Films*, **351**, 260–263.
34. Sirghi, L. and Hatanaka, Y. (2003) Hydrophilicity of amorphous TiO_2 ultra-thin films. *Surf. Sci.*, **530**, L323–L327.
35. Radecka, M., Rekas, M., Trenczek-Zajac, A., and Zakrzewska, K. (2008) Importance of the band gap energy and flat band potential for application of modified TiO_2 photoanodes in water photolysis. *J. Power. Sources*, **181**, 46–55.
36. Abou-Helal, M.O. and Seeber, W.T. (2002) Preparation of TiO2 thin films by spray pyrolysis to be used as a photocatalyst. *Appl. Surf. Sci.*, **195**, 53–62.
37. Conde-Gallardo, A., Guerrero, M., Castillo, N., Soto, A.B., Fragoso, R., and Cabañas-Moreno, J.G. (2005) TiO2 anatase thin films deposited by spray pyrolysis of an aerosol of titanium diisopropoxide. *Thin Solid Films*, **473**, 68–73.
38. Aarik, J., Aidla, A., Mändar, H., and Uustare, T. (2001) Atomic layer deposition of titanium dioxide from TiCl4 and H2O: investigation of growth mechanism. *Appl. Surf. Sci.*, **172**, 148–158.
39. Aarik, J., Aidla, A., Uustare, T., Ritala, M., and Leskelä, M. (2000) Titanium isopropoxide as a precursor for atomic layer deposition: characterization of titanium dioxide growth process. *Appl. Surf. Sci.*, **161**, 385–395.
40. Bessergenev, V.G., Pereira, R.J.F., Mateus, M.C., Khmelinskii, I.V., Vasconcelos, D.A., Nicula, R. et al. (2006) Study of physical and photocatalytic properties of titanium dioxide thin films prepared from complex precursors by chemical vapour deposition. *Thin Solid Films*, **503**, 29–39.
41. Mills, A., Elliott, N., Parkin, I.P., O'Neill, S.A., and Clark, R.J. (2002) Novel TiO_2 CVD films for semiconductor photocatalysis. *J. Photochem. Photobiol., A*, **151**, 171–179.
42. Karuppuchamy, S., Jeong, J.M., Amalnerkar, D.P., and Minoura, H. (2006) Photoinduced hydrophilicity of titanium dioxide thin films prepared by cathodic electrodeposition. *Vacuum*, **80**, 494–498.
43. Li, Y., Yu, X., and Yang, Q. (2009) Fabrication of TiO_2 nanotube thin films and their gas sensing properties. *J. Sens.* (Special issue: Nanomaterials for Chemical Sensing Technologies, (eds M. Penza et al.), Hindawi Publishing Corp) **2009**, 1–19.
44. Akpan, U.G. and Hameed, B.H. (2010) The advancements in sol-gel method of doped-TiO_2 photocatalysts. *Appl. Catal., A*, **375**, 1–11.
45. Dzik, P., Vesely, M., and Chomoucka, J. (2010) Thin layers of photocatalytic TiO_2 prepared by ink-jet printing of a sol-gel precursor. *J. Adv. Oxid. Technol.*, **13**, 172–183.
46. Yang, M., Li, L., Zhang, S., Li, G., and Zhao, H. (2010) Preparation, characterisation and sensing application of inkjet-printed nanostructured TiO2 photoanode. *Sens. Actuators, B*, **B147**, 622–628.
47. Sirés, I., Brillas, E., Cerisola, G., and Panizza, M. (2008) Comparative depollution of mecoprop aqueous solutions by electrochemical incineration using BDD and PbO2 as high oxidation power anodes. *J. Electroanal. Chem.*, **613**, 151–159.
48. Poh, W.C., Loh, K.P., Zhang, W.D., Triparthy, S., Ye, J.-S., and Sheu, F.-S. (2004) Biosensing properties of diamond and carbon nanotubes. *Langmuir*, **20**, 5484–5492.
49. Mitani, N. and Einaga, Y. (2009) The simple voltammetric analysis of acids using highly boron-doped diamond macroelectrodes and microelectrodes. *J. Electroanal. Chem.*, **626**, 156–160.
50. Holt, K.B., Bard, A.J., Show, Y., and Swain, G.M. (2004) Scanning electrochemical microscopy and conductive probe atomic force microscopy studies of hydrogen-terminated boron-doped diamond electrodes with different doping levels. *J. Phys. Chem. B*, **108**, 15117–15127.

51. Kalish, R. (1999) Doping of diamond. *Carbon*, **37**, 781–785.
52. Yu, H., Chen, S., Quan, X., Zhao, H., and Zhang, Y. (2008) Fabrication of a TiO_2-BDD heterojunction and its application as a photocatalyst for the simultaneous oxidation of an azo dye and reduction of Cr(VI). *Environ. Sci. Technol.*, **42**, 3791–3796.
53. Manivannan, A., Spataru, N., Arihara, K., and Fujishima, A. (2005) Electrochemical deposition of titanium oxide on boron-doped diamond electrodes. *Electrochem. Solid-State Lett.*, **8**, C138–C140.
54. Qu, J. and Zhao, X. (2008) Design of BDD-TiO_2 hybrid electrode with p-n function for photoelectrocatalytic degradation of organic contaminants. *Environ. Sci. Technol.*, **42**, 4934–4939.
55. Han, Y., Zhang, S., Zhao, H., Wen, W., Zhang, H., Wang, H. *et al.* (2009) Photoelectrochemical characterization of a robust TiO_2/BDD heterojunction electrode for sensing application in aqueous solutions. *Langmuir*, **26**, 6033–6040.
56. Han, Y., Qiu, J., Miao, Y., Han, J., Zhang, S., Zhang, H. *et al.* (2011) Robust TiO_2/BDD heterojunction photoanodes for determination of chemical oxygen demand in wastewaters. *Anal. Methods.* doi: 10.1039/c1ay05193h
57. Jiang, D., Zhang, S., and Zhao, H. (2007) Photocatalytic degradation characteristics of different organic compounds at TiO_2 nanoporous film electrodes with mixed anatase/rutile phases. *Environ. Sci. Technol.*, **41**, 303–308.
58. Chen, Y. and Dionysiou, D.D. (2006) Effect of calcination temperature on the photocatalytic activity and adhesion of TiO_2 films prepared by the P-25 powder-modified sol-gel method. *J. Mol. Catal. A: Chem.*, **244**, 73–82.
59. Kim, D.J., Hahn, S.H., Oh, S.H., and Kim, E.J. (2002) Influence of calcination temperature on structural and optical properties of TiO_2 thin films prepared by sol-gel dip coating. *Mater. Lett.*, **57**, 355–360.
60. Dunlop, P.S.M., Byrne, J.A., Manga, N., and Eggins, B.R. (2002) The photocatalytic removal of bacterial pollutants from drinking water. *J. Photochem. Photobiol. A: Chem.*, **148**, 355–363.
61. Jiang, D. (2004) Fundamental studies of photocatalytic processes at TiO_2 nanoporous electrodes by photoelectrochemical techniques. PhD thesis. Griffith University, Gold Coast, AustraliaGold Coast, Australia.
62. Kawahara, T., Konishi, Y., Tada, H., Tohge, N., Nishii, J., and Ito, S. (2002) A patterned TiO_2(anatase)/TiO_2(rutile) bilayer-type photocatalyst: Effect of the anatase/rutile junction on the photocatalytic activity. *Angew. Chem. Int. Ed.*, **41**, 2811–2813.
63. Jiang, L.-C. and Zhang, W.-D. (2009) Electrodeposition of TiO_2 nanoparticles on multiwalled carbon nanotube arrays for hydrogen peroxide sensing. *Electroanalysis*, **21**, 988–993.
64. Llobet, E., Espinosa, E.H., Sotter, E., Ionescu, R., Vilanova, X., Torres, J. *et al.* (2008) Carbon nanotube-TiO_2 hybrid films for detecting traces of O_2. *Nanotechnology*, **19**, 375501.
65. Sánchez, M. and Rincón, M.E. (2009) Sensor response of sol-gel multiwalled carbon nanotubes-TiO2 composites deposited by screen-printing and dip-coating techniques. *Sens. Actuators, B*, **140**, 17–23.
66. Shen, Q., You, S.-K., Park, S.-G., Jiang, H., Guo, D., Chen, B. *et al.* (2008) Electrochemical biosensing for cancer cells based on TiO_2/CNT nanocomposites modified electrodes. *Electroanalysis*, **20**, 2526–2530.
67. Ueda, T., Takahashi, K., Mitsugi, F., and Ikegami, T. (2009) Preparation of single-walled carbon nanotube/TiO_2 hybrid atmospheric gas sensor operated at ambient temperature. *Diam. Relat. Mater.*, **18**, 493–496.
68. Gao, B., Chen, G.Z., and Li Puma, G. (2009) Carbon nanotubes/titanium dioxide (CNTs/TiO2) nanocomposites prepared by conventional and novel surfactant wrapping sol-gel methods exhibiting enhanced photocatalytic activity. *Appl. Catal., B*, **89**, 503–509.

69. Woan, K., Pyrgiotakis, G., and Sigmund, W. (2009) Photocatalytic carbon-nanotube-TiO$_2$ composites. *Adv. Mater.*, **21**, 2233–2239.
70. Yu, Y., Yu, J.C., Yu, J.-G., Kwok, Y.-C., Che, Y.-K., Zhao, J.-C. et al. (2005) Enhancement of photocatalytic activity of mesoporous TiO$_2$ by using carbon nanotubes. *Appl. Catal. A-Gen.*, **289**, 186–196.
71. Li, L., Yang, M., Zhang, S., Liu, P., Li, G., Wen, W. et al. (2010) The fabrication of CNTs/TiO$_2$ photoanodes for sensitive determination of organic compounds. *Nanotechnology*, **21**. doi: 10.1088/0957-4484/21/48/485503
72. Zhang, S., Wen, W., Jiang, D., Zhao, H., John, R., Wilson, G.J. et al. (2006) Photoelectrochemical characterisation of TiO$_2$ thin films derived from microwave hydrothermally processed nanocrystalline colloids. *J. Photochem. Photobiol. A-Chem.*, **179**, 305–313.
73. American Public Health Association, American Water Works Association, and Water Environment Federation (1995) Standard Methods for the Examination of Water and Wastewater, 19th edn, 1 v. (various paging), Apha-Awwa-Wef, Washington, DC.
74. Taschuk, M.T., Steele, J.J., van Popta, A.C., and Brett, M.J. (2008) Photocatalytic regeneration of interdigitated capacitor relative humidity sensors fabricated by glancing angle deposition. *Sens. Actuators, B*, **B134**, 666–671.
75. Matthews, R.W., Abdullah, M., and Low, G.K.C. (1990) Photocatalytic oxidation for total organic carbon analysis. *Anal. Chim. Acta*, **233**, 171–179.
76. Lee, K.-H., Kim, Y.-C., Suzuki, H., Ikebukuro, K., Hashimoto, K., and Karube, I. (2000) Disposable chemical oxygen demand sensor using a microfabricated clark-type oxygen electrode with a TiO$_2$ suspension solution. *Electroanalysis*, **12**, 1334–1338.
77. Kim, Y.C., Sasaki, S., Yano, K., Ikebukuro, K., Hashimoto, K., and Karube, I. (2001) Photocatalytic sensor for the determination of chemical oxygen demand using flow injection analysis. *Anal. Chim. Acta*, **432**, 59–66.
78. Kim, Y.-C., Sasaki, S., Yano, K., Ikebukuro, K., Hashimoto, K., and Karube, I. (2002) A flow Method with photocatalytic oxidation of dissolved organic matter using a solid-phase (TiO$_2$) reactor followed by amperometric detection of consumed oxygen. *Anal. Chem.*, **74**, 3858–3864.
79. Ai, S., Li, J., Yang, Y., Gao, M., Pan, Z., and Jin, L. (2004) Study on photocatalytic oxidation for determination of chemical oxygen demand using a nano-TiO$_2$-K$_2$Cr$_2$O$_7$ system. *Anal. Chim. Acta*, **509**, 237–241.
80. Chai, Y., Ding, H., Zhang, Z., Xian, Y., Pan, Z., and Jin, L. (2006) Study on photocatalytic oxidation for determination of the low chemical oxygen demand using a nano-TiO$_2$-Ce(SO$_4$)$_2$ coexisted system. *Talanta*, **68**, 610–615.
81. Li, J., Ai, S., Yang, Y., Gu, F., Cao, Y., and Jin, L. (2003) Study on photocatalytic detection of groundwater chemical oxygen demand values based on nanometer TiO$_2$/KMnO$_4$ synergic system. *Huaxue Chuanganqi*, **23**, 49–54.
82. Zhang, Z., Yuan, Y., Fang, Y., Liang, L., Ding, H., and Jin, L. (2007) Preparation of photocatalytic nano-ZnO/TiO$_2$ film and application for determination of chemical oxygen demand. *Talanta*, **73**, 523–528.
83. Li, J., Li, L., Zheng, L., Xian, Y., and Jin, L. (2006) Determination of chemical oxygen demand values by a photocatalytic oxidation method using nano-TiO$_2$ film on quartz. *Talanta*, **68**, 765–770.
84. Li, C. and Song, G. (2009) Photocatalytic degradation of organic pollutants and detection of chemical oxygen demand by fluorescence methods. *Sens. Actuators B*, **B137**, 432–436.
85. Zhu, L., Chen, Y.E., Wu, Y., Li, X., and Tang, H. (2006) A surface-fluorinated-TiO$_2$-KMnO$_4$ photocatalytic system for determination of chemical oxygen demand. *Anal. Chim. Acta*, **571**, 242–247.
86. Frank, S.N. and Bard, A.J. (1977) Semiconductor electrodes. 12. Photoassisted oxidations and photoelectrosynthesis at polycrystalline titanium dioxide

electrodes. *J. Am. Chem. Soc.*, **99**, 4667–4675.

87. Jiang, D., Zhao, H., Zhang, S., and John, R. (2006) Comparison of photocatalytic degradation kinetic characteristics of different organic compounds at anatase TiO2 nanoporous film electrodes. *J. Photochem. Photobiol., A: Chem.*, **177**, 253–260.

88. Zhang, S., Li, L., Zhao, H., and Li, G. (2009) A portable miniature UV-LED-based photoelectrochemical system for determination of chemical oxygen demand in wastewater. *Sensor Actuators, B*, **141**, 634–640.

89. Ruzicka, J. and Hansen, E.H. (1981) *Flow Injection Analysis*, Chemical Analysis, Vol. 62, John Wiley & Sons, Inc., New York, p. xi, 207.

90. Zhao, H. (2004) Photoelectrochemical determination of chemical oxygen demand. Intl. Pat. WO 2004088305.

91. Li, J., Zheng, L., Li, L., Shi, G., Xian, Y., and Jin, L. (2007) Determination of chemical oxygen demand using flow injection with Ti/TiO2 electrode prepared by laser anneal. *Meas. Sci. Technol.*, **18**, 945–951.

92. Zhang, B., Li, X., and Yan, B. (2008) Advances in HPLC detection – towards universal detection. *Anal. Bioanal. Chem.*, **390**, 299–301.

93. Ivandini, T.A., Sarada, B.V., Terashima, C., Rao, T.N., Tryk, D.A., Ishiguro, H. et al. (2002) Electrochemical detection of tricyclic antidepressant drugs by HPLC using highly boron-doped diamond electrodes. *J. Electroanal. Chem.*, **521**, 117–126.

94. Rao, T.N., Sarada, B.V., Tryk, D.A., and Fujishima, A. (2000) Electroanalytical study of sulfa drugs at diamond electrodes and their determination by HPLC with amperometric detection. *J. Electroanal. Chem.*, **491**, 175–181.

95. Trojanowicz, M. (2000) *Flow Injection Analysis: Instrumentation and Applications*, 1st edn, World Scientific Publishing Co. Pte. Ltd, Singapore.

96. Skoog, D.A., Holler, F.J., and Nieman, T.A. (1998) *Principles of Instrumental Analysis*, 5th edn, Thomson Learning, California.

97. Stulík, K. and Pacáková, V. (1981) Electrochemical detection techniques in high-performance liquid chromatography. *J. Electroanal. Chem.*, **129**, 1–24.

98. Fox, M.A. and Tien, T.P. (1988) A photoelectrochemical detector for high-pressure liquid chromatography. *Anal. Chem.*, **60**, 2278–2282.

99. Li, L., Zhang, S., and Zhao, H. (2011) A low cost universal photoelectrochemical detector for organic compounds based on photoelectrocatalytic oxidation at a nanostructured TiO2 photoanode. *J. Electroanal. Chem.*, **15**, 211–217.

100. Micheli, A.L. (1984) Fabrication and performance evaluation of a titania automotive exhaust gas sensor. *Am. Ceram. Soc. Bull.*, **63**, 694–698.

101. Gouma, P.I., Mills, M.J., and Sandhage, K.H. (2000) Fabrication of free-standing titania-based gas sensors by the oxidation of metallic titanium foils. *J. Am. Ceram. Soc.*, **83**, 1007–1009.

102. Harris, L.A. (1980) A titanium dioxide hydrogen detector. *J. Electrochem. Soc.*, **127**, 2657–2662.

103. Satake, K., Katayama, A., Ohkoshi, H., Nakahara, T., and Takeuchi, T. (1994) Titania NO_x sensors for exhaust monitoring. *Sens. Actuators, B*, **20**, 111–117.

104. Obee, T.N. and Brown, R.T. (1995) TiO_2 photocatalysis for indoor air applications: effects of humidity and trace contaminant levels on the oxidation rates of formaldehyde, toluene, and 1,3-butadiene. *Environ. Sci. Technol.*, **29**, 1223–1231.

105. Traversa, E., Di Vona, M.L., Licoccia, S., Sacerdoti, M., Carotta, M.C., Gallana, M. et al. (2000) Sol-gel nano-sized semiconducting titania-based powders for thick-film gas sensors. *J. Sol-Gel Sci. Technol.*, **19**, 193–196.

106. Yang, T.-Y., Lin, H.-M., Wei, B.-Y., Wu, C.-Y., and Lin, C.-K. (2003) UV enhancement of the gas sensing properties of nano-TiO_2. *Rev. Adv. Mater. Sci.*, **4**, 48–54.

107. Radecka, M., Lyson, B., Lubecka, M., Czapla, A., and Zakrzewska, K. (2009) Photocatalytical decomposition of

contaminants on thin film gas sensors. *Acta Physica Polonica*, **A 117**, 415–419.
108. Si, S.H., Fung, Y.S., and Zhu, D.R. (2005) Improvement of piezoelectric crystal sensor for the detection of organic vapors using nanocrystalline TiO_2 films. *Sens. Actuators, B*, **108**, 165–171.
109. Mills, A. and Le Hunte, S. (1997) An overview of semiconductor photocatalysis. *J. Photochem. Photobiol., A*, **108**, 1–35.
110. Skubal, L.R., Meshkov, N.K., and Vogt, M.C. (2002) Detection and identification of gaseous organics using a TiO_2 sensor. *J. Photochem. Photobiol.A: Chem.*, **148**, 103–108.
111. Noguchi, T., Fujishima, A., Sawunyama, P., and Hashimoto, K. (1998) Photocatalytic degradation of gaseous formaldehyde using TiO_2 film. *Environ. Sci. Technol.*, **32**, 3831–3833.
112. Chen, Z. and Lu, C. (2005) Humidity sensors: a review of materials and mechanisms. *Sens. Lett.*, **3**, 274–295.

4
Mesoporous Materials for Water Treatment

Yonghui Deng and Dongyuan Zhao

In the past decade, rapid development of global economics and industry, environmental pollution, energy crisis, food safety, and biotechnology have aroused more and more public concerns. Water pollution is a major global environmental problem. It has been suggested that it is one of the leading worldwide causes of deaths and diseases [1]. Thus, efficient control of the discharge of industrial wastewater is in emergent demand before its release to the environment. Wastewater generated from industrial activities can contain pollutants such as suspended solids, nutrients, heavy metals, oils and greases, and other toxic organic and inorganic chemicals. If this untreated water is discharged directly, pollutants can cause serious harm to the environment and human health. Significant progress has been made in removing toxic substances, such as heavy metal ions, polyaromatic organics, and microcystins (MCs) through various methods, including sedimentation, membrane separation, adsorption, and oxidation. Among these strategies, adsorption-based processes may lead to one of the most efficient routes for removal of toxic substances from wastewater [2].

Traditionally, silica gels and activated carbons were widely adopted as adsorbents for water treatment. These materials are mainly microporous and thus not suitable for the adsorption of large molecules because of the restriction of pore size. However, the development of advanced adsorbents with improved performance toward large molecules and bearing multifunctions (e.g., magnetic separation, molecular sieves, and recognition) has become a growing objective of research and practical application. In this regard, materials with regular and large mesopores tend to become indispensable. As a result, ordered mesoporous materials, especially those composed of silica and carbon frameworks, have attracted global interests. Ordered mesoporous materials, as an intriguing porous material, have attracted considerable attention because of their high specific surface areas, regular and tunable large pore sizes, and high pore volumes, as well as stable and interconnected frameworks with active pore surface for easy modification or functionalization, meeting the requirements as promising adsorbents. In this chapter, we focus on the removal of various pollutants, such as metal ions, anion, and organic compounds from industrial wastewater by using the unique mesoporous materials as sorbents.

Functional Nanostructured Materials and Membranes for Water Treatment, First Edition.
Edited by Mikel Duke, Dongyuan Zhao, and Raphael Semiat.
© 2013 Wiley-VCH Verlag GmbH & Co. KGaA. Published 2013 by Wiley-VCH Verlag GmbH & Co. KGaA.

4.1
Adsorption of Heavy Metal Ions

Many industries, such as electroplating, battery, mining, and ceramics, may discharge wastewater-containing heavy metals that are harmful to water environment and living organisms. These metal ions include cadmium, lead, and mercury. Since their discovery in 1992 [3], ordered mesoporous silicas have been widely adopted as adsorbents for trapping heavy metal ions. However, most as-prepared ordered mesoporous silicas show quite limited adsorption capacity. It is almost a prerequisite to attach functional groups on the pore surface to generate specific binding sites. Fortunately, the hydroxyl-group-enriched surface of ordered mesoporous silicas makes it easy for functionalization. Two main approaches have been developed to chemically bind functional groups, namely, postgrafting and co-condensation (Figure 4.1) [4]. Both the methods adopt organosilanes carrying different terminal groups. The former utilizes cross-linking reaction between surface silanol groups and organosilanes, whereas the latter relates to the coassembly between organosilanes and surfactants. In order to specifically bind different heavy metal ions, the basic principle lies on the different behaviors of complex chemistry, that is, soft metal ions (e.g., Hg^{2+}) are more likely to form stable complexes with ligands carrying soft electron donor atoms (e.g., thiol) and vice versa. The most successful development is the fabrication of silica-based hybrids bearing thiol groups on ordered mesoporous silicas, which have exceptional binding affinity toward Hg^{2+}, which is one of the most threatening metal ions [5].

Recently, ordered mesoporous silica materials containing functionalized organic monolayers have been synthesized. Solid-state nuclear magnetic resonance (NMR) spectra show that a cross-linked monolayer of mercaptopropylsilane is covalently bound to the mesoporous silica and closely packed on the pore surface. The relative surface coverage of the monolayers can be systematically varied up to 76%. These materials are extremely efficient in removing mercury and other heavy metals from both aqueous and nonaqueous waste streams. These materials can be regenerated and possess good reusability [6]. Bibby and Mercier [7] reported the synthesis of the microspherical adsorbents with uniform mesopore channels by fluoride-catalyzed surfactant-directed co-condensation of tetraethoxysilane (TEOS) and 3-mercaptopropyltrimethoxysilane (MPTMS), using mildly acidic nonionic surfactant solutions. The adsorption kinetics of the adsorbents shows that the uptake of mercury ions is rather slow, with diffusion coefficients ranging between 10^{-14} and 10^{-15} $m^2 s^{-1}$. The coefficients were found to increase as a function of the thiol group density of the adsorbents. The diffusion coefficients of the mercury (II) adsorption process also increased as a function of time, indicating synergistic acceleration of the uptake rate with increasing mercury ion loading in the materials. Zhang et al. [8] synthesized a thioether functionalized organic–inorganic mesoporous silica composite for the first time by a one-step co-condensation of TEOS and (1, 4)-bis(triethoxysilyl)propane tetrasulfide $(CH_3CH_2O)_3Si(CH_2)_3S-S-S-S(CH_2)_3Si(OCH_2CH_3)_3$ (BTESPTS). The new material showed highly selective and capacious adsorption for Hg^{2+}. The adsorbent is

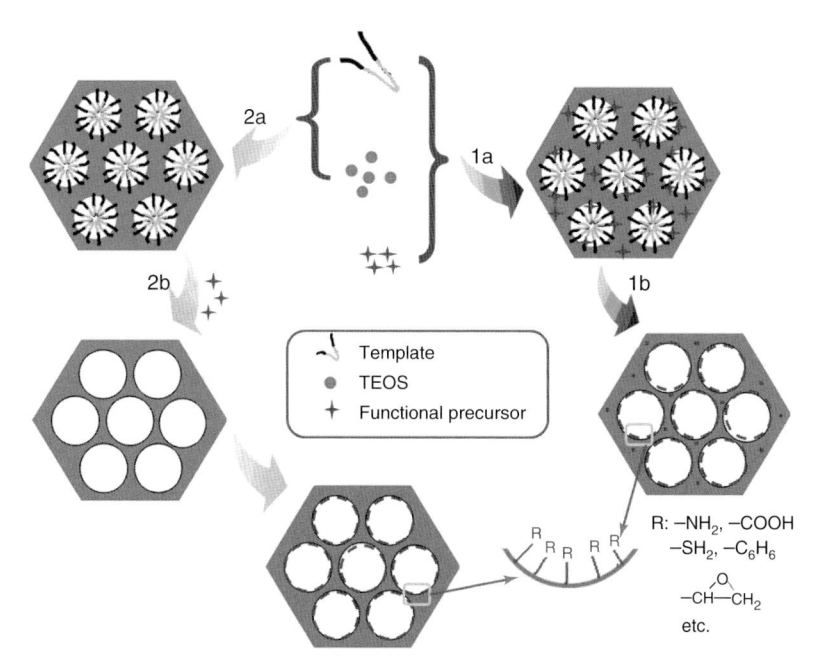

Figure 4.1 Schematic illustration of the two approaches to mesoporous silica materials with different surface functionalities: (1) co-condensation, wherein functional organosilanes and TEOS are simultaneously used as the silica sources for templating synthesis of ordered mesoporous silicas by using Pluronic P123 as a template (1a), and mesoporous silica bearing functional groups can be obtained after the removal of the templates through solvent extraction (1b); (2) the postgrafting, wherein mesoporous silicas are synthesized through soft templating by using TEOS as a silica source and P123 as a template (2a), followed by the removal of templates through calcination. Then, organosilanes are used to react with the silanol groups on the pore wall, creating mesoporous silicas with functionalize pore surfaces (2b).

environmentally friendly because BTESPTS is much less toxic than MPTMS, and the highly selective adsorption for Hg^{2+} also makes the recycling of Hg^{2+} possible.

In summary, ligands bearing thiol-based groups (thiol, thiourea, thioether, etc.) [9–11] have been extensively grafted to mesoporous silica MCM-41 and SBA-15 with high surface coverage and easy accessibility. These hybrids show excellent binding affinity and large capacity (up to a few thousand of milligrams per gram of sorbent) to Hg^{2+}, as well as good stability and high selectivity over other metal ions. They can efficiently remove Hg^{2+} in waste streams from ~10 ppm down to undetectable levels (the inductively coupled plasma spectroscopy with limit of detection of 1 ppb). Moreover, such binding effect of Hg^{2+} ions in mesopores could prevent bacteria (much larger in size than the mesopores) from forming deadly methyl mercury. Factors influencing Hg^{2+} uptake include pH value, porosity, and structure of sorbents. It is also reported that well-ordered mesostructures favor fast mass transport than disordered structures, and functional groups obtained by postgraft are more easily accessible than those by co-condensation.

Similarly, analogous materials grafted with terminal amino-based (amine, urea, polyamidoamine, etc.) or carboxylic groups have shown efficient binding properties to many hard heavy metal ions (such as Cu^{2+}, Zn^{2+}, Cr^{3+}, Fe^{3+}, Cd^{2+}, and Ni^{2+}). Algarra et al. [12] have evaluated the potential of removing nickel and copper ions from industrial electroplating wastewaters using MCM-41 type mesoporous silica materials functionalized with different ratios of aminopropyl groups. Preliminarily, they determined the optimal experimental conditions for the retention of heavy ions. Results show that the mesoporous silica material has a great selectivity against sodium, indicating that ionic strength does not affect the extraction. Benitez et al. [13] successfully synthesized urea-containing mesoporous silica with intraframework urea groups (UreaMS) via a surfactant-templated route by co-condensation of an organosilane precursor and TEOS. A detailed characterization by chemical analysis, ^{29}Si-MAS-NMR, X-ray photoelectron spectroscopy, thermogravimetric analysis, and FT-IR spectroscopy confirmed the integrity of urea groups inside the pore walls of the material. The results revealed a density of one urea group per 13–16 silicon atoms in the bulk material. The adsorption capacity for Fe(III) cations can reach $0.19\,mmol\,g^{-1}$. Heidari et al. [14] systematically investigated the removal of Ni(II), Cd(II), and Pb(II) ions from aqueous solution by using mesoporous silica materials, namely, bulk MCM-41, and their nanoparticles – NH_2-MCM-41 (amino functionalized MCM-41), and nano NH_2-MCM-41. The authors found that NH_2-MCM-41 sample showed the highest uptake for metal ions in aqueous solution. The results indicate that the adsorption of Ni(II), Cd(II), and Pb(II) ions on the surface of the adsorbent increased with the increase of pH values. The maximum adsorption capacity of NH_2-MCM-41 for Ni(II), Cd(II), and Pb(II) was found to be 12.36, 18.25, and $57.74\,mg\,g^{-1}$, respectively. Therefore, the main adsorption process of the heavy metal ions is related with surface complexation. In the same case, the precipitation of metal ions, that is, forming insoluble metal hydroxides on the surface, could also happen under certain condition (at increased pH values) [15]. Compared with other sorbents, such as activated carbons (with a Pb^{2+} adsorption capacity of $38.5\,mg\,g^{-1}$) [16], the mesoporous NH_2-MCM-41 sorbents exhibit much higher adsorption capacity because of the high density of their functional groups.

Ordered mesoporous silicas possessing dual functional groups capable of both detection and uptake of metal ions are a recent development. For example, a mesoporous silica-based chemodosimeter could specifically react with Hg^{2+} ions with the release of some dyes into the solution and the binding of Hg^{2+} from solution at the same time, which changes the color to deep blue and removal of Hg^{2+}. Thus, the detection of Hg^{2+} with naked eye can be achieved. Ros-Lis et al. [17] designed a dual-function hybrid material (U1) for simultaneous chromofluorogenic detection and removal of Hg^{2+} in an aqueous environment (Figure 4.2). In this strategy, the mesoporous solid (UVM-7, University of Valencia Materials-7) is decorated with thiol groups that were treated with squaraine dye III to give a 2,4-bis(4-dialkylaminop henyl)-3-hydroxy-4-alkylsulfanylcyclobut-2-enone (APC) derivative. The derivative is covalently anchored to the inorganic silica matrix. Hg^{2+} ions can react with the APC fragment in sample U1 with release of the

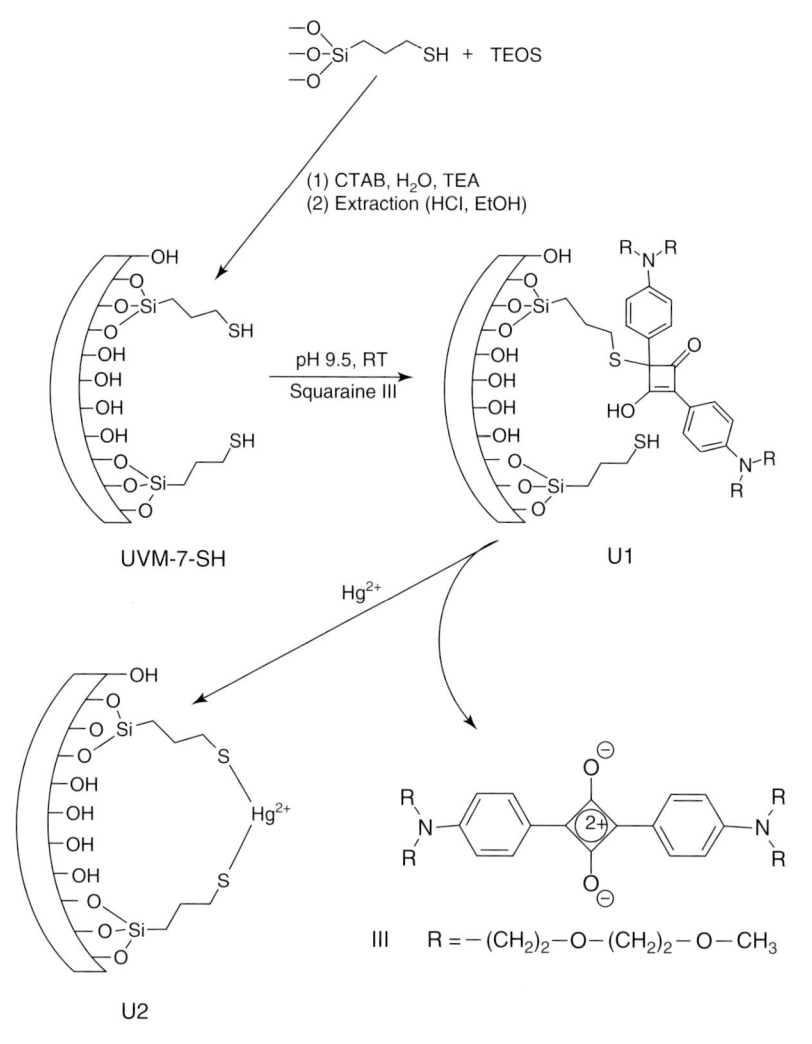

Figure 4.2 Synthesis route and chemosensor design protocol of mesoporous adsorbents for selective indication of Hg^{2+} ions. The thiol groups modified mesoporous silica (UVM-7-SH) was synthesized through co-condensation by using CTAB as the template, mercaptopropyltriethoxysilane and TEOS as the silica sources, and triethylamine (TEA) as a basic catalyst. After the removal of CTAB templates by ethanol (containing HCl) extraction method, the obtained UVM-7-SH sample was treated with dye squaraine III (the indicator), resulting in dye-immobilized mesoporous silica (U1). On contact with Hg^{2+}, the materials can convert into U2 by complexing with Hg^{2+} and release the blue, fluorescent squaraine dye III to the solution, indicating the Hg^{2+} concentration. (Source: Reproduced from Ref. [17].)

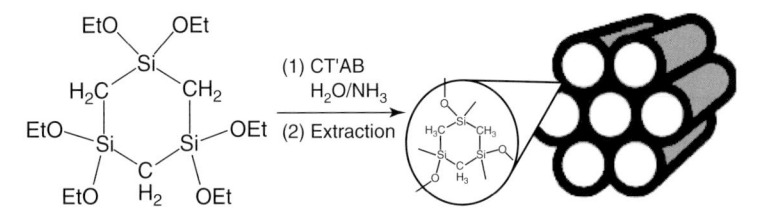

Figure 4.3 The synthesis of periodic mesoporous organosilicas (PMOs). Step 1, formation of mesostructured CTAB/organosilica composite through the ammonia catalyzed hydrolysis and condensation of a cyclic trimer $[SiCH_2(OEt)_2]_3$ precursor in the presence of CTAB as a structure-directing agent; Step 2, removal of CTAB template through solvent extraction. (Source: Reproduced from Ref. [18].)

squaraine dye into the solution, which turns deep blue and fluoresces strongly. Naked-eye Hg^{2+}-detection is thus accomplished in an easy-to-use procedure. The sample U1 not only acts as a chemodosimeter that signals the presence of Hg^{2+} down to parts-per-billion concentrations but at the same time is also an excellent adsorbent for the removal of mercury cations from aqueous solution. The amount of adsorbed mercury ranges from 0.7 to 1.7 mmol g^{-1}, depending on the degree of the functionalization.

Functional mesoporous organosilica and carbon materials are also capable of capacious adsorption of metals. Especially for the well-developed ordered mesoporous carbons (OMCs), they hold many attractive features such as high porosity, tunable pore structures and sizes, good stability, and easy surface modification, which may render them as the next generation of adsorbents. Periodic mesoporous organosilicas (PMOs) are a new class of organic–inorganic hybrid nanomaterials, which were developed right after the discovery of ordered mesoporous silicas (Figure 4.3) [18–21]. In PMO materials, the organic bridging groups are incorporated into the pore channel walls, which can tune the chemical and physical properties of the materials and enable a broad range of applications such as adsorption, catalysis and sensing, and separations. Wu et al. [22] synthesized well-ordered cubic benzene-bridged PMOs functionalized with variable contents of thiol groups via a one-step co-condensation of 1,4-bis(triethoxysilyl)benzene (BTEB) and MPTMS in a highly acidic medium with the use of cetyltrimethylammonium bromide (CTAB) as a structure-directing agent. The thiol-functionalized benzene-silica materials with different molar percentage ratio (x) of MPTMS/(BTEB+MPTMS) were investigated for metal ion adsorption (Table 4.1). These materials exhibited high adsorption affinity toward metal ions such as Hg^{2+} and Ag^+, but not for Cd^{2+}, Co^{2+}, and Pb^{2+}. The evidence that there are strong interactions between the thiol groups and Hg^{2+} and Ag^+ ions is provided by NMR. Zhao and coworkers synthesized OMC FDU-15 functionalized with carboxyl group. This material exhibits high affinity toward metal ions, and the adsorption capacity can be achieved at 1.0 mmol g^{-1} [23].

Table 4.1 Sulfur contents and Hg^{2+} and Ag^+ adsorption capacities of the thiol-functionalized benzene-silica materials with different thiol contents in the metal ion mixture (SH-Bz-x).[a]

x	S content (mmol g^{-1})	Hg (mg g^{-1})	Ag (mg g^{-1})	Hg/S	Ag/S
10.0	0.68	112	140	0.8	1.9
12.5	0.79	157	172	1.0	2.0
16.7	1.09	221	216	1.0	1.8
25.0	1.33	320	296	1.2	2.1

[a] The adsorption capacities for Cd^{2+}, Pb^{2+}, and Co^{2+} ions are 26, 38, and 36 mg g^{-1}, respectively.
Source: Reproduced from Ref. [22].

4.2 Adsorption of Anions

Many anions, such as arsenate, chromate, selenate, perrhenate, and phosphate, can be present in wastewater as pollutant sources. Waters contaminated by these anions have raised widespread public attention. Specific binding of these anions is one of the most challenging problems in chemistry, biology, materials, and environmental science. The basic strategy to trap such toxic anions by ordered mesoporous silicas is to immobilize amino groups or metal-chelated ligands on the mesopore surface. The former could directly bind anions through positive amino ligands, whereas the latter usually requires sequential grafting of organic ligands and immobilization of metal ions. On the basis of the former method, Hamoudi et al. [24] prepared mesoporous silica SBA-15-based sorbents via a postsynthesis grafting method, using 3-aminopropyltrimethoxysilane (N-silane), [1-(2-aminoethyl)-3-aminopropyl] trimethoysilane (NN-silane), and 1-[3-(trimethoxysilyl)-propyl]-diethylenetriamine (NNN-silane), followed by acidification in HCl solution to convert the attached surface amino groups to positively charged ammonium moieties. For the same organoalkoxysilane/silica molar ratio, the adsorption capacity increased markedly with increase in the number of protonated amines in the functional groups. Therefore, maximum adsorption capacities of 1.07, 1.70, and 2.46 mmol H_2PO_4 g^{-1} adsorbent were obtained using mono-, di-, and triammonium-functionalized SBA-15, respectively. For the metal-chelated ligands-grafted material, proper ligands are pregrafted on silica surface, while metal ions can chelate with the ligands as well as targeted anions. The binding to anions involves direct coordination or displacement of one ligand to the metal center. For example, ethylenediamine-modified mesoporous silicas can chelate Cu^{2+}, further binding arsenate and chromate with high capacity and selectivity [25]. Yoshitake and coworkers [26] reported that Fe^{3+}, Co^{2+}, Ni^{2+}, and Cu^{2+} ions fixed by diamino-functionalized hexagonal mesoporous silica MCM-41 and cubic MCM-48 work as adsorption centers for arsenate ions. The effectiveness of the removal of arsenate from a dilute solution is of the order $Fe^{3+} > H^+ \sim Co^{2+} > Ni^{2+} \sim Cu^{2+}$, which is slightly different from the adsorption capacity: $Fe^{3+} > Co^{2+} \sim H^+ > Ni^{2+} > Cu^{2+}$. The Fe center binds nearly three arsenate ions, which allows the adsorption capacity of Fe/NN-MCM-48 to be the largest one at 2.5 mmol g^{-1} (of adsorbent) among the adsorbents of arsenate. Therefore, Fe^{3+} ions coordinated to amino ligands fixed on mesoporous silica MCM-41 work as a strong adsorbent for toxic oxyanions. The maximum adsorption amounts can be achieved to 1.56, 0.99, 0.81, and 1.29 mmol g^{-1} for arsenate, chromate, selenate, and molybdate, respectively [27].

Compared to heavy metal ions, the removal of anions by ordered mesoporous materials based on the adsorption process has been rarely studied, probably because of both negatively charged silica surface and anions, resulting in the difficulty to graft stable connections capable of binding both silica surface and targeted anions. Future studies may try to synthesize novel mesoporous composites with certain species (e.g., iron oxides) of high affinity being exposed to targeted anions (e.g., arsenate).

4.3
Adsorption of Organic Pollutants

The effluent from many industries, such as printing and dying and textiles, contains a wide variety of organic pollutants. Ordered mesoporous materials are widely used as adsorbents to remove many organic compounds from aqueous streams. One of the most studied topics is the adsorption of dyes. Similar to binding heavy metal ions, ordered mesoporous silicas modified with carboxylic and amino groups have been fabricated for capturing basic and acidic dyes, respectively. It shows excellent performance (good selectivity, large capacity, and extremely rapid adsorption rate) because of the designed strong electrostatic interactions and high surface areas. Both the adsorbent and the dye can be easily recovered by simple washing with an alkaline or acid solution. Ho *et al.* [28] prepared ordered mesoporous silica MCM-41 adsorbents by grafting amino- and carboxylic-containing functional groups for the removal of Acid blue 25 and methylene blue (MB) dyes from wastewater. The amino-containing mesoporous silica adsorbents have a large adsorption capacity and a strong affinity for the Acid blue 25. It can selectively remove Acid blue 25 from a mixture of dyes (i.e., Acid blue 25 and MB). The ordered mesoporous silica functionalized with -COOH group is a good adsorbent for MB, displaying excellent adsorption capacity and selectivity. Furthermore, these adsorbents can be regenerated by simple washing with alkaline or acid solution to recover both the adsorbents and the adsorbed dyes. Yan *et al.* [29] synthesized anhydride-group-grafted mesoporous silica by direct co-condensation of silanes with TEOS in the presence of different surfactant templates, which led to silica materials with different pore sizes and a high density of carboxylic acid groups. These materials were used as adsorbents for the removal of three basic dyestuffs (MB, phenosafranine, and night blue) from wastewater. The measurements performed showed that, probably because of their high surface area, good affinity of carboxylic groups and large number of binding sites, the mesoporous materials obtained exhibit a high adsorption capacity and an extremely rapid adsorption rate. Furthermore, these carboxylic-functionalized adsorbents can be regenerated by simply washing with acid solution to recover both the adsorbents and the adsorbed dyes.

Without surface modification, surfactants in as-made ordered mesoporous silicas are essential for removing organic molecules. Zhao et al. [30] synthesized mesoporous silica materials by using self-assembling micellar aggregates cetylpyridinium bromide (CPB) and CTAB, respectively. The authors systemically investigated the effect of the composition (the presence and absence of surfactants, different kinds of templates) on the sorption performance. The as-prepared materials showed excellent retention performance toward chloroacetic acids, toluene, naphthalene, and methyl orange, and the materials without surfactants do not show such performance.

In contrast, as organic pollutants are always more or less hydrophobic, one can envisage that OMCs would be better adsorbents than silica. Carbon is inert, stable, light, and mainly hydrophobic, presenting high affinity toward organic pollutants. Moreover, unlike the prerequisite of generating binding sites in the ordered mesoporous silica case, only physisorption is enough to efficiently adsorb many organics on carbon materials. Yuan et al. [31] used MB and neutral red (NR) as probe molecules to investigate adsorption behaviors of OMCs synthesized by using mesoporous silica SBA-15 as the hard templates. The results showed that the volume of the mesopores with size larger than 3.5 nm is a crucial factor for the adsorption capacity and rate of MB on mesoporous carbons. However, the most probable pore diameter of mesoporous carbons was found to be vital to the adsorption capacity and rate of NR. Wu et al. [32] synthesized highly ordered mesoporous carbonaceous phenol-formaldehyde resins with two-dimensionally (2D) hexagonal and three-dimensionally (3D) bicontinuous cubic mesostructures by nanocasting process using SBA-15 and KIT-6 as hard templates. These ordered mesoporous carbonaceous phenolic resins exhibit excellent performances in removing toxic basic organic compounds from wastewater, with fast adsorption kinetics, high adsorption affinity, and large adsorption capacities. Asouhidou et al. [33] investigated OMC CMK-3 samples with hexagonal structure and a disordered mesoporous carbon for the sorption of Remazol Red 3BS (C.I. 239) dye in comparison to three commercial activated carbons (Takeda 5A, Calgon carbon, and Norit SAE-2) and mesoporous silica with a wormhole pore structure. The results showed that the CMK-3 sample exhibits higher sorption capacities (576 mg g^{-1}) in comparison to the commercial activated carbons (181 mg g^{-1}), which possess either microporous (Takeda 5A and Calgon carbon) or combined micro-/mesoporous (Norit SAE-2) structures and the mesoporous silica.

Indeed, OMCs templated from mesoporous silicas by using nanocasting approach are widely adopted for removing dyes and other organic compounds with high adsorption capacities and fast diffusion kinetics [34]. With the recent advances in the preparation of OMCs by organic–organic self-assembly, these materials are more likely to be the next cheap generation but highly efficient adsorbents for removing organic pollutants. Zhuang et al. [35] synthesized OMCs through

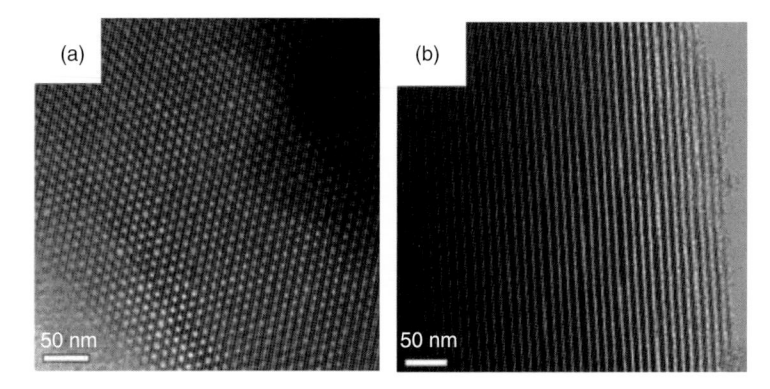

Figure 4.4 TEM images for the mesoporous carbons (MPSC/C) derived from the mesoporous silica–carbon (MPSC) composites by the silica using HF solution, viewed from the directions parallel (a) and perpendicular (b) to the pore channels. (Source: Reproduced from Ref. [35].)

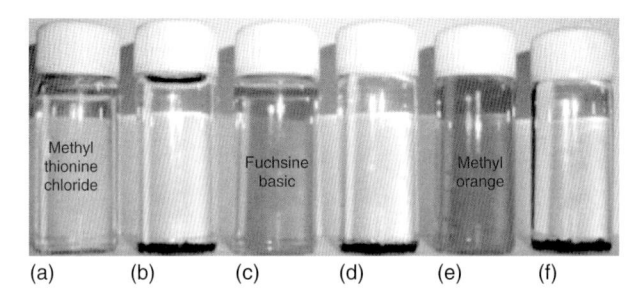

Figure 4.5 Optical photographs of the dye-polluted water (a,c,e) before and (b,d,f) after the adsorption by the mesoporous carbons (MPSC/C) derived from the mesoporous silica–carbon (MPSC) composites by dissolving the silica using HF solution. (a,b) Basic dye methylthionine chloride (MC), (c,d) basic dye fuchsine basic (FB), and (e,f) acidic and azo dye methyl orange (MO). (Source: Reproduced from Ref. [35].)

a tricomponent self-assembly method. In this method, triblock copolymer F127 is used as the template, TEOS and resol are used as the silica and carbon sources, respectively, and ethanol is used as the solvent. After thermosetting the F127/resol/silica mesostructured composite and prolysis treatment, ordered mesoporous silica–carbon (MPSC) hybrid materials are obtained. Removal of the silica component with HF solution can give rise to mesoporous carbon (MPSC/C) with high surface area, large pore volume, and bimodal pore size distribution (Figures 4.4 and 4.5). The authors demonstrated the application of the novel OMCs in efficient disposal of wastewater containing bulky dye molecules. The adsorption amount for the bulky dye (methylthionine chloride, fuchsine basic, rhodamine B, brilliant yellow, methyl orange, or Sudan G) is almost twice that of the activated carbon in which mesopores contribute almost 100% to the total surface area and volume. The OMC adsorbent has a high adsorption rate (>99.9%) for low-concentration dyes; good performance in decoloration regardless of the dye nature, including basic, acidic, or azo dyes; and high stability after dye elution. The authors found that the spatial effect of dye molecules is the determinative factor for adsorption in OMCs with various pore textural properties. The mesoporous carbon with an extremely high surface area (~2580 m² g^{-1}), a large pore volume (2.16 cm³ g^{-1}), and bimodal pores (6.4 and 1.7 nm) prepared from the silica–carbon composite shows the highest adsorption capacities for bulky basic dyes.

4.4 Multifunctional Modification of Sorbents

Besides the effort to create high porosity and specific surface functionalities, considerate work has been done to endow mesoporous materials with more sophisticated functions, such as magnetic properties, to achieve fast and effective adsorption and separation processes. MCs, a family of cyclic heptapeptide toxins with over 80 analogs, have a large molecular dimension (2–3 nm) (Figure 4.6). They are produced by the cyanobacterial cells when they die. Microcystins are very toxic for plants and animals as well as humans. They may cause serious damage to the liver and strongly inhibit protein phosphatases. Unfortunately, microcystins are extremely stable and resist common chemical breakdown such

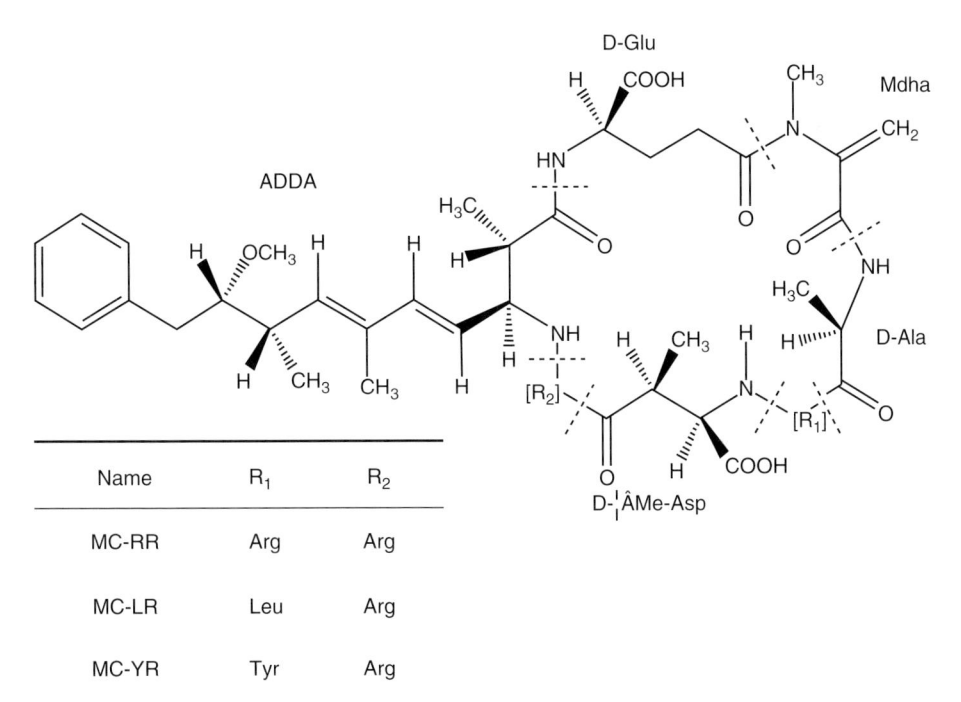

Name	R$_1$	R$_2$
MC-RR	Arg	Arg
MC-LR	Leu	Arg
MC-YR	Tyr	Arg

Figure 4.6 Chemical structure of microcystins (Wherein, ADDA, D-Glu, Mdha, D-Ala, and D-U-Me-Asp refer to 3-amino-9-methoxy-2,6,8-trimethyl-10-phenyldeca-4, 6-dienoic acid, D-glutamic acid, N-methyldehydroalanine, D-Alanine, and D-erythro-U-methylaspartic acid, respectively.). (Source: Reproduced from Ref. [36].)

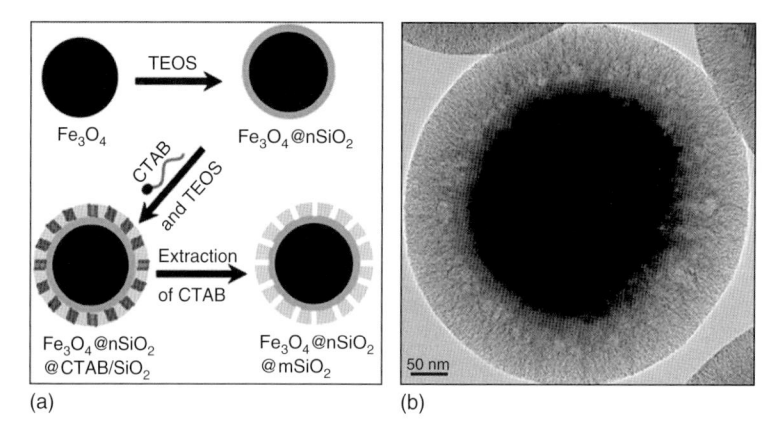

Figure 4.7 (a) Synthesis route: first, magnetic Fe_3O_4 particles are coated with a nonporous SiO_2 ($nSiO_2$) layer by a sol–gel method. Then, the obtained $Fe_3O_4@nSiO_2$ microspheres are coated with a layer of mesostructured $CTAB/SiO_2$ composite by using CTAB as a template and TEOS as a silica source. Finally, CTAB templates are removed through refluxing in acetone, resulting in the core-shell magnetic mesoporous silica microspheres. (b) TEM image of the core-shell magnetic mesoporous silica microspheres. (Source: Reproduced from Ref. [36].)

as hydrolysis or oxidation. Therefore, the adsorption of microcystins by using high surface area porous materials is promising for the removal of microcystins in waters. Deng et al. [36] synthesized superparamagnetic microspheres with a $Fe_3O_4@SiO_2$ core and a perpendicularly aligned mesoporous silica shell through a surfactant-templating sol–gel approach (Figure 4.7a). The magnetic mesoporous microspheres (Figure 4.7b) possess high magnetization (53.3 emu g^{-1}), high surface area (365 m^2 g^{-1}), large pore volume (0.29 cm^3 g^{-1}), and uniform mesopore (2.3 nm). Using the unique core–shell microspheres with accessible large pores and excellent magnetic responsiveness, a fast removal of microcystins with high efficiency (>95%) can be achieved. Remarkably, these novel magnetic mesoporous microspheres can be conveniently recycled with a magnet and exhibit excellent reusability.

Zhai et al. [37] synthesized OMC materials with magnetic frameworks via a "one-pot" block-copolymer self-assembly strategy associated with a direct carbonization process from resol, ferric citrate, and the triblock copolymer Pluronic F127. The mesoporous nanocomposites obtained possess nanosized γ-Fe_2O_3 confined in amorphous carbon frameworks, an ordered 2D hexagonal (space group $P6mm$) mesostructure, uniform mesopores (~4.0 nm), and high surface areas (up to 590 m^2 g^{-1}) and pore volumes (up to 0.48 cm^3 g^{-1}). Maghemite nanocrystals with a small particle size (<9.3 nm) are confined in the matrix of amorphous carbon frameworks. The maghemite/carbon nanocomposites exhibit excellent superparamagnetic behaviors that are favorable for their application, such as magnetic separation. Further H_2O_2 oxidation treatment of the magnetic nanocomposites endows plenty of oxygen-containing functional groups on the carbon surface, which improves their hydrophilic properties significantly. Such modified

Figure 4.8 (a) SEM and (b) TEM images of the hierarchically macroporous/mesoporous silica monoliths obtained by using commercial polyurethane foams as the scaffold. The inset in (a) shows the photograph of the hierarchically porous silica monoliths. (Source: Reproduced from Ref. [38].)

nanocomposites with hydrophilic and magnetic framework show evidently improved adsorption properties of water and fuchsine base dye molecules in water and an easy separation procedure. Xue et al. [38] fabricated hierarchically porous silica monoliths through a confinement self-assembly approach in the skeletal scaffolds of commercial polyurethane (PU) foams (Figure 4.8). The porous silica monoliths exhibit large macropores (100–500 μm in diameter), adjustable uniform mesopores (6.5–9.3 nm in diameter), high surface areas (340–780 m² g⁻¹), and large pore volumes (0.48–1.16 cm³ g⁻¹). The resulting macroporous/mesoporous silica monoliths show an excellent capability in adsorption and removal of microcystin-LR in wastewater. The hierarchically porous silica monoliths provide a fast mass transfer during adsorption of microcystins and attained a fast and efficient removal of microcystins (96% in 10 min).

4.5
Photocatalytic Degradation of Organic Pollutants

The photocatalytic decomposition of organic pollutants in water has been widely studied by using semiconductor nanomaterials as the photocatalysts. A semiconductor nanoparticle, on irradiation with light of certain wavelength corresponding to its bandgap, can be excited to generate numerous electron–hole pairs. The electron and hole can migrate to particles' surface and initiate the reductive and oxidative reaction, respectively. The hole can oxidize H_2O and/or hydroxyl groups to form hydroxyl radicals, while the electron can reduce dissolved O_2 molecules to form various species, such as O_2^- and H_2O_2 [39, 40]. During the photocatalytic degradation of organic pollutant in water solution, the compounds can be adsorbed on the semiconductor and oxidized and decomposed directly. Many metal oxides, such as TiO_2, ZnO, WO_3, $SrTiO_3$, and $BiVO_4$, have been explored as photocatalysts.

Among them, TiO$_2$ is the most popular one because of its unique characteristics, such as chemical inertness, photostability, and biocompatibility [41]. It is well known that, as a heterogeneous process, photocatalysis is a surface-dependent reaction; therefore, metal oxide catalysts with high surface area are highly desirable for photocatalytic degradation in water purification and remediation.

Zhao and coworkers [42], using the triblock copolymer Pluronic P123 (EO$_{20}$PO$_{70}$EO$_{20}$) as a template, synthesized ordered mesoporous titania phosphate materials (TiPO$_4$) by incorporating phosphorus from phosphoric acid directly into the framework of TiO$_2$. The materials obtained possess high surface area up to 300 m^2 g^{-1}, crystallized frameworks with crystalline size of ~6.0 nm, and bandgap of ~3.17 eV. The mesoporous TiPO$_4$ materials show an excellent activity for the degradation of n-pentane with the rate two times higher than that of commercial titania nanoparticles P25. The high photocatalytic activity is speculated to be related to its extended bandgap energy, high surface area, and the existence of Ti ions in the tetrahedral coordination of the mesoporous TiPO$_4$ materials.

It is commonly accepted in photocatalysis based on semiconductor nanomaterials that small crystalline size can lead to powerful redox ability because of the quantum size effect. Moreover, the small crystalline sizes are also beneficial for the separation of the photogenerated electron–hole pairs. These features should slow the rate of electron–hole recombination and increase the photocatalytic activity. In this regard, Hirao and coworkers [43] synthesized nanosized mesoporous titania powder with high surface area and anatase crystalline walls via hydrothermal process by using CTAB as a surfactant-directing agent. The mesoporous TiO$_2$ nanoparticles obtained after calcination at 400 °C have a mean diameter of ~17 nm, pore size of ~2.5 nm, and surface area up to 320 m^2 g^{-1}. Photocatalytic degradation of Rhodamine B shows that the TiO$_2$ nanoparticles obtained are more active than both commercial P25 powder and mesoporous TiO$_2$ nanoparticles obtained under the same condition but after being calcined at a higher temperature (600 °C). The lower photocatalytic activity of the sample obtained at 600 °C is attributed to its reduced specific surface area (221 m^2 g^{-1}).

Although titania is one of the most studied semiconductor photocatalysts, it can be activated only under UV light irradiation of wavelength <387 nm owing to its large bandgap of 3.2 eV. Moreover, the photogenerated electron–hole pairs of titania are easy to recombine, which greatly influences their photocatalytic efficiency. Therefore, much work has recently been done to improve the photocatalytic efficiency of TiO$_2$ either by shifting its optical response to the visible range so as to make good use of solar energy or by deposition of metal nanoparticles (e.g., Pt) to inhibit the electron–hole recombination. Lin et al. [44] synthesized visible light responsive carbon-doped mesoporous titania films by using glucose as both a carbon source and a pore-forming agent. The films obtained have a high surface area of ~283 m^2 g^{-1} and a thickness of about 10 μm and are composed of anatase crystalline TiO$_2$ particles. The prepared C-doped TiO$_2$ films exhibited excellent photocatalytic activity in the degradation of the dye Reactive Brilliant Red X-3B under UV and visible light irradiation compared with that of the smooth TiO$_2$ and P25 films. Sanchez and coworkers [45] synthesized mesoporous

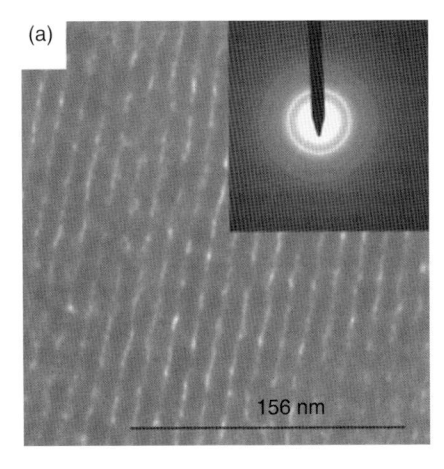

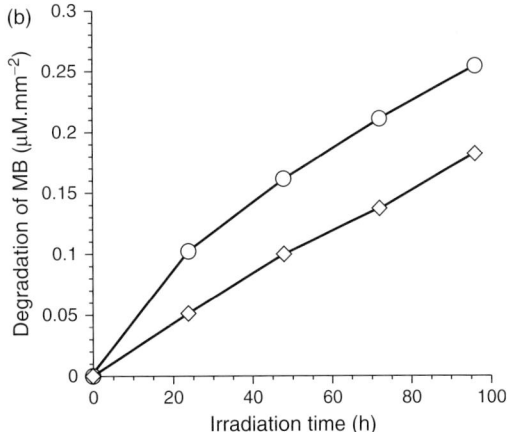

Figure 4.9 (a) TEM image of the N-doped mesoporous titania obtained by annealing at 500 °C in ammonia vapor and (b) comparison of the methylen blue (MB) photodegradation activity between N-doped mesoporous titania (circle) and conventional mesoporous TiO_2 without nitrogen-doping (rectangle) under visible light. The inset in (a) shows the selected area electronic diffraction assigned to polycrystalline anatase TiO_2 [45].

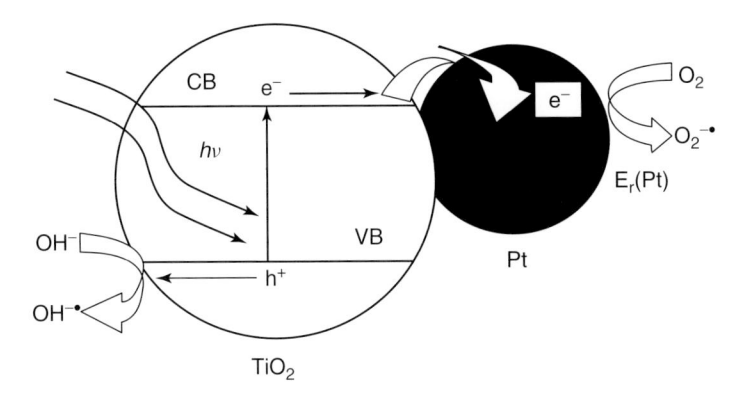

Figure 4.10 Promoted charge-carrier separation on mesoporous Pt-containing titania photocatalyst by trapping a photoinduced electron on the Pt/TiO_2 junction [46].

N-doped nanocrystalline titania films by thermal treatment of mesoporous titania templated from Pluronic P123 with ammonia vapor. Photocatalytic degradation results indicate that the N-doped mesoporous titania has a much faster degradation rate than the conventional sample without nitrogen under visible light (Figure 4.9).

Noble metals such as Pt, Pd, Ag, and Au deposited on TiO_2 can act as sinks for photoexcited electrons, hindering the recombination of charge carriers, that is, electrons and holes (Figure 4.10). On the basis of this principle, Wang et al. [46] synthesized mesoporous titania films via a soft-templated method by using titanium tetraisopropoxide as an inorganic source and Pluronic P123 as a template. Pt nanoparticles were loaded onto the calcined mesoporous titania via a

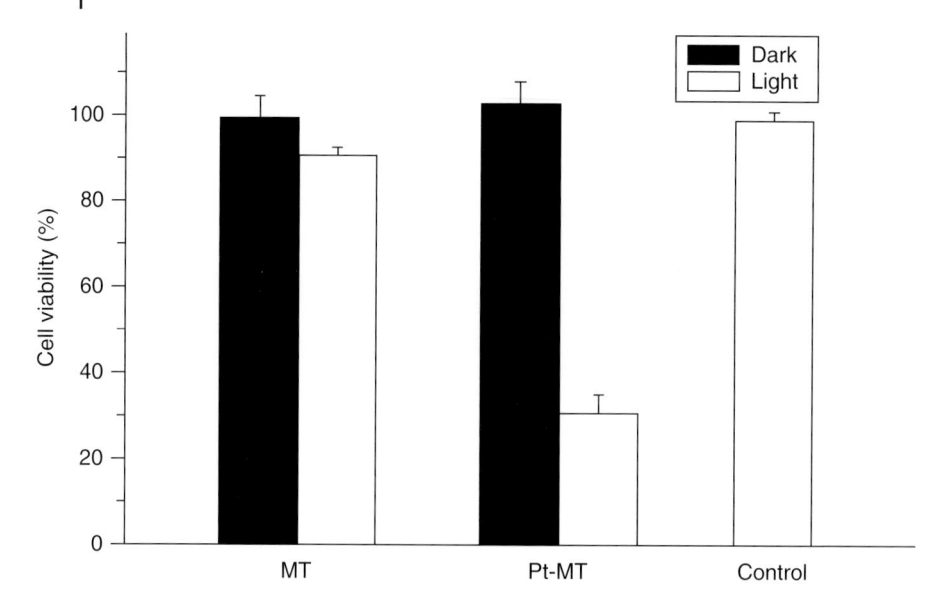

Figure 4.11 Percentage of cell survival of micrococcus lylae on the mesoporous titania (MT) and Pt-containing mesoporous titania (Pt-MT) films after irradiation with 365 nm UV light. The control experiment was carried out without the films [46].

photodeposition method, resulting in a Pt-containing film. Under 365 nm UV light, the Pt-containing mesoporous titania films exhibit an excellent performance in destroying *Micrococcus lylae*. Significant cell death was observed only for the photoactivated mesoporous Pt-TiO$_2$ film, with which 70% of the *M. lylae* cells were killed within 60 min (Figure 4.11). The significant difference between normal mesoporous titania and Pt-TiO$_2$ is due to the interaction of the trapped Pt nanoclusters with the nanoanatase-TiO$_2$ host.

4.6
Conclusions and Outlook

The rapid industrialization, urbanization, and technological development have, to some extent, resulted in severe environmental pollution, especially water pollution. In this chapter, from the viewpoint of fundamental study, we summarized the main approaches based on ordered mesoporous materials as the absorbents for wastewater treatment. The basic principle for removing hazardous compounds in wastewater is the specific and relative strong interactions between the absorbents and target compounds. Therefore, great efforts have been devoted to creating mesoporous materials with desired surface functionalities, high-density anchoring sites, and high adsorption capacity. Various synthesis methods such as postgrafting and co-condensation have been proposed to generate mesoporous materials with high-density functional moieties for specific and efficient removal of toxic species, including heavy metals, organic compounds (dyes and aromatics), and anions. Novel mesoporous materials with multifunctionalities, such as magnetic separability and photocatalytic activity, have been created and have attracted great interest because of the ease of manipulation of the absorbents and the ability to decompose the toxic organics *in situ*.

From the point of view of practical application and production cost, among various mesoporous materials, silica-based ones should be an ideal candidate absorbent for water treatment, and postgrafting method should be a more suitable strategy to create high-density surface functionalities for high-efficiency wastewater treatment. Currently, the big challenge is that the cost of silica-based mesoporous materials is still higher than that of traditional absorbents such as activated carbon and porous polymer resin (fibers or beads). Therefore, further work needs to be done to reduce the cost through innovative synthesis strategies, such as recycling the block copolymer templates during synthesis. In addition, although extensive research on water treatment using mesoporous materials has been done, systematic investigations on the synthesis, functionalization (or surface modification), adsorption, and reusability are yet to be carried out so as to build up a comprehensive guideline for mass production, functionalization, and application of mesoporous materials. Considering the excellent properties of ordered mesoporous materials in the uptake of hazardous compounds, the commercialization of mesoporous materials and even large-scale application in wastewater treatment will come into being sooner or later given that the production cost is further reduced to that of activated carbons.

Acknowledgments

This work was supported by the NSF of China (20890123, and 21073040), the State Key Basic Research Program of the PRC (2009AA033701 and 2009CB930400), the Innovation Program of Shanghai Municipal Education Commission (13ZZ004), Shanghai Rising Star Project of STCSM (12QH1400300) and Science and Technology Commission of Shanghai Municipality (08DZ2270500), and Shanghai Leading Academic Discipline Project (B108).

References

1. Zhang, J.F., Mauzerall, D.L., Zhu, T., Liang, S., Ezzati, M., and Remais, J.V. (2010) *Lancet*, **375**, 1110.
2. Wu, S.J., Zhang, L.F., and Zhang, B.R. (2010) *Ind. Water Treat.*, **30**, 1.
3. Dresge, C.T., Roth, W.J., Vartuli, J.C., and Beck, J.S. (1992) *Nature*, **359**, 710.
4. Hoffmann, F., Cornelius, M., Morell, J., and Fröba, M. (2006) *Angew. Chem. Int. Ed.*, **45**, 3216.
5. Wu, Z. and Zhao, D. (2011) *Chem. Commun.*, **47**, 3332.
6. Feng, X., Fryxell, G.E., Wang, L.Q., Kim, A.Y., Liu, J., and Kemner, K.M. (1997) *Science*, **276**, 923.
7. Bibby, A. and Mercier, L. (2002) *Chem. Mater.*, **14**, 1591.
8. Zhang, L., Zhang, W., Shi, J., Hua, Z., Li, Y., and Yan, J. (2003) *Chem. Commun.*, 210.
9. Liu, A.M., Hidajat, K., Kawi, S., and Zhao, D.Y. (2000) *Chem. Commun.*, 1145.

10. Antochshuk, V., Olkhovyk, O., Jaroniec, M., Park, I.-S., and Ryoo, R. (2003) *Langmuir*, **19**, 3031.
11. Delace, C., Gaslain, F.O.M., Lebeau, B., and Walcarius, A. (2009) *Talanta*, **79**, 877.
12. Algarra, M., Jimez, M.V., Rodruez-Castell, E., Jimez-Lez, A., and Jimez-Jimez, J. (2005) *Chemosphere*, **59**, 779.
13. Benitez, M., Das, D., Ferreira, R., Pischel, U., and Garca, H. (2006) *Chem. Mater.*, **18**, 5597.
14. Heidari, A., Younesi, H., and Mehraban, Z. (2009) *Chem. Eng. J.*, **153**, 70.
15. Jiang, Y., Gao, Q., Yu, H., Chen, Y., and Deng, F. (2007) *Microporous Mesoporous Mater.*, **103**, 316.
16. Sekar, M., Sakthi, V., and Rengaraj, S. (2004) *J. Colloid Interface Sci.*, **279**, 307.
17. Ros-Lis, J.V., Casasús, R., Comes, M., Coll, C., Marcos, M.D., Martínez-Máñez, R., Sancenón, F., Soto, J., Amorós, P., Haskouri, J.E., Garró, N., and Rurack, K. (2008) *Chem.—Eur. J.*, **14**, 8267.
18. Hatton, B., Landskron, K., Whitnall, W., Perovic, D., and Ozin, G.A. (2005) *Acc. Chem. Res.*, **38**, 305.
19. Asefa, T., MacLachlan, M.J., Coombs, N., and Ozin, G.A. (1999) *Nature*, **402**, 867.
20. Inagaki, S., Guan, S., Fukushima, Y., Ohsuna, T., and Terasaki, O. (1999) *J. Am. Chem. Soc.*, **121**, 9611.
21. Melde, B.J., Holland, B.T., Blanford, C.F., and Stein, A. (1999) *Chem. Mater.*, **11**, 3302.
22. Wu, H.-Y., Chen, C.-T., Hung, I.M., Liao, C.-H., Vetrivel, S., and Kao, H.-M. (2010) *J. Phys. Chem. C*, **114**, 7021.
23. Wu, Z.X., Webley, P.A., and Zhao, D.Y. (2010) *Langmuir*, **26**, 10277.
24. Hamoudi, S., El-Nemr, A., and Belkacemi, K. (2010) *J. Colloid Interface Sci.*, **343**, 615.
25. Fryxell, G.E., Liu, J., Hauser, T.A., Nie, Z., Ferris, K.F., Mattigod, S., Gong, M., and Hallen, R.T. (1999) *Chem. Mater.*, **11**, 2148.
26. Yoshitake, H., Yokoi, T., and Tatsumi, T. (2003) *Chem. Mater.*, **15**, 1713.
27. Yokoi, T., Tatsumi, T., and Yoshitake, H. (2004) *J. Colloid Interface Sci.*, **274**, 451.
28. Ho, K.Y., McKay, G., and Yeung, K.L. (2003) *Langmuir*, **19**, 3019.
29. Yan, Z., Tao, S., Yin, J., and Li, G. (2006) *J. Mater. Chem.*, **16**, 2347.
30. Zhao, Y.X., Ding, M.Y., and Chen, D.P. (2005) *Anal. Chim. Acta*, **542**, 193.
31. Yuan, X., Zhuo, S.-P., Xing, W., Cui, H.-Y., Dai, X.-D., Liu, X.-M., and Yan, Z.-F. (2007) *J. Colloid Interface Sci.*, **310**, 83.
32. Wu, Z.X., Meng, Y., and Zhao, D.Y. (2010) *Microporous Mesoporous Mater.*, **128**, 165.
33. Asouhidou, D.D., Triantafyllidis, K.S., Lazaridis, N.K., Matis, K.A., Kim, S.-S., and Pinnavaia, T.J. (2009) *Microporous Mesoporous Mater.*, **117**, 257.
34. Kim, D.J., Lee, H.I., Yie, J.E., Kim, S.-J., and Kim, J.M. (2005) *Carbon*, **43**, 1868.
35. Zhuang, X., Wan, Y., Feng, C.M., Shen, Y., and Zhao, D.Y. (2009) *Chem. Mater.*, **21**, 706.
36. Deng, Y., Qi, D., Deng, C., Zhang, X., and Zhao, D.Y. (2008) *J. Am. Chem. Soc.*, **130**, 28.
37. Zhai, Y.P., Dou, Y.Q., Liu, X.X., Tu, B., and Zhao, D.Y. (2009) *J. Mater. Chem.*, **19**, 3292.
38. Xue, C., Wang, J., Tu, B., and Zhao, D.Y. (2009) *Chem. Mater.*, **22**, 494.
39. Yamashita, H., Honda, M., Harada, M., Ichihashi, Y., Anpo, M., Hirao, T., Itoh, N., and Iwamoto, N. (1998) *J. Phys. Chem. B*, **102**, 10707.
40. Pan, J.H., Dou, H.Q., Xiong, Z.G., Xu, C., Ma, J.Z., and Zhao, X.S. (2010) *J. Mater. Chem.*, **20**, 4512.
41. Kuwahara, Y. and Yamashita, H. (2011) *J. Mater. Chem.*, **21**, 2407.
42. Yu, J.C., Zhang, L.Z., Zheng, Z., and Zhao, J.C. (2003) *Chem. Mater.*, **15**, 2280.
43. Peng, T.Y., Zhao, D., Dai, K., Shi, W., and Hirao, K. (2005) *J. Phys. Chem. B*, **109**, 4947.
44. Lin, X.X., Rong, F., Ji, X., and Fu, D.G. (2011) *Microporous Mesoporous Mater.*, **142**, 276.
45. Martínez-Ferrero, E., Sakatani, Y., Boissière, C., Grosso, D., Fuertes, A., Fraxedas, J., and Sanchez, C. (2007) *Adv. Funct. Mater.*, **17**, 3348.
46. Wang, X., Yu, J.C., Yip, H.Y., Wu, L., Wong, P.K., and Lai, S.Y. (2005) *Chem.—Eur. J.*, **11**, 2997.

5
Membrane Surface Nanostructuring with Terminally Anchored Polymer Chains

Yoram Cohen, Nancy Lin, Kari J. Varin, Diana Chien, and Robert F. Hicks

5.1
Introduction

Membrane separations utilize dense or porous membranes as a semipermeable barrier to selectively separate solution or suspension components. Membrane separation processes that make use of microfiltration (MF) and ultrafiltration (UF) are governed by a transmembrane pressure driving force whereas nanofiltration (NF) and reverse osmosis (RO) separations are governed by both applied pressure and osmotic pressure. In contrast, in gas separations [1, 2] and pervaporation [3, 4] processes, a transmembrane chemical potential difference is the main driving force for solute separation. Transport in dense polymeric membranes is via the solution-diffusion mechanism [5–7], whereas transport in porous membranes [8, 9] is governed by differences in the characteristic mean free path, as well as size and geometry for the gaseous molecules. MF and UF processes are common in the biotechnology industry for protein filtration and bacteria/virus removal [10, 11] and in the water treatment and desalination industry for colloidal filtration and removal of microorganisms [12–14]. RO and NF membranes are widely used in water treatment [13, 15–17] and water desalination [18–21], dairy and beverages [22–26], and wastewater reclamation/water reuse [27–29]. Gas separation membranes are common in industrial gas purification/production [1, 2, 30, 31], and pervaporation membranes are utilized in solvent recovery, as well as in dehydration and aroma and fragrance extraction [3, 4, 32, 33].

Membranes are generally classified on the basis of the size of the species being rejected by the membrane (Figure 5.1). Commercial MF and UF membranes are typically based on polysulfone (PSf) or polyether sulfone (PES) with a pore size in the range of ∼100–0.1 μm and 0.1 μm–5 nm for MF and UF membranes, respectively [34, 35]. Alternative polymers include polypropylene (PP) or polyethylene for MF and polyvinylidene fluoride (PVDF) or PES for UF membranes. NF membranes with a pore size range of ∼8–0.5 nm are typically used in water treatment [34, 35] for removal of polyvalent ions (i.e., water softening) and natural organic matter removal (for water desalination) purposes. RO membranes that have the smallest pore size (∼<0.5 nm) are typically used for water desalination given their high

Functional Nanostructured Materials and Membranes for Water Treatment, First Edition.
Edited by Mikel Duke, Dongyuan Zhao, and Raphael Semiat.
© 2013 Wiley-VCH Verlag GmbH & Co. KGaA. Published 2013 by Wiley-VCH Verlag GmbH & Co. KGaA.

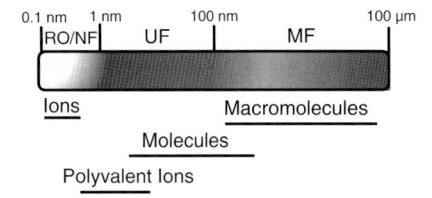

Figure 5.1 Schematic illustration of pore size ranges for different membranes.

rejection of both monovalent and divalent mineral ions. NF and RO membranes are typically thin-film-composite (TFC) membranes in which the active layer consists of either an aromatic polyamide layer of ~150–200 nm thickness or are cellulose acetate membranes, the former being more widely used. Such TFC membranes consist of a PES layer sandwiched between an aromatic polyamide layer and a nonwoven sheet [36, 37]. TFC NF membranes are often classified as "loose" RO membranes, while NF membranes are composed of the same materials, their pore size is larger than that in RO membranes, and they separate primary polyvalent ions [38]. It is noted that polypiperazine-, PSf-, and PES-based NF membranes are also commercially available (2008).

Despite the rapid growth of the membrane industry and increasing acceptability of membrane processes in various industry sectors (particularly in water treatment and desalination applications), membrane fouling by colloids, microorganisms, and organics as well as mineral scaling present an operational challenge for MF, UF, NF, and RO membrane-based processes [11, 15–17, 29]. Accordingly, this chapter first provides a brief overview of membrane fouling and scaling, followed by a review of surface structuring of MF/UF/NF/RO membranes focusing on the assessment of surface nanostructuring via graft polymerization methods for membrane performance improvements.

5.2
Membrane Fouling

The majority of industrial membrane processes operate in the cross-flow configuration (Figure 5.2). As the fluid passes through the membrane feed channel, permeate is produced under the action of applied transmembrane pressure with the target species being rejected by the membrane. As a result, species concentrations at the membrane surface increase downstream along the membrane channel (a phenomenon known as *concentration polarization*) to an extent that is dictated largely by the permeate flux and the cross-flow velocity. The cross-flow velocity governs the degree of back diffusion from the membrane surface to the bulk of the channel (by either Fickian or hydrodynamic transport) which in turns affects the development of the concentration boundary layer. The degree of CP at the membrane surface is typically estimated via the simple film model [39]:

$$\text{CP} = \frac{c_m}{c_b} = 1 - R_o + R_o \exp\left(\frac{J}{k}\right) \tag{5.1}$$

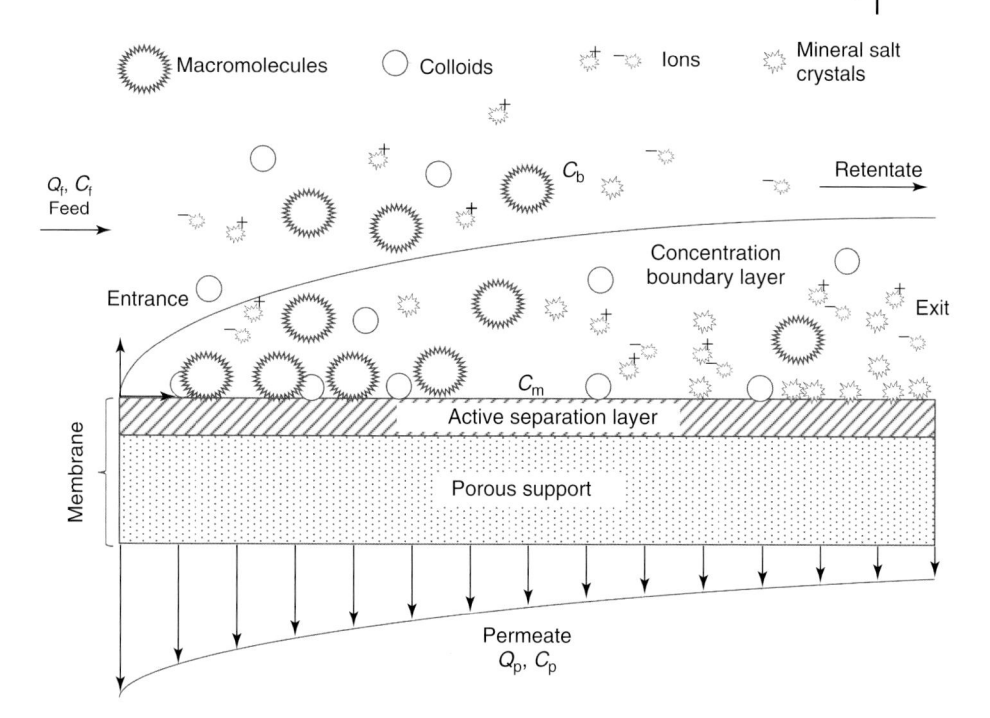

Figure 5.2 Illustration of membrane process in cross-flow configuration with foulants in the bulk and on the membrane surface (e.g., proteins, microorganisms, humic acids, polysaccharides, inorganic colloids, ions, and mineral salt crystals). Q_f: volumetric flow rate of the feed solution; C_f: concentration of the feed solution; C_b: concentration in the bulk solution; C_m: concentration at the membrane surface; Q_p: volumetric flow rate of the permeate solution; and C_p: concentration of the permeate flow.

in which C_m and C_b are solute concentrations (mg l^{-1}) near the membrane surface and in the bulk solution, respectively. The observed (nominal) species rejection is given as $R_o = 1 - C_p/C_b$, where J is the permeate flux (l m^2 s^{-1}), C_p is the permeate concentration, and k (m s^{-1}) is the feed-side solute mass transfer coefficient. Increased concentration of solution species (dissolved, colloidal, or particulate) can result in decreased permeate flux and can further lead to membrane surface fouling because of surface deposition, adsorption, or pore blockage [40, 41], all of which can constitute either reversible (i.e., easily cleaned with a solvent) or irreversible fouling where aggressive membrane cleaning may be needed to restore membrane permeability [1, 15, 42–45]. In NF and RO processes, CP can, under certain conditions, lead to significant supersaturation of the solution in the membrane feed channel, thereby leading to precipitation of mineral salts (when the concentration of the salt ions exceeds their solubility limit) and thus membrane mineral scaling [46–50].

Biofouling, organic/colloid, and inorganic/scaling are the major membrane modes of fouling resulting in permeate flux decline (Figure 5.3). Biofouling occurs when microorganisms attach to and colonize the membrane surface,

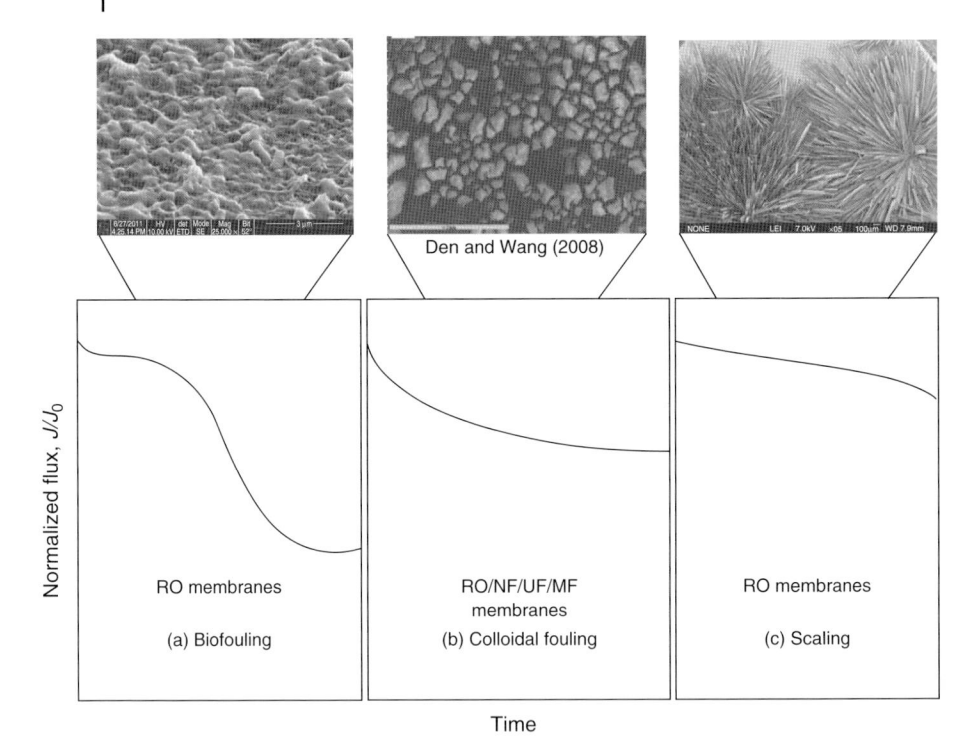

Figure 5.3 Illustration of types of membrane fouling (source of images provided below) and associated permeate flux. The flux is normalized with respect to the initial flux with typical foulant images: (a) biofouling, (b) colloidal organic fouling [64], and (c) gypsum scaling.

forming a biofilm. The initial stages of microorganisms' attachment onto the membrane surface (e.g., via extracellular polysaccharides [51]) generally causes a minor decrease in permeate flux. However, drastic flux decline can occur as microorganisms grow and colonize the membrane surface [52–55]. Various natural organics common in surface waters, such as humic acids, polysaccharides, lipids, and proteins, as well as a variety of microorganisms, can interact with the membrane surface (e.g., via van der Waals, electrostatics, and hydrophobic interactions, as well as hydrogen bonding) such that the surface attachments result in membrane fouling [56, 57]. Humic acids, polysaccharides, and biological foulants (such as proteins and bacteria/microorganisms) can interact with the membrane surface and further adsorb irreversibly onto the membrane [58, 59]. Organic, biological, and colloidal fouling are more severe at the entrance region of the membrane channel owing to both the higher permeate flux in this region and the higher mass transfer coefficient/deposition velocity at the entrance region [48]. In RO and NF membrane systems, in particular, macromolecular fouling and interaction of macromolecules such as humic acids with calcium, for example, have been shown to result in severe fouling and consequently flux decline [60–63].

Although membranes can be cleaned to restore their performance, membrane fouling and mineral scaling can damage the membrane and shorten its useful service life [65]. The aforementioned facts have motivated intensive efforts to develop various operational process mitigation strategies. However, given that membranes used for water treatment and desalination, which are the focus of this chapter, are based on a limited number of base polymers, the available range of surface chemistries is limited. Therefore, the options available for mitigating membrane fouling via the optimal choice of surface chemistry are limited. Accordingly, significant efforts have been devoted over the last three decades to develop both new membrane materials and methods of membrane surface modifications to increase membrane fouling resistance.

5.3
Strategies for Mitigation of Membrane Fouling and Scaling

Over the past few decades, various approaches have been advanced for improving membrane surface properties, such as hydrophilicity, surface topography, and surface functionality in order to decrease membrane fouling/scaling propensity. The major methods of mitigating membrane fouling include (i) surface modification with suitable hydrophilic and charged functional groups, which are generally expected to retard the adsorption of microorganisms and other hydrophobic solutes [57, 66–77]; (ii) feed pretreatment, including the use of chemical additives (e.g., dispersants and antiscalants), UF, and dissolved air floatation [18, 78–80]; and (iii) periodic membrane cleaning [81–85]. Even with feed pretreatment, membrane fouling cannot be completely eliminated. Periodic membrane cleaning is costly, time consuming, and can also accelerate the need for membrane replacement [86]. Of the aforementioned three categories of fouling mitigation, membrane surface modification (or structuring) has received significant attention and has been regarded as most critical for developing membranes of low fouling propensity [87].

Membrane surface modification is typically achieved by either physical or chemical methods. Physical modification methods to improve membrane fouling resistance and/or increase permeability include the synthesis of RO/NF membranes in which nanoparticles are embedded in the active polyamide layer of the membrane [71, 74] and physical coating of the membrane surface with polymers or surfactants [57]. Studies have also shown that adsorption of certain polymers (e.g., methylcellulose, poly(vinyl alcohol) and poly(vinylpyrrolidone) (PVP)) and surfactants (e.g., polyethyoxylates and polysorbate) onto the surface of MF/UF membranes can increase membrane permeability (for a limited period), lower fouling propensity, and/or lessen the degree of irreversible fouling [88–90]. Coating with hydrophilic polymers (i.e., polyethyleneimine, phospholipid) has been reported to decrease membrane fouling [91–93]. However, a major deficiency with physical coating of the membrane surface (e.g., with polymers, surfactants, or other chemical modifiers) is the lack of long-term stability of such coatings. Although chemical coatings (or chemical adsorption) have been shown to increase membrane permeability, while maintaining the target membrane rejection, permeability improvement is often temporary because of gradual attrition (or leaching) of the surface coating additives [94]. Alternatively, surfactants can also be used during membrane formulation to enhance membrane properties. For example, it has been shown that the incorporation of an amphiphilic copolymer such as polyacrylonitrile-*graft*-polyethylene oxide (PAN) and PSf-*graft*-polyethylene glycol to the PAN and PSf membrane casting solutions, respectively, improved the surface hydrophilicity and pure water permeability [95, 96]. Chemical surface treatment of RO/NF membranes via exposure to acids (e.g., hydrofluoric acid) [73, 76] has been shown to improve the membrane flux, albeit often at the expense of reduced salt rejection. Surface plasma treatment is another surface modification method that has been proposed in order to functionalize (or hydrophilize) membrane surfaces to improve membrane properties [68, 97–103].

Chemical surface modifications by covalently attaching preformed hydrophilic polymer chains onto the membrane surface activated using chemical initiators [67, 70, 104, 105], gamma [69, 106–108], ultraviolet (UV) [106, 109–111] irradiation, or via plasma-based polymerization [77, 112–114] have been proposed as a means of reducing membrane fouling propensity. In polymer grafting (or "grafting to"), however, the ability to attain high surface graft density is limited because of steric hindrance [115, 116]. In contrast, surface graft polymerization (or "grafting from"), whereby polymer chains are "grown" from surface active sites, enables one to attain a higher surface chain density and better control over the properties of the resulting polymer layer (Sections 5.4.1 and 5.4.2). It has been argued that surface screening by end-grafted polymer chains, partial mobility of hydrophilic end-attached polymer chains [70, 117–119], as well as reduced chain surface propensity for foulant adhesion all contribute to increased membrane fouling/scaling resistance (Figure 5.4).

Graft polymerization is typically achieved by surface activation via grafting a surface reactive initiator [66, 121] or monomer [122–124] or by using plasma [77, 125] or irradiation [67, 69, 107] to form surface reactive groups for subsequent polymerization using a suitable vinyl monomer. It is noted that when an initiator

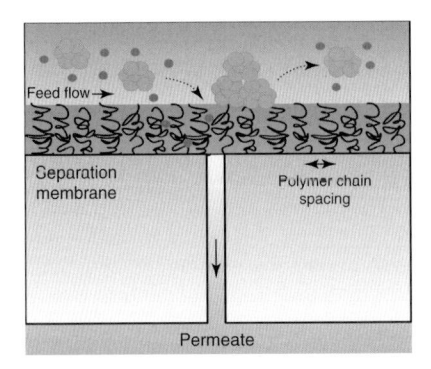

Figure 5.4 Schematic illustration of a grafted polymer layer for imparting fouling resistance to a membrane surface [120].

is used in the bulk solution, polymer chains that form in the bulk solution can lead to significant polymer grafting and thus a grafted polymer layer with chains of a broad molecular size distribution (i.e., high polydispersity) [125]. Therefore, growth of surface chains from active surface sites via addition polymerization, where the polymerization reaction is surface initiated and with limited introduction of radicals in the monomer solution, is preferred as it enables the formation of a grafted layer of higher surface chain density and lower polydispersity [126–128]. Indeed, various studies have demonstrated that surface graft polymerization of hydrophilic polymers onto MF [129–131], UF [111, 112, 131], and RO/NF [66, 70, 120, 132] can be effective in reducing the extent of biofouling, protein adhesion, as well as mineral scaling of RO membranes. The major approaches to membrane surface graft polymerization are reviewed in the subsequent sections summarizing the level of improvements in membrane performance achieved by the various major proposed methods.

5.4
Membrane Surface Structuring via Graft Polymerization

5.4.1
Overview

Surface free radical graft polymerization (FRGP) [133, 134] is a classical approach of surface modification [135, 136]. For FRGP to occur sufficiently reactive surface sites must be present on the membrane surface [134]. Active surface sites are typically reactive surface groups such as initiators [136], free radicals [77, 111, 112, 137–139], or vinyl groups [120, 125, 140, 141] that serve as anchoring sites for subsequent polymer chain growth. Surface activation methods include exposure of the membrane surface to plasma [97, 100, 112, 142], gamma [107, 108, 143] or UV irradiation [106, 110, 111, 114, 145], or a solution containing a suitable initiator [67, 70, 104, 105]. The preferred mode of chain growth is via propagation [146] (or "grafting from"), although attachment of growing chain in solution via termination with surface chains (i.e., polymer grafting or "grafting to"; Figure 5.5) can also occur in some cases [147].

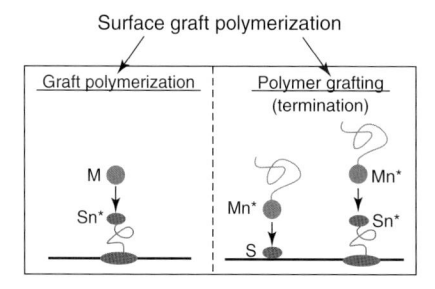

Figure 5.5 Graft polymerization and polymer grafting.

Graft polymerization from surface active sites without resorting to the use of bulk phase initiators (in the monomer solution) is preferable in order to ensure chain growth that is dominated by monomer addition (i.e., propagation reaction). Such an approach minimizes the possibility of bulk homopolymer formation and grafting onto the surface; the latter "grafting to" process leading to a broader molecular weight (MW) distribution. For example, surface grafting of an initiator onto the membrane surface [148, 149] or surface activation using a bulk initiator [67, 150–152] enables surface chain growth via thermal or chemical initiation of chain growth. Similarly, surface grafting of vinyl groups [124, 153] can serve as anchoring points. Surface activation using vinyl groups is particularly suitable for graft polymerization modification of ceramic membranes as a variety of vinyl silanes can be grafted onto inorganic oxide surfaces. For polymeric membranes, plasma and irradiation (gamma and UV) surface activation methods have been particularly popular research laboratory methods for both surface and pore-filling modification of polymeric membranes. A review of the aforementioned major membrane modification approaches is presented in Sections 5.4.2–5.4.6.

5.4.2
Reaction Schemes for Graft Polymerization

FRGP requires the presence of surface active species such as grafted vinyl monomer (Figure 5.6a), free radicals, or other surface-active species created via surface exposure to an activating solution, gas, or plasma (Figure 5.6b), or grafted initiator (Figure 5.6c). The active surface species provide anchoring sites for subsequent graft polymerization using a suitable vinyl monomer and if required additional bulk initiator.

The general reaction scheme for free radical polymerization [147] consists of surface initiation, followed by chain growth by propagation, chain transfer, and chain termination (Figure 5.7). In this reaction scheme, I_2, M, and S are the initiator, monomer, and reactive surface sites, respectively, and I· and M· are the free initiator and monomer radicals, respectively. $M_n·$ and H_n are growing and terminated homopolymer chains, respectively, consisting of n monomers (formed in the bulk as a result of homopolymerization and are not covalently bonded to the surface). $S_n·$ and G_n are the growing and terminated surface grafted chains consisting of n monomers, respectively. A bulk initiator (I_2) can undergo a series of decomposition reactions (Figure 5.7, Equations 1–3) and react with surface species (S) to create a surface bound radical (S·) that serves as the anchoring site for chain growth (Figure 5.7, Equation 5). In certain surface activation approaches, S· may be created directly through surface activation with plasma or irradiation (e.g., UV or gamma radiation) or through direct grafting of a vinyl monomer that is initiated with a bulk initiator to form a surface radical. Surface (Figure 5.7, Equation 7) and bulk chain (Figure 5.7, Equation 6) polymerizations then proceed by sequential monomer (M) addition from solution, as well as reaction of homopolymer chains (S·) in solution with surface active sites (Figure 5.7, Equations 10, 14, 15). The FRGP method results in polymer chains that are terminally and covalently anchored to the

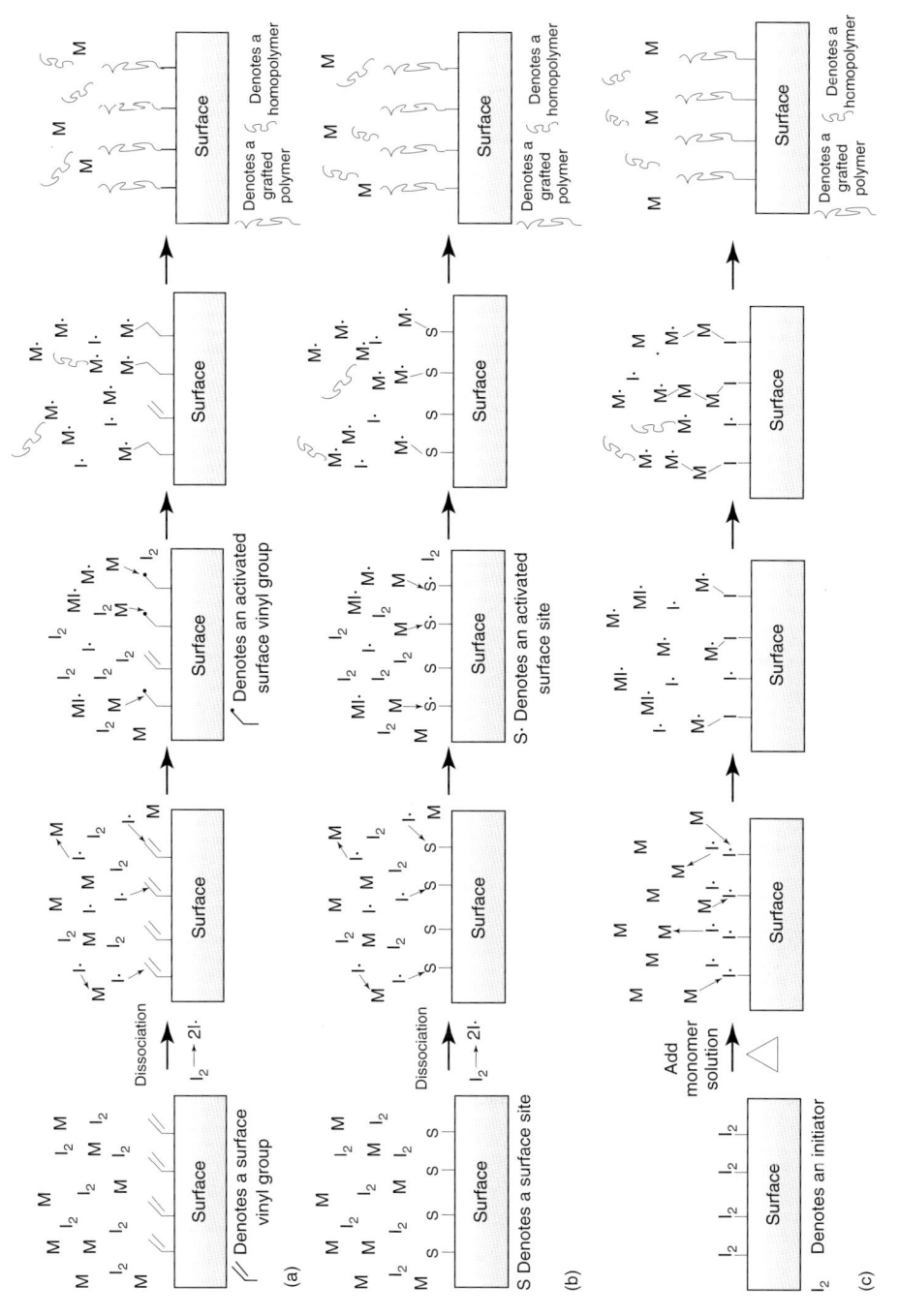

Figure 5.6 Schematic illustration of three main approaches to surface graft polymerization with surface activation using: (a) grafted vinyl monomer, (b) bulk initiator, and (c) grafted surface initiator.

Initiation		Chain transfer	
$I_2 \rightleftharpoons (I_2^{\bullet}) \rightarrow 2I\bullet$	(1)	$M_n\bullet + M \rightarrow H_n + M_1\bullet$	(8)
$I_2^S + M \rightarrow IMI$	(2)	$S_n\bullet + M \rightarrow G_n + M_1\bullet$	(9)
$IMI \rightarrow I\bullet + M_1\bullet$	(3)	$M_n\bullet + S \rightarrow H_n + S_1\bullet$	(10)
$I\bullet + M \rightarrow M_1\bullet$	(4)	$S_n\bullet + S \rightarrow G_n + S_1\bullet$	(11)
$I\bullet + S \rightarrow S_1\bullet$	(5)		
Propagation		**Termination**	
$M_n\bullet + M \rightarrow M_{n+1}\bullet$	(6)	$M_m\bullet + M_n\bullet \rightarrow H_{m+n}$	(12)
$S_n\bullet + M \rightarrow S_{n+1}\bullet$	(7)	$M_m\bullet + M_n\bullet \rightarrow H_m + H_n$	(13)
		$S_m\bullet + M_n\bullet \rightarrow G_{m+n}$	(14)
		$S_m\bullet + M_n\bullet \rightarrow G_m + H_n$	(15)
		$S_m\bullet + S_n\bullet \rightarrow G_{m+n}$	(16)
		$S_m\bullet + S_n\bullet \rightarrow G_m + G_n$	(17)

Figure 5.7 Reaction scheme for free radical graft polymerization [147]. (The subscripts n and m refer to differing number of monomers in growing or terminated chains).

surface (G_n) as well as in homopolymer chains in solution (H_n). Chain growth is also impacted by chain transfer (Figure 5.7, Equations 8–11) and chain termination (Figure 5.7, Equations 12–17). As growing chains in solution also contribute to the formation of the grafted polymer surface layer, they lead to broadening of the surface chain size distribution [116, 154].

5.4.3
Surface Activation with Vinyl Monomers

Surface graft polymerization via activation with vinyl monomers (Figure 5.6a) has been shown to be effective for developing dense polymeric brush layers [116, 120, 123, 155] suitable for the synthesis of low fouling membranes. This approach is particularly suitable for inorganic membranes where vinyl group attachments can be easily accomplished via surface silylation [116, 123, 156]. For example, end-attached PVP chains (via surface graft polymerization), forming a robust brush polymer layer of terminally and chemically anchored macromolecular chains, were shown to be effective in reducing fouling in MF and UF treatment of oil-in-water microemulsions [157] and reduction of protein adsorption [158] without attrition of the grafted modification layer as is commonly encountered in physical membrane surface coating methods.

Graft polymerization kinetic studies [147] have suggested that the contribution of polymer grafting (Figure 5.7, Equations 10, 14, 15) to the total polymer graft yield is more important at high reaction temperatures and/or low initial monomer concentrations. This has been demonstrated for vinylpyrrolidone [147] and vinylpyridine [149] graft polymerization. Membrane graft polymerization studies [116, 123, 124, 156] have shown that graft polymer yield (mg m^{-2}) generally increases with increasing temperature and initial monomer concentration. Surface chain propagation was reported to be the dominant grafting process with higher reaction temperature promoting the formation of a denser layer of shorter grafted chains, whereas a higher initial monomer concentration enhances both graft density and average grafted chain length. Although the above conclusions were derived on the basis of graft polymerization onto surface anchored vinyl silanes, the general reaction scheme for graft polymerization (Figure 5.7, Equations 6–17) should be applicable, without a loss of generality, to other graft polymerization systems irrespective of the surface anchoring scheme.

5.4.4
Surface Activation with Chemical Initiators

Membrane surface activation via exposure of the surface to a solution containing chemical initiators (Figure 5.6b) or grafting or chemisorption of an initiator (Figure 5.6c) have been proposed specifically for modification of polymeric membranes. In the first approach (Figure 5.6b), the initiator reacts with certain surface functional groups (depicted as S in Figure 5.6b) to create reactive surface radicals, whereas in the second approach (Figure 5.6c) an initiator is first physisorbed, chemisorbed, or grafted onto the membrane surface (Figure 5.6c). The decomposition of the bulk initiator or fragments of surface-bound (or sorbed) initiator can promote polymer initiation in the solution phase; thus, formed homopolymers can graft onto the surface thereby increasing surface chain polydispersity, as well as grafted polymer yield (Figures 5.6b and 5.7, Equations 14, 15) [159]. Other important considerations also include initiator-membrane-solvent compatibility and temperature required to carry out radical-forming reactions, especially when using thermal initiators.

The aforementioned approaches have been reported primarily for peroxide and azo compounds for surface nanostructuring of RO polyamide membranes [152], UF PES membranes [151], and polyolefinic and PP hollow fiber MF membranes [141, 150, 160] at a temperature range of ∼60–85 °C. Such initiators generally produce two radicals per initiator due to thermal homolytic dissociation (Figure 5.7, Equation 1) [161]. The homolytic decomposition of peroxide initiators involves the thermal cleavage of the weak O–O bond (bond dissociation energy = 154 kJ mol^{-1}) to produce radical species [162]. While the dissociation energy of the C–N bond is high (∼290 kJ mol^{-1} [161]) for azo compounds, which are typically used for activation of relatively hydrophobic polymer membrane surfaces (e.g., aromatic polyamide RO [152] membranes). Peroxide initiators have been used for graft polymerization of hydrophobic PP MF [141, 160] and PES UF [151] membranes with acrylic acid monomers. It has been proposed that for the aforementioned membranes, the initiator decomposes to form radicals, which participate in hydrogen abstraction with the surface [151, 152], thereby creating activated surface macroradical sites ([151, 163], Figure 5.6c). However, it is noted that mere exposure of the membrane surface to an initiator does not ensure that the reactive initiator species would provide covalent anchoring of the grafted polymers with the membrane surface. Graft polymerization studies on modification of aromatic polyamide RO membranes with 3-allyl-5,5,-dimethylhydantoin using 2,2′-azobisisobutyramidine dihydrochloride (AIBA) [152], reported that the positively charged AIBA in aqueous solutions [164] strongly associated (through electrostatic interactions) with the negatively charged membrane surface. Polyamide membranes modified with 3-allyl-5,5-dimethylhydantoin demonstrated reduction of microbial fouling as quantified by about a factor of three reduction in flux decline relative to the unmodified membrane [152]. MF and UF membranes modified with polyacrylic acid have demonstrated reduced protein surface adsorption (up to ∼20% [160]) and improved resistance to protein biofouling. Membrane graft polymerization via simultaneous membrane surface exposure to a reaction mixture containing the initiator, monomer, and solvent has also been reported with the use of redox initiators [66, 132, 165–167]. Such approaches have been reported, for example, for modification of aromatic polyamide RO membranes with acrylic acids, acrylates, and acrylamide type monomers [66, 104, 165]. Studies have also been reported on the modification of polyamide and polypiperazine-based NF membranes via graft polymerization of acrylic acids and methacrylates [132], PES NF membranes with acrylamides, acrylic acids, and acrylate monomers [86, 166, 167].

Persulfate initiators and oxyacids of sulfur form efficient redox-systems [168], such as the commonly used potassium/sodium persulfate and potassium/sodium metabisulfite redox initiators ($K_2S_2O_8/K_2S_2O_8$ and $K_2S_2O_5/Na_2S_2O_5$), which have been used for surface structuring of polyamide RO/NF, polypiperazine-based NF, and PES UF membranes. Graft polymerization has been achieved with acrylic acids, acrylate, and acrylamide type monomers [86, 132, 165]. In the presence of $S_2O_5^{2-}$, the persulfate $S_2O_8^{2-}$ decomposes into two free radical species SO_4^{-} [169]. In the ongoing debate [67] regarding the activity of the radical SO_4^{-}, some claim that SO_4 initially reacts with water to form an OH radical, whereas others report that the SO_4^{-} reacts directly with the surface to produce surface radicals through hydrogen abstraction. An advantage of using redox initiators is that the rate of radical productions are reasonable at moderate temperatures [161] and these initiators can be used at room temperature. As a result, the rate of polymer termination is reduced [162]. Also, reported reaction times to achieve a reasonable level of surface polymer coverage have been achieved over periods in the range of ∼15 min to as long as 6 h [132, 165, 167].

The performance of membranes modified via surface initiation-induced graft polymerization has been shown to depend on the specific monomer used (e.g., monomer containing charged and/or hydrophilic groups) and its initial concentration, molar ratio of initiator components (oxidant : reductant), reaction conditions (e.g., temperature and pH), and reaction time [66, 132, 167]. Longer reaction time and higher initial monomer concentration are likely to reduce membrane permeability, although fouling resistance would generally increase with improved grafted surface polymer coverage. Membranes modified via graft polymerization with acrylic acid and acrylate monomers have typically demonstrated better fouling resistance characteristics relative to unmodified membranes in terms of a lower flux decline. For example, ∼32% flux decline (in the course of 30 day desalting of brackish water) was reported for 3-sulfoproyl methacrylate grafted PES NF membrane compared to greater flux decline (~54%) for the unmodified base PES NF membrane [166]. Fouling resistance in desalting of brackish river water using polyamide RO membranes graft polymerized using the same monomer also demonstrated increased fouling resistance as verified by surface analysis via ATR-FTIR [165].The aforementioned studies and others have shown that in general, graft polymerized RO membranes exhibit reduction in permeability (Section 5.5).

Although chemically initiated free radical graft polymerization can be applied to different membrane surfaces activated with a variety of chemical initiators, there are several limitations to this approach. Chemical initiators can also initiate polymerization in the bulk solution (homopolymerization), thus resulting in polymer grafting (or "grafting to"). A more serious concern is the relatively long reaction periods (as long as 18 h) reported in most studies, which is infeasible for large-scale commercial deployment. Membrane modification using the aforementioned approaches in-place (i.e., with prefabricated membrane elements in their pressure vessels) is feasible for specialized applications even with such long reaction periods; however, reproducibility of the approach would be important to demonstrate for wide-scale acceptance of this membrane surface structuring technology.

5.4.5
Irradiation-Induced Graft Polymerization

Membranes' surfaces can be directly modified via irradiation or activated by irradiation for subsequent graft polymerization. Gamma and UV radiation are the two most popular methods of surface irradiation for membrane surface modifications and accordingly are discussed in Sections 5.4.5.1 and 5.4.5.2.

5.4.5.1 Gamma-Induced Graft Polymerization

Gamma irradiation is typically employed using cobalt 60 (Co^{60}). This irradiation method has been reported for surface modifications of MF and UF membranes [138, 170–172]. A variety of membranes have been modified via gamma radiation, including PP [108, 138, 171, 172], PES [170], and fluorinated Teflon [173]. Also, a variety of different monomers have been used for membrane graft polymerization via gamma-initiation such as styrene [170], acrylic acid [138, 173], acrylonitrile [171, 172], and 2-hydroxyethyl methacrylate (HEMA) [108].

Irradiation of surface molecules causes homolytic fission [67], which results in the formation of surface free radicals on the backbone of the polymeric chains of the membrane. Membrane surface irradiation, in the presence of oxygen, can lead to the formation of peroxides on the membrane surface [67]. Graft polymerization can be carried out during [170, 173–175] or directly after [138, 171, 172] surface activation by irradiation. The gamma dose is typically in the range of 0.5–20 Mrad [138, 170–172, 174] with an exposure of 0.25–0.45 Mrad h^{-1} for 0.5–75 h [138, 170–172, 174]. Polymer membrane surface activation via gamma irradiation is typically accomplished in three different ways in order to create radicals to initiate graft polymerization: (i) preirradiation, (ii) peroxidation, and (iii) mutual irradiation [67] (Figure 5.8).

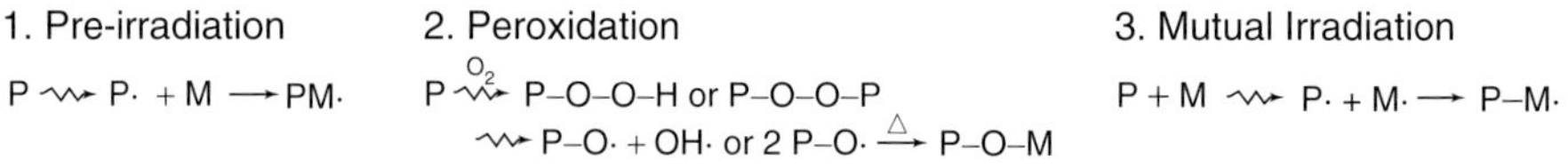

1. Pre-irradiation

P $\leadsto$ P· + M $\longrightarrow$ PM·

2. Peroxidation

P $\xrightarrow{O_2}$ P–O–O–H or P–O–O–P
$\leadsto$ P–O· + OH· or 2 P–O· $\xrightarrow{\Delta}$ P–O–M

3. Mutual Irradiation

P + M $\leadsto$ P· + M· $\longrightarrow$ P–M·

Figure 5.8 Gamma-assisted grafted polymerization mechanisms where · is a free radical, P is the polymeric membrane, M is the monomer, and $\leadsto$ indicates gamma irradiation.

In preirradiation, the polymeric membrane is exposed to gamma rays in an inert environment, and the free radicals are formed on the membrane backbone [67]. Subsequently, graft polymerization proceeds by exposing the activated surface to the desired monomer. The monomer can be in either a liquid (with or without solvent) or vapor phase. An advantage of the preirradiation method is that homopolymerization is limited as polymerization proceeds directly from surface active sites without the presence of initiators in solution. However, the above approach can result in polymer chain scission as well as surface etching and alteration of the surface pore structure [67]. In peroxidation, the membrane surface is gamma irradiated in the presence of gaseous oxygen leading to the formation of hydroperoxides or diperoxides on the membrane surface [67]. The surface peroxy groups are then exposed to the target monomer solution at an elevated temperature (60–100 °C) in order to decompose the peroxy groups to form surface radicals, which serve to initiate FRGP with a suitable vinyl monomer. An advantage of the peroxidation method is that the intermediate peroxy groups are relatively stable over several hours at 20 °C [67, 176]. Finally, in mutual irradiation, the membrane is simultaneously exposed to the monomer and gamma radiation [67]. Free radicals are formed on both the membrane surface and the monomer solution (or gas phase). This allows for graft polymerization via monomer addition to surface chains (i.e., "grafting from") and also results in the formation of polymer chains in solution (or gas phase); the latter polymerization reaction leads to increased polydispersity of surface chains as well as potentially cross-linked surface chains.

The degree of grafting (i.e., percent of the monomer in solution added to the surface) is reported to generally increase with increasing gamma irradiation dose up to a plateau region [171, 172]. The degree of grafting has also been reported to increase with initial monomer concentration and/or reaction time up to a maximum and then decreases [171, 172] as expected when the rate of chain termination and homopolymer formation increase faster than the rate of surface graft polymerization. Although the degree of grafting achieved via gamma surface activation has been reported in various studies [138, 170–173], limited characterizations have been provided regarding membrane permeability, rejection, and fouling resistance. Some partial success has been reported in imparting fouling resistance to PP (MF/UF) membranes modified via gamma irradiation/graft polymerization; however, in most cases, gain in membrane rejection was at the expense of reduced permeability [108] (Figure 5.11). For example, work [108] on modifying PP membranes (MF/UF) with HEMA (in solution) via the preirradiation method demonstrated a slight increase in membrane rejection of the protein bovine serum albumin (BSA) (from 28% up to ∼34% for $1\,g\,l^{-1}$ solution) with increased fouling resistance as indicated by over a factor of 2 flux decline reduction for the modified membrane.

Although moderate success has been achieved with surface modification via gamma radiation, this is not a low-cost process and safety considerations are paramount [67]. Also, gamma-ray irradiation typically requires long exposure times (1–24 h) to generate sufficient radical concentration on the membrane surfaces [138, 171–175], which are not feasible for commercial applications.

5.4.5.2 UV-Induced Graft Polymerization

UV-induced graft polymerization, otherwise known as *photo-induced polymerization*, has been primarily used for surface modification of MF and UF membranes [144, 177–181]. UV-assisted graft polymerization can be accomplished via direct exposure of a UV photosensitive membrane surfaces (e.g., PSf, PES) to UV light [145, 177, 178, 182–185] or with the use of a photoinitiator [181, 186–189]. In UV surface activation, the photoreactive chromospores (photosensitive sites that allow for bond rupture) enable the formation of free radical sites on the membrane backbone via photolysis [185]. Membrane surfaces containing phenoxy-phenyl sulfone sites [185] such as polyaryl sulfone [183, 185], PSf [178], and PES [145, 177, 182, 184] have been the primary candidates for UV-induced graft polymerization. Accordingly, UF and MF membranes have been the popular target of surface modification by researchers, via the above approach. Exposure of the membrane surface to UV light causes bond cleavage at the photoreactive chromospore sites, which results in the creation of free radical formation for subsequent initiation of graft polymerization. For example, UV exposure of the phenoxy-phenol sulfone group on the backbone of PES leads to chain cleavage of the carbon–sulfur bond at the sulfone linkage (Figure 5.9, [185]). This results in the formation of two radicals (aryl radical and

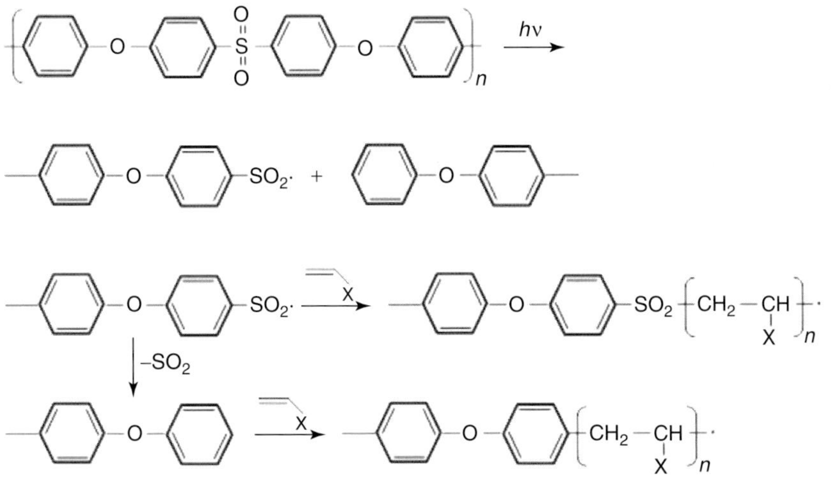

Figure 5.9 UV-assisted graft polymerization mechanism of PES [67].

sulfonyl radical) at each end of the PES chain [185]. Also, an additional radical may be formed with the loss of the sulfur dioxide on the sulfonyl radical site. The created surface radicals serve as both initiation and anchoring sites for vinyl monomer graft polymerization. UV surface exposure in the wavelength range of 200–500 nm is typically around 20–70 mW cm^{-2} for a period from 1 min to as long as 1 h [177, 181, 182, 187–189]. The degree of graft polymerization typically increases with increasing initial monomer concentration and increasing irradiation dose [144, 178, 179].

If the membrane support is not photosensitive, a photoinitiator has to be first adsorbed onto the membrane surface or absorbed into the surface region. Upon exposure to UV, the photoinitiator dissociates into free radicals that can then abstract hydrogen from the membrane polymer backbone, forming polymeric radicals that can then be used to carry out polymerization (Figure 5.3) [67, 190]. This approach is, therefore, similar to the one described in Section 5.4.2 but with UV instead of thermal initiator activation. Common photoinitiators reported for PP and poly(vinylidene fluoride) (MF/UF) membrane surface modification with a variety of monomers (e.g., HEMA, acrylamide, and 2-aminoethyl methacrylate (AEMA)) include benzoin ethyl ether [67, 186], xanthone [67], and benzophenone (BPO) [181, 186–189], the latter being particularly popular because of its peroxy group that can readily form free radicals [187–189, 181]. As described in Section 5.4.4, abstraction of hydrogen from the membrane surface polymer backbone also results in introduction of free radicals to the bulk monomer solution (during the graft polymerization process), and thus decreased grafting efficiency and grafted polydispersity of the grafted chains [190].

UV-induced graft polymerization process can be used with either photoactive membrane materials or with the use of photoinitiators. In the former case, prolonged surface exposure to UV can lead to membrane degradation [144, 179, 182] that can also be manifested by pore enlargement [182] due to polymer chain scission [144, 145, 179, 182]. Generally, polymer membrane surface structuring via UV-assisted graft polymerization, as with other techniques relying on chemical surface activation, leads to increased solute rejection at the expense of decreased permeability, in part owing to the pore blockage by the grafted polymer chains [144, 178, 179] (Figure 5.11). In some cases, however, reported slight increase in permeability (~3–5%) has been attributed to increasing surface hydrophilicity or pore widening as in the case of UF PES membrane graft polymerization with poly(ethylene glycol) methacrylate (PEGMA) [144]. It has also been shown [178] that hollow fiber PSf UF membranes can be upgraded to NF membrane performance (e.g., with $CaCl_2$ rejection of 40–90%) via diallyl dimethyl ammonium chloride (DADMAC) surface nanostructuring. Modification of MF/UF PP membranes with AEMA improved fouling resistance (quantified by ~7% lower flux decline and about 20% greater initial flux recovery (after DI water cleaning of the fouled membranes)), relative to the unmodified membranes, as quantified by filtration of 200× diluted activated sludge (350 mg l^{-1} COD) [188]. In general, fouling tests with BSA and model bacteria have shown that MF and UF membranes nanostructured with hydrophilic polymers demonstrated increased fouling resistance [42, 144, 177, 179,

182–184, 186, 191]. The fouling resistance is attributed to the increased negative surface charge imparted to the otherwise typically hydrophobic unmodified membranes enabling repulsion of negatively charged colloids and reduction of bacterial adhesion [144, 183, 186, 192, 193]. It has also been suggested that neutral surface charge, attained via surface modification, can be effective in promoting biofouling resistance [186].

5.4.6
Plasma-Initiated Graft Polymerization

Plasma-induced or assisted graft polymerization relies on the use of plasma to activate a monomer (in the gas phase) and/or the membrane surface. Plasma is an ionized gas containing positively and negatively charged species, mix with electrons and the (gas) neutral state [137]. Plasma can be formed, for example, by flowing an inert source gas, such as oxygen, nitrogen, or hydrogen, through a cloud of electrons generated by an external electric energy source [125, 137]. Membrane surface modification with plasma can be accomplished in two distinct approaches. In the first method, the membrane surface is exposed to plasma along with a monomer in the gas phase in order to create monomer radicals and active surface sites, resulting in the formation of a surface polymer layer that is partially cross-linked bonded to the substrate (Figure 5.10a) [137, 194, 195]. In the second method (Figure 5.10b) [77, 100, 112, 131, 142], the surface is first activated via plasma treatment and subsequently exposed to the monomer either in a gas or in a liquid solution (with bulk initiator if desired).

Increased membrane surface hydrophilicity via plasma (CO_2 or N_2) surface treatment alone was reported to increase the hydrophilicity of PSf UF membranes, as well as increased membrane resistance to protein fouling, attributed increased peroxide, acid, and ester groups [99, 196]. PSf UF membrane surface treatment with N_2 plasma resulted in both acidic and basic surface functionalities [99]. Plasma treatment time, power level, and the plasma gas can have a significant effect on the degree of surface activation and thus the resulting membrane performance. For example, treatment with oxygen plasma was shown to increase PAN UF membrane surface hydrophilicity [139]; however, prolonged plasma treatment time can result in membrane etching (degradation) enlarged membrane pores and thus increased membrane permeability and decreased solute rejection [68, 112, 197]. Polymer degradation (measured via polymer weight loss over time) has also been reported for glassy polymer films (e.g., polycarbonate) due to excessive exposure to oxygen plasma [198].

Plasma surface activation can be achieved with essentially most types of polymer. However, the types of surface radicals or active species depend on the plasma type. For example, poly(tetrafluoroethylene) surface treatment with argon plasma was shown (via electron spin resonance) to result in radical formation at different locations (mid-chain or end-chain) along the carbon [199]. The plasma-induced surface radicals rapidly converted to peroxy radicals when the surface was exposed to air [200]. However, there was a gradual loss of the positive effect of plasma

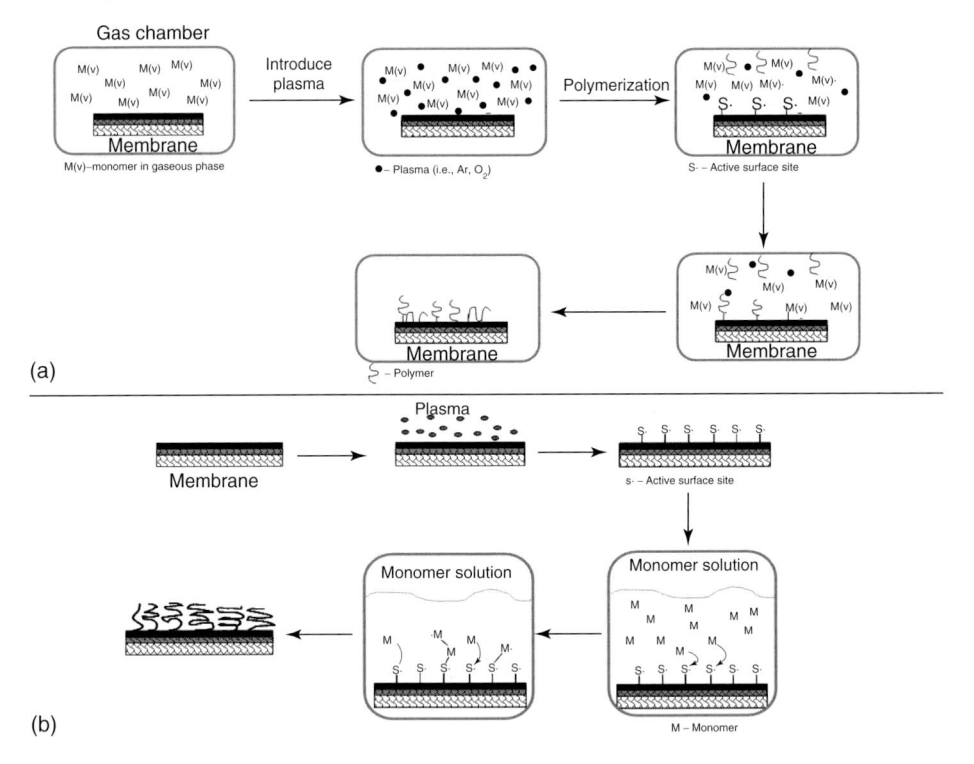

Figure 5.10 Schematic of plasma-induced graft polymerization (a) monomer in gaseous phase and plasma are introduced to initiate graft polymerization and (b) surface activation by plasma followed by graft polymerization via monomer introduction.

surface treatment [201–205] due to both reactivity with gaseous species in air or liquid solution to be treated, as well as due to chain and polar group reorientation in the surface region. Therefore, surface modification by plasma treatment alone is insufficient as the stability of the created active surface functional groups is limited.

In plasma-induced graft polymerization, plasma species such as electrons, ions, and radicals can abstract hydrogen from a carbon backbone (hydrogen abstraction) or break the C–C bond to produce surface radicals that serve as the anchoring locations for graft polymerization typically using a suitable vinyl monomer. The use of plasma has been reported, for example, for graft polymerization of acrylamide onto polyamide RO membrane [206], 2,3-epoxypropylmethacrylate onto polyurethane pervaporation membrane [207], acrylic acid onto poly(4-methyl-1-pentene) [208], n-vinylpyrrolidone onto PAN UF membrane [209], and vinyl acetate onto nylon-4 pervaporation membranes [210]. Poly(ether sulfone) UF membranes were among the first to be surface structured via plasma-induced graft polymerization [112, 137, 196] owing to the excellent thermal and chemical stability of these membranes and their widespread industrial use. For example, graft polymerized *n*-vinyl-2-pyrrolidone onto PES membranes [112], activated by low-pressure (0.2 Torr) helium plasma followed by air exposure and increased membrane permeability by ~25%. Membrane protein fouling resistance increased with initial permeability nearly completely recovered via water and caustic solution cleaning. Similar improvements, with respect to the initial permeability and in membrane BSA fouling tests, have also been reported for PAN and PSf membranes surface structured with acrylic acid, HEMA, and methacrylic acid monomers [211–213]. It is noted, that as long as active plasma species are energetic they can penetrate the subsurface and thus lead to etching [112, 197], pore enlargement [68, 139, 196], and graft polymerization within the pore space [139, 214].

Plasma surface activation before graft polymerization with a gas-phase monomer, or simultaneously with the introduction of monomer in the gas phase can also be achieved as reported in a number of studies on the modification of PAN UF, PES, and PSf MF/UF membranes with acrylamide, acrylic acid, and styrenemonomer vapors [77, 131, 139, 196, 215]. For example, PES membranes modified with low-pressure argon plasma followed by polymerization with acrylamide vapor [77] or acrylic acid [77, 131] demonstrated increased polymer grafting with increased plasma exposure time (up to 30 min) and plasma power (from 25 to 40 W). It was also shown that acrylic acid graft polymerized PES MF membranes exhibited reasonable protein BSA fouling resistance with caustic solution postfiltration, enabling up to 95% initial flux recovery for the polymer-grafted membranes relative to 80 and 64% flux recovery for Ar-plasma treated and unmodified membrane, respectively [77, 131].

Recent studies have also shown that atmospheric pressure hydrogen plasma can be successfully utilized for polyamide RO/NF membrane surface activation followed by nanostructuring via graft polymerization with hydrophilic vinyl monomers such as methacrylic acid and acrylamide [120]. Starting with high permeability membranes, it was possible to synthesize RO membranes of permeability up to approximately two to three times higher than for a commercial RO membrane (Figure 5.12) of the same level of salt rejection and surface roughness (~70 nm), while also being of lower propensity for nucleation mineral salts (demonstrated for calcium sulfate dihydrate). Simple cleaning of the membranes with water enabled complete recovery of the original membrane permeability and salt rejection. These membranes also demonstrated fouling resistance as demonstrated with alginic acid and BSA fouling tests.

Surface structuring assisted by the use of plasma has received increasing attention for use in graft polymerization methods, as surface activation is achieved without the need for chemical initiators and generation of a high density of surface active sites. Methods that rely on the use of low-pressure plasma membrane surface modification or activation and/or initiation of polymerization with monomers in the gas phase have provided a significant body of knowledge regarding the effectiveness of surface nanostructuring for improving membrane performance. However, low-pressure (vacuum) operation is unattractive for large-scale adaptation [125] due to the need for using vacuum chambers. The recent development of atmospheric pressure plasma sources [216–218] has made plasma-assisted graft polymerization possible under atmospheric pressure conditions. Therefore, membrane surface activation with atmospheric plasma followed by graft polymerization has the potential for being adapted in continuous large-scale [125] commercial membrane fabrication operations by optimizing the available plasma source area and treatment time. With plasma surface treatment, it is possible to achieve a high density of surface active sites for subsequent graft polymerization.

5.5
Summary

The majority of MF, UF, and RO membranes are made of either apolar or somewhat charged polymers. However, adequate water surface wettability and fouling resistance requires the membrane surface to be hydrophilic. Therefore, membrane surface structuring via graft polymerization has been the focus of numerous studies over the last three decades (Appendix A) focusing on improving membrane performance for water treatment, purification, and desalting applications with respect to flux, rejection, and fouling resistance. Although various approaches to surface graft polymerization have been proposed, large-scale commercial adaption of such methods are yet to emerge.

In general, surface membrane surface structuring with hydrophilic polymers, while improving membrane wettability and fouling resistance, results in reduced permeability (Figures 5.11 and 5.12) because of surface or pore blockage. A survey of published data on graft polymerization of MF and UF membranes (Figure 5.11) and NF and RO membranes (Figure 5.12) reveals that, in general, the permeability of the modified membranes is lower than that of the original membrane. In certain cases, increased permeability was observed in surface modification, but this is generally at the expense of decreased solute rejection. Fouling resistance of modified membranes is generally quantified by observation of lower flux decline when exposing the membrane to water sources (model solutions or field water) containing organic or bacterial foulants or under conditions that can lead to membrane mineral salt scaling. As the fouling resistance of the membrane is increased, it is possible that continued operation, when compared at the same initial flux and cross-flow velocity (in the membrane module), the permeability of the unmodified membrane would fall below that of the modified membrane to which fouling resistance has been imparted by graft polymerization. This improved performance is contingent on the level of fouling resistance and the degree by which the modified membrane permeability has decreased as a consequence of membrane modification. Unfortunately, the selection of commercial membranes with respect to performance and polymeric material is limited, and thus, it is often difficult to optimize the graft polymerization process to arrive at the target performance. An approach to overcome the above dilemma is to optimize the permeability and rejection of the starting membrane material such that on surface nanostructuring membrane performance is upgraded to the target level.

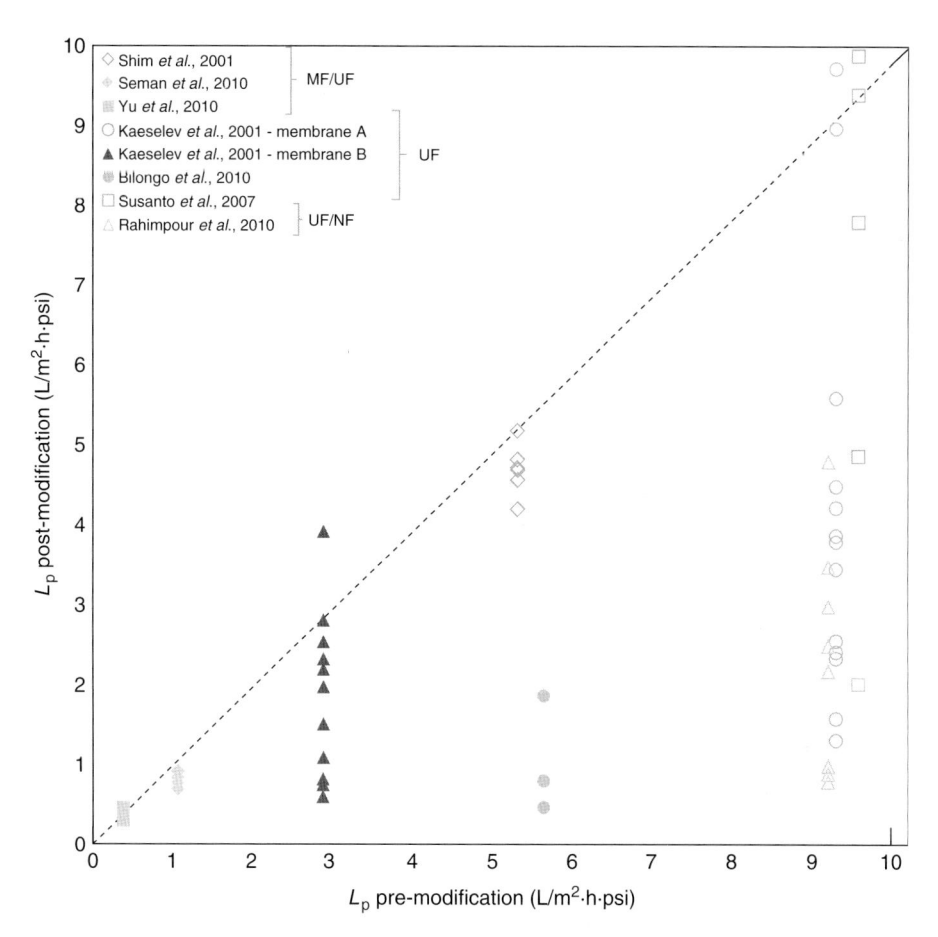

Figure 5.11 MF and UF membrane permeabilities before and after surface nanostructuring with different hydrophilic polymers. HEMA-grafted polypropylene [108], NVP-grafted polyethersulfone [177], HEA-grafted polypropylene [181], NVP, AAG, AAP-grafted polysulfone and polyethersulfone [179], DADMAC-grafted polysulfone [178], PEGMA-grafted polyethersulfone [144], and AA, HMA, mPDA-grafted polyimide [180]. (Membrane materials: PP = polypropylene, PS = polysulfone, PES = polyethersulfone, PI = polyimide, monomer for graft polymerization: HEMA = 2-hydroxyethyl methacrylate, NVP = N-vinylpyrrolidone, HEA = 2-hydroxylethyl acrylate, AAG = 2-acrylamidoglycolic acid, AAP = 2-acrylamido-2-methyl-1-propanesulfonic acid, DADMAC = diallyl dimethyl ammonium chloride, PEGMA = polyethyleneglycol methacrylate, AA = acrylic acid, and mPDA = 1,3-phenylene-diamine.) Note: Two membranes in Kaeselev et al. [179] showed decreased BSA rejection of ∼87% (△) and 0.6% (⌃), respectively, relative to the unmodified membranes, while one of the membrane in Susanto et al. [144] (□) demonstrated increased rejection (from 56 to 62% for BSA) and a slight increase (∼3%) in permeability.

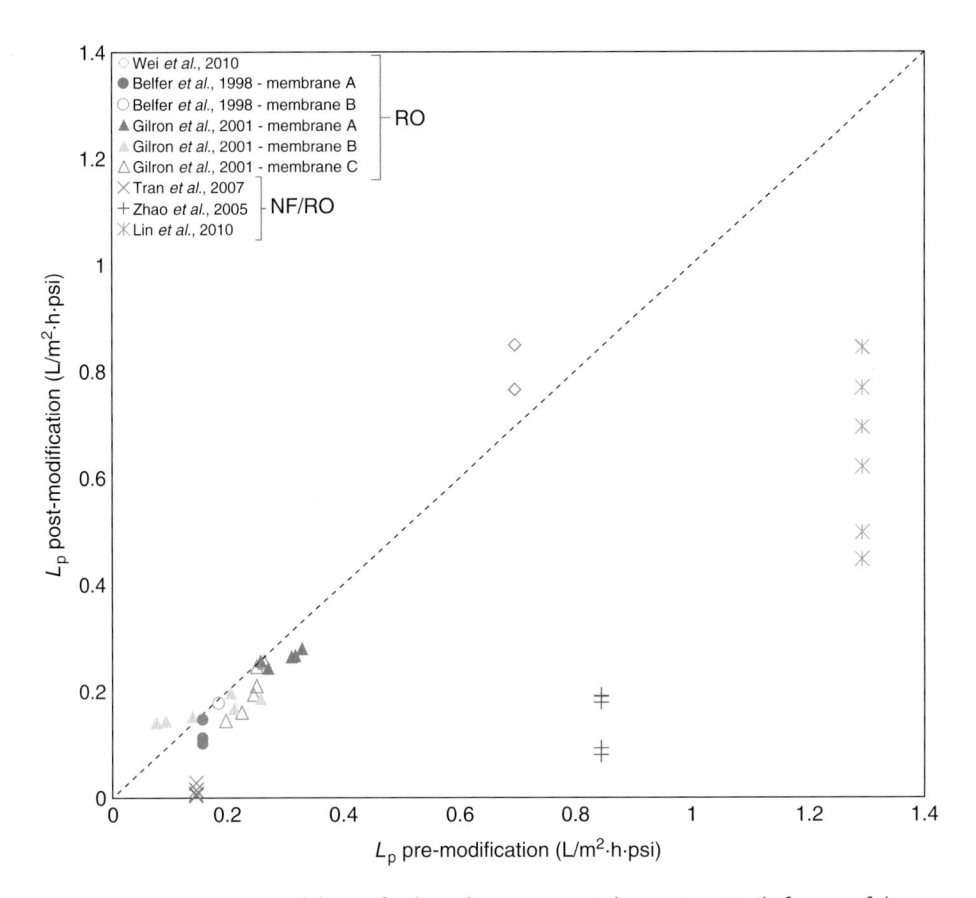

Figure 5.12 Permeabilities of polyamide RO and NF membranes before and after surface nanostructuring with different hydrophilic polymers. ADMH-grafted polyamide [152], MAA, PEGMA-grafted polyamide [66], PEGMA, AMPS, SPM-grafted polyamide [165], AA-grafted polyacrylonitrile [139], NVP-grafted polyacrylonitrile [219], and MAA, and AAm-grafted polyamide [120]. (Membrane materials: polyamide and polyacrylonitrile; monomer for graft polymerization: ADMH = 3-allyl-5,5-dimethylhydantoin, MAA = methacrylic acid, PEGMA = polyethyleneglycol methacrylate, AMPS = 2-acrylamido-2-methyl propane sulfonate, SPM = sulfopropylmethacrylate, AA = acrylic acid, NVP = N-vinylpyrrolidone, and AAm = acrylamide.)

Note: Salt rejection (NaCl) for two of the membranes in Wei et al. [152] (◊), which showed improve permeability relative to the unmodified membrane, decreased from 96.6 to 91.8%. Improved membrane permeability in Gilron et al. [165] (Δ) was accompanied by increase of ∼1–3% in salt rejection (NaCl). Also, the membranes in Lin et al. [120] were synthesized in an integrated process (interfacial polymerization followed by graft polymerization) starting with a loose NF membrane (salt rejection of ∼30%) that was nanostructured to result in RO membrane performance with sodium chloride rejection of ∼95% and permeability of up to about twice that of commercial RO membranes of similar salt rejection).

References

1. Baker, R.W. (2002). Future directions of membrane gas separation technology. *Ind. Eng. Chem. Res.*, **46**, 1393–1411.
2. Koros, W.J. and Fleming, G.K. (1993). Membrane-based gas separation. *J. Membr. Sci.*, **83**, 1–80.
3. Feng, X. and Huang, R.Y.M. (1997). Liquid separation by membrane pervaporation: a review. *Ind. Eng. Chem. Res.*, **36**, 1048–1066.
4. Smitha, B., Suhanya, D., Sridhar, S., and Ramakrishna, M. (2004). Separation of organic-organic mixtures by pervaporation – a review. *J. Membr. Sci.*, **241**, 1–21.
5. Lonsdale, H.K., Merten, U., and Riley, R.L. (2003). Transport properties of cellulose acetate osmotic membranes. *J. Appl. Polym. Sci.*, **9**, 1341–1362.
6. Robinson, J.P., Tarleton, E.S., Millington, C.R., and Nijmeijer, A. (2004). Solvent flux through dense polymeric nanofiltration membranes. *J. Membr. Sci.*, **230**, 29–37.
7. Wijmans, J.G. and Baker, R.W. (1995). The solution-diffusion model: a review. *J. Membr. Sci.*, **107**, 1–21.
8. Pandey, P. and Chauhan, R.S. (2001). Membranes for gas separation. *J. Membr. Sci.*, **26**, 853–893.
9. Schofield, R.W., Fane, A.G., and Fell, C.J.D. (1990). Gas and vapour transport through microporous membranes. I. Knudsen-poiseuille transition. *J. Membr. Sci.*, **53**, 159–171.
10. Nagata, N., Herouvis, K.J., Dziewulski, D.M., and Belfort, G. (2004). Cross-flow membrane microfiltration of a bacteriol fermentation broth. *Biotechnol. Bioeng.*, **34**, 447–466.
11. Reis, R. and Zydney, A. (2001). Membrane separation in biotechnology. *Curr. Opin. Biotechnol.*, **12**, 208–211.
12. Jacangelo, J.G. (1990). Efficacy of Membrane Processes for Microbial Control in Drinking Water. Paper presented at National Conference on Environmental Engineering (ASCE).
13. Madaeni, S.S. (1999). The application of membrane technology for water disinfection. *Water Res.*, **33**, 301–308.
14. Madsen, R.F. (1987). Membrane technology as a tool to prevent dangers to human health by water-reuse. *Desalination*, **67**, 381–393.
15. Goosen, M.F.A., Sablani, S.S., Al-Hinai, H., Al-Obeidani, S., Al-Belushi, R., and Jackson, D. (2004). Fouling of reverse osmosis and ultrafiltration membranes: a critical review. *Sep. Sci. Technol.*, **39**, 2261–2297.
16. Hilal, N., Al-Zoubi, H., Darwish, N.A., Mohamma, A.W., and Abu Ababi, M. (2004). A comprehensive review of nanofiltration membranes: treatment, pretreatment, modeling, and atomic force microscopy. *Desalination*, **170**, 281–308.
17. Jacangelo, J.G., Trussell, R.R., and Watson, M. (1997). Role of membrane technology in drinking water treatment in the United States. *Desalination*, **113**, 119–127.
18. Fritzmann, C., Lowenberg, J., Wintgens, T., and Melin, T. (2007). State-of-the-art of reverse osmosis desalination. *Desalination*, **216**, 1–76.
19. Karagiannis, I.C. and Soldatos, P.G. (2008). Water desalination cost literature: review and assessment. *Desalination*, **223**, 448–456.
20. Khawaji, A.D., Kutubkhanah, I.K., and Wei, J.-M. (2008). Advances in seawater desalination technologies. *Desalination*, **221**, 47–69.
21. Redondo, J.A. (2001). Brackish-, sea-, and wastewater desalination. *Desalination*, **138**, 29–40.
22. Akoum, O., Jaffrin, M.Y., Ding, L.H., and Frappart, M. (2004). Treatment of dairy process waters using a vibrating filtrating system adn NF and RO membranes. *J. Membr. Sci.*, **235**, 111–122.
23. Girard, B. and Fukumoto, L.R. (2000). Membrane processing of fruit juices and beverages: a review. *Food Sci. Nutr.*, **40**, 91–157.
24. Kiss, I., Vatai, G., and Bekassy-Molnar, E. (2004). Must concentrate using membrane technology. *Desalination*, **162**, 295–300.

25. Rosenberg, M. (1995). Current and future applications for membrane processes in the dairy industry. *Trends Food Sci. Technol.*, **6**, 12–19.
26. Vourch, M., Balannec, B., Chaufer, B., and Dorange, G. (2005). Nanofiltration and reverse osmosis of model process waters from the dairy industry to produce water for reuse. *Desalination*, **172**, 245–256.
27. Abu Qdais, H. and Moussa, H. (2004). Removal of heavy metals from wastewater by membrane processes: a comparative study. *Desalination*, **164**, 105–110.
28. Simon, A., Nghiem, L.D., Le-clech, P., Khan, S.J., and Drewes, J.E. (2009). Effects of membrane degradation on the removal of pharmaceutically active compounds (PhACs) by NF/RO filtration processes. *J. Membr. Sci.*, **340**, 16–25.
29. Van der Bruggen, B., Cornelis, G., Vandecasteele, C., and Devreese, I. (2005). Fouling of nanofiltration and ultrafiltration membranes applied for wastewater regeneration in the textile industry. *Desalination*, **175**, 111–119.
30. Dyer, P.N., Richards, R.E., Russek, S.L., and Taylor, D.M. (2000). Ion transport membrane technology for oxygen separation and syngas production. *Solid State Ionics*, **134**, 21–33.
31. Strathmann, H. (2004). Membrane separation process: current relevance and future opportunities. *AICHE J.*, **47**, 1077–1087.
32. Baker, R.W., Cussler, E.L., Eykamp, W., Koros, W.J., Riley, R.L., and Strathmann, H. (1991). *Membrane Separation Systems: Recent Developments and Future Directions* (Noyes Data Corporation, Park Ridge, NJ).
33. Karlsson, H.O.E. and Tragardh, G. (1993). Pervaporation of dilute organic-waters mixtures. A literature review on modelling studies and applications to aroma compound recovery. *J. Membr. Sci.*, **76**, 121–146.
34. Perry, R.H. and Green, D.W. (1997) *Perry's Chemical Engineers's Handbook*, 7th edn, McGraw-Hill, New York.
35. Van der Bruggen, B., Vandecasteele, C., Van Gestel, T., Doyen, W., and Leysen, R. (2003) A review of pressure-driven membrane processes in wastewater treatment and drinking water production. *Environ. Prog.*, **22**, 46–56.
36. Cadotte, J.E., Petersen, R.J., Larson, R.E., and Erickson, E.E. (1980) A new thin-film composite seawater reverse osmosis membrane. *Desalination*, **32**, 25–31.
37. Singh, P.S., Joshi, S.V., Trivedi, J.J., Devmurari, C.V., Rao, A.P., and Ghosh, P.K. (2006) Probing the structural variations of thin film composite RO membranes obtained by coating polyamide over polysulfone membranes of different pore dimensions. *J. Membr. Sci.*, **278**, 19–25.
38. Baker, R.W. (2004) *Membrane Technology and Applications* 2nd edn, (John Wiley & Sons, Ltd, Chichester).
39. Zydney, A. (1997) Stagnant film model for concentration polarization in membrane systems. *J. Membr. Sci.*, **130**, 275–281.
40. Mulder, M. (1996) *Basic Principles of Membrane Technology* (Kluwer Academic Publishers, Dordrecht).
41. Oliviear, R. (1997) Understanding adhesion: a means for preventing fouling. *Exp. Therm Fluid Sci.*, **14**, 316–322.
42. Hilal, N., Ogunbiyi, O.O., Miles, N.J., and Nigmatullin, R. (2005) Methods employed for control of fouling in MF and UF membranes: a comprehensive review. *Sep. Sci. Technol.*, **40**, 1957–2005.
43. Kimura, K., Hane, Y., Watanabe, Y., Amy, G., and Ohkuma, N. (2004) Irreversible membrane fouling during ultrafiltration of surface water. *Water Res.*, **38**, 3431–3441.
44. Potts, D.E., Ahlert, R.C., and Wang, S.S. (1981) A critical review of fouling of reverse osmosis membranes. *Desalination*, **36**, 235–264.
45. Yamamura, H., Kimura, K., and Watanabe, Y. (2007) Mechanism involved in the evolution of physically irreversible fouling in microfiltration and ultrafiltration membranes used for drinking water treatment. *Environ. Sci. Technol.*, **41**, 6789–6794.

46. Gouellec, Y.A.L. and Elimelech, M. (2002) Calcium sulfate (gypsum) scaling in nanofiltration of agricultural drainage water. *J. Membr. Sci.*, **205**, 279–291.
47. Hasson, D., Drak, A., and Semiat, R. (2001) Inception of $CaSO_4$ scaling on RO membranes at various water recovery levels. *Desalination*, **139**, 73–81.
48. Lyster, E. and Cohen, Y. (2007) Numerical study of concentration polarization in a rectangular reverse osmosis membrane channel: permeate flux variation and hydrodynamic end effects. *J. Membr. Sci.*, **303**, 140–153.
49. Tzotzi, C., Pahiadaki, T., Yiantsios, S.G., Karabelas, A.J., and Andritsos, N. (2007) A study of $CaCO_3$ scale formation and inhibition in RO and NF membrane processes. *J. of Membr. Sci.*, **296**, 171–184.
50. Uchymiak, M., Rahardianto, A., Lyster, E., Glater, J., and Cohen, Y. (2007) A novel RO Ex situ scale observation detector (EXSOD) for mineral scale characterization and early detection. *J. Membr. Sci.*, **291**, 86–95.
51. Kim, A.S., Chen, H., and Yuan, R. (2006) EPS biofouling in membrane filtration: an analytic modeling study. *J. Colloid Interface Sci.*, **303**, 243–249.
52. Baker, J.S. and Dudley, L.Y. (1998) Biofouling in membrane systems- a review. *Desalination*, **118**, 81–90.
53. Flemming, H.-C., Schaule, G., Griebe, T., Schmitt, J., and Tamachkiarowa, A. (1997) Biofouling- the achilles heel of membrane processes. *Desalination*, **113**, 215–225.
54. Flemming, H.C. and Schaule, G. (1988) Biofouling on membranes: a microbiological approach. *Desalination*, **70**, 95–119.
55. Sheikholeslami, R. (1999) Composite fouling- inorganic and biological. *Environ. Prog.*, **18**, 113–122.
56. Chen, V., Fane, A.G., and Fell, C.J.D. (1992) The use of anionic surfactants for reducing fouling of ultrafiltration membranes: their effects and optimization. *J. Membr. Sci.*, **67**, 249–261.
57. Wilbert, M.C., Pellegrino, J., and Zydney, A. (1998) Bench-scale of surfactant-modified reverse osmosis/nanofiltration membranes. *Desalination*, **115**, 15–32.
58. Bessiere, Y., Jefferson, B., Goslan, E., and Bacchin, P. (2009) Effect of hydrophilic/hydrophobic fractions of natural organic matter on irreversible fouling of membranes. *Desalination*, **249**, 182–187.
59. Kanani, D.M., Sun, X., and Ghosh, R. (2008) Reversible and irreversible membrane fouling during in-line microfiltration of concentrated protein solutions. *J. Membr. Sci.*, **315**, 1–10.
60. Jin, X., Huang, X., and Hoek, E.M.V. (2009) Role of specific ion interaction in seawater RO membrane fouling by alginic acid. *Environ. Sci. Technol.*, **43**, 3580–3587.
61. Schafer, A.I., Fane, A.G., and Waite, T.D. (1998) Nanofiltration of natural organic matter: removal, fouling and the influence of multivalent ions. *Desalination*, **118**, 109–122.
62. Tang, C.Y., Kwon, Y.-N., and Leckie, J.O. (2007) Fouling of reverse osmosis and nanofiltration membranes by humic acid—effects of solution composition and hydrodynamic conditions. *J. Membr. Sci.*, **290**, 86–94.
63. Yoon, S.-H., Lee, C.-H., Kim, K.-J., and Fane, A.G. (1998) Effect of calcium ion on teh fouling of nanofilter by humic acid in drinking water production. *Water Res.*, **7**, 2180–2186.
64. Den, W. and Wang, C.J. (2008) Removal of silica from brackish water by electrocoagulation pretreatment to prevent fouling of reverse osmosis membranes. *Sep. Purif. Technol.*, **59**, 318–325.
65. Tragardh, G. (1989) Membrane cleaning. *Desalination*, **71**, 325–335.
66. Belfer, S., Purinson, Y., Fainshtein, R., Radchenko, Y., and Kedem, O. (1998) Surface modification of commercial composite polyamide reverse osmosis membranes. *J. Membr. Sci.*, **139**, 175–181.
67. Bhattacharya, A. and Misra, B.N. (2004) Grafting: a versatile means to modify polymers-techniques, factors and applications. *Prog. Polym. Sci.*, **29**, 767–814.

68. Bryak, M., Gancarz, I., Pozniak, G., and Tylus, W. (2002) Modification of polysulfone membranes. 4. Ammonia plasma treatment. *Eur. Polym. J.*, **38**, 717–726.
69. Bucio, E., Arenas, E., and Burillo, G. (2006) Radiation grafting of N-isopropylacrylamide onto polypropylene films by preirradiation method. *Mol. Cryst. Liq. Cryst.*, **447**, 203–213.
70. Freger, V., Gilron, J., and Belfer, S. (2002) TFC polyamide membrane modified by grafting of hydrophilic polymers: an FT-IR/AFM/TEM study. *J. Membr. Sci.*, **209**, 283–292.
71. Jeong, B.-H., Hoek, E.M.V., Yan, Y., Subramani, A., Huang, X., Hurwitz, G., Ghosh, A.K., and Jawor, A. (2007) Interfacial polymerization of thin film nanocomposites: a New concept for reverse osmosis membranes. *J. Membr. Sci.*, **294**, 1–7.
72. Kim, S.H., Kwak, S.-Y., Sohn, B.-H., and Park, T.H. (2003) Design of TiO_2 nanoparticles self-assembled aromatic polyamide thin-film composite (TFC) membranes as an approach to solve biofouling problem. *J. Membr. Sci.*, **211**, 157–165.
73. Kulkarni, A., Mukherjee, D., and Gill, W.N. (1996) Flux enhancement by hydrophilization of thin film composite reverse osmosis membranes. *J. Membr. Sci.*, **114**, 39–50.
74. Lee, H.S., Im, S.J., Kim, J.H., Kim, H.J., Kim, J.P., and Min, B.R. (2008) Polyamide thin-film nanofiltration membranes containing TiO_2 nanoparticles. *Desalination*, **219**, 48–56.
75. Liu, L.-F., Yu, S.-C., Zhou, Y., and Gao, C.-J. (2006) Study on a novel polyamide-urea reverse osmosis composite membrane (ICIC-MPD) II. Analysis of membrane antifouling performance. *J. Membr. Sci.*, **283**, 133–146.
76. Mukherjee, D., Kulkarni, A., and Gill, W.N. (1994) Flux enhancement of thin film composite reverse osmosis membranes by chemical surface modification. *J. Membr. Sci.*, **97**, 231–249.
77. Wavhal, D.S. and Fisher, E.R. (2003) Membrane surface modification by plasma-induced polymerization of acrylamide for improved surface properties and reduced protein fouling. *Langmuir*, **19**, 79–85.
78. Brehant, A., Bonnelye, V., and Perez, M. (2002) Comparison of MF/UF pre-treatment with conventional filtration prior to RO membranes for surface seawater desalination. *Desalination*, **144**, 353–360.
79. Múñoz Elguera, A. and Pérez Báez, S.O. (2005) Development of the most adequate pre-treatment for high capacity seawater desalination plants with open intake. *Desalination*, **184**, 173–183.
80. Van Hoof, S.C.J.M., Hashimb, A., and Kordes, A.J. (1999) The effect of ultrafiltration as pretreatment to reverse osmosis in wastewater reuse and seawater desalination applications. *Desalination*, **124**, 231–242.
81. Cornelissen, E.R., Vrouwenvelder, J.S., Heijman, S.G.J., Viallefont, X.D., Van Der Kooij, D., and Wessels, L.P. (2007) Periodic air/water cleaning for control of biofouling in spiral wound membrane elements. *J. Membr. Sci.*, **287**, 94–101.
82. Lee, H., Amy, G., Cho, J., Yoon, Y., Moon, S.-H., and Kim, I.S. (2001) Cleaning strategies for flux recovery of an ultrafiltration membrane fouled by natural organic matter. *Water Res.*, **35**, 3301–3308.
83. Madaeni, S.S. and Mansourpanah, Y. (2004) Chemical cleaning of reverse osmosis membranes fouled by whey. *Desalination*, **161**, 13–24.
84. Madaeni, S.S., Mohamamdi, T., and Moghadam, M.K. (2001) Chemical cleaning of reverse osmosis membranes. *Desalination*, **134**, 77–82.
85. Muthukumaran, S., Kentish, S., Lalchandani, S., and Askokkumar, M. (2005) The optimisation of ultrasonic cleaning procedures for dairy fouled ultrafiltration membranes. *Ultrason. Sonochem.*, **12**, 29–35.
86. Belfer, S., Fainchtain, R., Purinson, Y., and Kedem, O. (2000) Surface characterization by FTIR-ATR spectroscopy

of polyethersulfone membranes-unmodified, modified and protein fouled. *J. Membr. Sci.*, **172**, 113–124.
87. Louie, J.S., Pinnau, I., Ciobanu, I., Ishida, K.P., Ng, A., and Reinhard, M. (2006) Effects of polyether-polyamide block copolymer coating on performance and fouling of reverse osmosis membranes. *J. Membr. Sci.*, **280**, 762–770.
88. Fane, A.G., Fell, C.J.D., and Kim, K.J. (1985) The effect of surfactant pretreatment on ultrafiltration of proteins. *Desalination*, **53**, 37–55.
89. Kim, A.J., Fane, A.G., and Fane, A.G. (1988) The performance of ultrafiltration membrane pretreated by polymers. *Desalination*, **70**, 229–249.
90. Xie, Y.-J., Yu, H.-Y., Wang, S.-Y., and Xu, Z.-K. (2007) Improvement of antifouling characteristics in a bioreactor of polypropylene microporous membrane by the adsorption of Tween 20. *J. Environ. Sci.*, **19**, 1461–1465.
91. Akhtar, S., Hawes, C., Dudley, L., Reed, I., and Stratford, P. (1995) Coatings reduces the fouling of microfiltration membranes. *J. Membr. Sci.*, **107**, 209–218.
92. Nystrom, M. (1989) Fouling of unmodified and modified polysulfone ultrafiltration membranes by ovalbumin. *J. Membr. Sci.*, **44**, 183–196.
93. Reuben, B.G., Perl, O., and Morgan, N.L. (1995) Phospholipid coatings for the prevention of membrane fouling. *J. Chem. Technol. Biotechnol.*, **63**, 85–91.
94. Dimov, A. and Islam, M.A. (1990) Hydrophilization of polyethylene membranes. *J. Membr. Sci.*, **50**, 97–100.
95. Asatekin, A., Kang, S., Elimelech, M., and Mayes, A.M. (2007) Anti-fouling ultrafiltration membranes containing polyacrylonitrile-graft-poly(ethylene oxide) comb copolymer additives. *J. Membr. Sci.*, **298**, 136–146.
96. Park, J.Y., Acar, M.H., Akthakul, A., Kuhlman, W., and Mayes, A.M. (2006) Polysulfone-graft-poly(ethylene glycol) graft copolymers for surface modification of polysulfone membranes. *Biomaterials*, **27**, 856–865.
97. Chu, P.K., Chen, J.Y., Wang, L.P., and Huang, N. (2002) Plasma-surface modification of biomaterials. *Mater. Sci. Eng., R Rep.*, **36**, 143–206.
98. Gancarz, I., Pozniak, G., and Bryjak, M. (1999) Modification of polysulfone membranes. 1. CO_2 Plasma treatment. *Eur. Polym. J.*, **35**, 1419–1428.
99. Gancarz, I., Pozniak, G., and Bryjak, M. (2000) Modification of polysulfone membranes. 3. Effect of nitrogen plasma. *Eur. Polym. J.*, **36**, 1563–1569.
100. Kang, M.S., Chun, B., and Kim, S.S. (2001) Surface modification of polypropylene membrane by Low-temperature plasma treatment. *J. Appl. Polym. Sci.*, **81**, 1555–1566.
101. Kim, K.S., Lee, K.H., Cho, K., and Park, C.E. (2002) Surface modification of polysulfone ultrafiltration membrane by oxygen plasma treatment. *J. Membr. Sci.*, **199**, 135–145.
102. Steen, M.L., Hymas, L., Havey, E.D., Capps, N.E., Castner, D.G., and Fisher, E.R. (2001) Low temperature plasma treatment of asymmetric polysulfone membranes for permanent hydrophilic surface modification. *J. Membr. Sci.*, **188**, 97–114.
103. Yu, H.-Y., Xie, Y.-J., Hu, M.-X., Wang, J.-L., Wang, S.-Y., and Xu, Z.-K. (2005) Surface modification of polypropylene microporous membrane to improve its antifouling property in MBR: CO_2 plasma treatment. *J. Membr. Sci.*, **254**, 219–227.
104. Belfer, S., Purinson, Y., and Kedem, O. (1998) Surface modification of commercial polyamide reverse osmosis membranes by radical grafting: An ATR-FTIR study. *Acta Polym.*, **49**, 574–582.
105. Kim, J.-H., Park, P.-K., Lee, C.-H., and Kwon, H.-H. (2008) Surface modification of nanofiltration membranes to improve the removal of organic micro-pollutants (EDCs and PhACs) in drinking water treatment: graft polymerization and cross-linking followed by functional group substitution. *J. Membr. Sci.*, **321**, 190–198.
106. Deng, J., Wang, L.P., Liu, L., and Yang, W. (2009) Developments and new applications of UV-induced surface graft

polymerizations. *Prog. Polym. Sci.*, **34**, 156–193.

107. Dessouki, A.M., Taher, N.H., and El-Arnaouty, M.B. (1998) Gamma ray induced graft copolymerization of N-vinylpyrrolidone, acrylamide and their mixtures onto polypropylene films. *Polym. Int.*, **45**, 67–76.

108. Shim, J.K., Na, H.S., Lee, Y.M., Huh, H., and Nho, Y.C. (2001) Surface modification of polypropylene membranes by γ-ray induced graft copolymerization and their solute permeation characteristics. *J. Membr. Sci.*, **190**, 215–226.

109. Susanto, H. and Ulbricht, M. (2007) Photografted thin polymer hydrogel layer on PES ultrafiltration membranes: characterization, stability, and influence on separation performance. *Langmuir*, **23**, 7818–7830.

110. Taniguchi, M., Pieracci, J., Samsonoff, W.A., and Belfort, G. (2003) UV-assisted graft polymerization of synthetic membranes: mechanistics studies. *Chem. Mater.*, **15**, 3805–3812.

111. Ulbricht, M., Matuschewski, H., Oechel, A., and Hicke, H.-G. (1996) Photo-induced graft polymerization surface modifications for the preparation of hydrophilic and low-proton-adsorbing ultrafiltration membranes. *J. Membr. Sci.*, **115**, 31–47.

112. Chen, H. and Belfort, G. (1999) Surface modification of poly(ether sulfone) ultrafiltration membranes by Low-temperature plasma-induced graft polymerization. *J. Appl. Polym. Sci.*, **72**, 1699–1711.

113. Cohen, Y. and Lewis, G.T. (2006) Graft Polymerization Induced by Atmospheric Pressure Plasma (U.S.A.).

114. Yasuda, H., Bumgarner, M.O., Marsh, H.C., and Morosoff, N. (1976) Plasma polymerization of some organic compounds and properties of polymers. *J. Polym. Sci., Part A: Polym. Chem.*, **14**, 195–224.

115. El Harrak, A., Carrot, G., Oberdisse, J., Jestin, J., and Boue, F. (2005) Atom transfer radical polymerization from silica nanoparticles using the 'grafting from' method and structural study via small-angle neutron scattering. *Polymer*, **46**, 1095–1104.

116. Nguyen, V., Yoshida, W., and Cohen, Y. (2003) Graft polymerization of vinyl acetate onto silica. *J. Appl. Polym. Sci.*, **87**, 300–310.

117. Bernards, M.T., Zhang, Z., Chen, S., and Jiang, S. (2008) Nonfouling polymer brushes via surface-initiated, Two-component atom transfer radical polymerization. *Macromolecules*, **41**, 4216–4219.

118. Kang, G., Cao, Y., Zhao, H., and Yuan, Q. (2008) Preparation and characterization of crosslinked poly(ethylene glyco) diacrylate membranes with excellent antifouling and solvent-resistant properties. *J. Membr. Sci.*, **318**, 227–232.

119. Li, Q., Xu, Z., and Pinnau, I. (2007) Fouling of reverse osmosis membranes by biopolymers in wastewater secondary effluent: Role of membrane surface properties and initial permeate flux. *J. Membr. Sci.*, **290**, 173–181.

120. Lin, N.H., Kim, M.M., Lewis, G.T., and Cohen, Y. (2010) Polymer surface nano-structuring of reverse osmosis membranes for fouling resistance and improved flux performance. *J. Mater. Chem.*, **20**, 4642–4652.

121. Lewis, G.T., Nguyen, V., Shih, W.-Y., and Cohen, Y. (2009) Reverse atom transfer radical graft polymerization of 4-vinylpyrdine onto inorganic oxide surfaces. *J. Appl. Polym. Sci.*, **113**, 437–449.

122. Bialk, M., Prucker, O., and Ruhe, J. (2002) Grafting of polymers to silid surfaces by using immobilized methacrylates. *Colloids Surf. A Physicochem. Eng. Asp.*, **198**, 543–549.

123. Browne, T., Chaimberg, M., and Cohen, Y. (1992) Graft-polymerization of vinyl-acetate onto silica. *J. Appl. Polym. Sci.*, **44**, 671–677.

124. Chaimberg, M. and Cohen, Y. (1991) Free-radical graft-polymerization of vinylpyrrolidone onto silica. *Ind. Eng. Chem. Res.*, **30**, 2534–2542.

125. Lewis, G.T., Nowling, G.R., Hicks, R.F., and Cohen, Y. (2007) Inorganic surface nanostructuring by atmospheric pressure plasma-induced graft polymerization. *Langmuir*, **23**, 10756–10764.

126. Bouhacina, T., Aimes, J.P., Gauthier, S., and Michel, D. (1997) Tribological behavior of a polymer grafted on silanized silica probed with a nanotip. *Phys. Rev. B*, **56**, 7694–7703.
127. Drummond, C., In, M., and Richetti, P. (2004) Behavior of adhesive boundary lubricated surface under shear: effect of grafted diblock copolymers. *Eur. Phys. J. E*, **15**, 159–165.
128. Orefice, R.L., Clark, A.E., and Brennan, A.B. (2006) AFM study on the interactions across interfaces containing attached polymer chains. *Macromol. Mater. Eng.*, **291**, 377–386.
129. Chen, Y., Deng, Q., Xiao, J., Nie, H., Wu, L., Zhou, W., and Huang, B. (2007) Controlled grafting from poly(vinylidene fluoride) microfiltration membranes via reverse atom transfer radical polymerization and antifouling properties. *Polymer*, **48**, 7604–7613.
130. Liu, Z.-M., Xu, Z.-K., Wang, J.-Q., Wu, J., and Fu, J.-J. (2004) Surface modification of polypropylene microfiltration membranes by graft polymerization of N-vinyl-2-pyrrolidone. *Eur. Polym. J.*, **40**, 2077–2087.
131. Wavhal, D.S. and Fisher, E.R. (2002) Hydrophilic modification of polyethersulfone membranes by low temperature plasma-induced graft polymerization. *J. Membr. Sci.*, **209**, 255–269.
132. Belfer, S., Fainshtain, R., Purinson, Y., Gilron, J., Nystrom, M., and Manttari, M. (2004) Modification of NF membrane properties by in situ redox initiated graft polymerization with hydrophilic monomers. *J. Membr. Sci.*, **239**, 55–64.
133. Castro, R.P., Monbouquette, H.G., and Cohen, Y. (2000) Shear-induced permeability changes in a polymer grafted silica membrane. *J. Membr. Sci.*, **179**, 207–220.
134. Painter, P.C. and Coleman, M.M. (1997) *Fundamentals of Polymer Science: An Introductory Text*, 2nd edn (CRC Press).
135. Edmondson, S., Osborne, V.L., and Huck, W.T.S. (2004) Polymer brushes via surface-initiated polymerizations. *Chem. Soc. Rev.*, **33**, 14–22.
136. Lewis, G. (2008) PhD thesis: Surface Nano-Structuring of Materials by Atmospheric Pressure Plasma-Induced Graft Polymerization. Chemical and Biomolecular Engineering Department, University of California, Los Angeles.
137. Inagaki, N. (1996) *Plasma Surface Modification and Plasma Polymerization* (Lancaster, PA Technomic Publishing Company, Inc..)
138. Mehta, I.K., Kumar, S., Chauhan, G.S., and Misra, B.N. (1990) Grafting onto isotactic polypropylene. 3. Gamma-rays induced graft-copolymerization of water-soluble vinyl monomers. *J. Appl. Polym. Sci.*, **41**, 1171–1180.
139. Tran, T.D., Mori, S., and Suzuki, M. (2007) Plasma modification of polyacrylonitrile ultrafiltration membrane. *Thin Solid Films*, **515**, 4148–4152.
140. Suzuki, M., Kishida, A., Iwata, H., and Ikada, Y. (1986) Graft copolymerization of acrylamide onto a polyethylene surface pretreated with a glow discharge. *Macromolecules*, **19**, 1804–1808.
141. Xu, Z., Wang, J., Shen, L., Men, D., and Xu, Y. (2002) Microporous polypropylene hollow fiber membrane: part I. Surface modification by the graft polymerization of acrylic acid. *J. Membr. Sci.*, **196**, 221–229.
142. Kim, M., Lin, N.H., Lewis, G.T., and Cohen, Y. (2010) Surface nano-structuring of reverse osmosis membranes via atmospheric pressure plasma-induced graft polymerization for reduction of mineral scaling propensity. *J. Membr. Sci.*, **354**, 142–149.
143. Misra, B.N., Mehta, I.K., Kanthwal, M., and Panjloo, S. (1987) Grafting onto poly(vinyl alcohol). II. Graft copolymerization of methyl acrylate and vinyl acetate and their binary mixture using γ-rays as initiator. *J. Polym. Sci., Part A: Polym. Chem.*, **25**, 2117–2125.
144. Susanto, H., Balakrishnan, M., and Ulbricht, M. (2007) Via surface functionalization by photograft copolymerization to low-fouling polyethersulfone-based ultrafiltration membranes. *J. Membr. Sci.*, **288**, 157–167.

145. Taniguchi, M. and Belfort, G. (2004) Low protein fouling synthetic membranes by UV-assisted surface grafting modification: varying monomer type. *J. Membr. Sci.*, **231**, 147–157.
146. Braunecker, W.A. and Matyjaszewski, K. (2007) Controlled/living radical polymerization: Features, developments, and perspectives. *Prog. Polym. Sci.*, **32**, 93–146.
147. Nguyen, V., Yoshida, W., Jou, J.D., and Cohen, Y. (2002) Kinetics of free-radical graft polymerization of 1-vinyl-2-pyrrolidone onto silica. *J. Polym. Sci., Part A: Polym. Chem.*, **40**, 26–42.
148. Lewis, G.T. and Cohen, Y. (2008) Controlled nitroxide-mediated styrene surface graft polymerization with atmospheric plasma surface activation. *Langmuir*, **24**, 13102–13112.
149. Lewis, G.T., Nguyen, V., and Cohen, Y. (2007) Synthesis of poly(4-vinylpyridine) by reverse atom transfer radical polymerization. *J. Polym. Sci., Part A: Polym. Chem.*, **45**, 5748–5758.
150. Childs, R.F., Weng, J.F., Kim, M., and Dickson, J.M. (2002) Formation of pore-filled microfiltration membranes using a combination of modified interfacial polymerization and grafting. *J. Polym. Sci., Part A: Polym. Chem.*, **40**, 242–250.
151. Shi, Q., Su, Y.L., Ning, X., Chen, W.J., Peng, J.M., and Jiang, Z.Y. (2010) Graft polymerization of methacrylic acid onto polyethersulfone for potential pH-responsive membrane materials. *J. Membr. Sci.*, **347**, 62–68.
152. Wei, X.Y., Wang, Z., Zhang, Z., Wang, J.X., and Wang, S.C. (2010) Surface modification of commercial aromatic polyamide reverse osmosis membranes by graft polymerization of 3-allyl-5,5-dimethylhydantoin. *J. Membr. Sci.*, **351**, 222–233.
153. Cohen, Y., Jou, J.D., Yoshida, W., and Bei, N. (eds) (1999) Surface modification of oxide surfaces by graft polymerization, in *Oxide Surfaces* (ed.J. Wingrave), Marcel Dekker, New York.
154. Wingrave, J.A. (2001) *Oxide Surfaces*, Marcel Dekker, New York.
155. Yoshida, W., Castro, R.P., Jou, J.D., and Cohen, Y. (2001) Multilayer alkoxysilane silylation of oxide surfaces. *Langmuir*, **17**, 5882–5888.
156. Chaimberg, M., Parnas, R., and Cohen, Y. (1989) Graft-polymerization of polyvinylpyrrolidone onto silica. *J. Appl. Polym. Sci.*, **37**, 2921–2931.
157. Faibish, R.S. and Cohen, Y. (2001) Fouling and rejection behavior of ceramic and polymer-modified ceramic membranes for ultrafiltration of Oil-in-water emulsions and microemulsions. *Colloids Surf.*, **191**, 27–40.
158. Rovira-Bru, M., Giralt, F., and Cohen, Y. (2001) Protein adsorption onto zirconia modified with terminally grafted polyvinylpyrrolidone. *J. Colloid Interface Sci.*, **235**, 70–79.
159. Sanli, O. and Pulat, E. (1993) Solvent-assisted graft-copolymerization of acrylamide on poly(ethylene-terephthalate) films using benzoyl peroxide initiator. *J. Appl. Polym. Sci.*, **47**, 1–6.
160. Wang, H., Yin, Y.H., Yang, S.T., and Li, C.B. (2009) Surface modification of polypropylene microporous membrane by grafting acrylic acid using physisorbed initiators method. *J. Appl. Polym. Sci.*, **112**, 3728–3735.
161. Odian, G.G. (1970) *Principles of Polymerization*, McGraw-Hill, New York.
162. Mishra, M.K. and Yagci, Y. (2009) *Handbook of Vinyl Polymers: Radical Polymerization, Process, and Technology*, 2nd edn, Boca Raton, FL., CRC Press).
163. Gaylord, N.G. and Mishra, M.K. (1983) Nondegradative reaction of maleic-anhydride and molten polypropylene in the presence of peroxides. *J. Polym. Sci., Part C: Polym. Lett.*, **21**, 23–30.
164. Dougherty, T. (1961) Chemistry of 2,2′-azobisisobutyramidine hydrochloride in aqueous solution-a water-soluble Azo initiator. *J. Am. Chem. Soc.*, **83**, 4849–4853.
165. Gilron, J., Belfer, S., Vaisanen, P., and Nystrom, M. (2001) Effects of surface modification on antifouling and performance properties of reverse osmosis membranes. *Desalination*, **140**, 167–179.
166. Reddy, A.V.R., Trivedi, J.J., Devmurari, C.V., Mohan, D.J., Singh, P., Rao, A.P.,

Joshi, S.V., and Ghosh, P.K. (2005) Fouling resistant membranes in desalination and water recovery. *Desalination*, **183**, 301–306.
167. Wang, M., An, Q.F., Wu, L.G., Mo, J.X., and Gao, C.J. (2007) Preparation of pH-responsive phenolphthalein poly(ether sulfone) membrane by redox-graft pore-filling polymerization technique. *J. Membr. Sci.*, **287**, 257–263.
168. Sarac, A.S. (1999) Redox polymerization. *Prog. Polym. Sci.*, **24**, 1149–1204.
169. Murugan, R. and Rao, K.P. (2003) Graft polymerization of glycidylmethacrylate onto coralline hydroxyapatite. *J. Biomater. Sci. Polym. Ed.*, **14**, 457–468.
170. Filho, A.A.M.F., and Gomes, A.S. (2006). Copolymerization of styrene onto polyethersulfone films induced by gamma ray irradiation. *Polym. Bull.*, **57**, 415–421.
171. Mehta, I.K., Sood, D.S., and Misra, B.N. (1989). Grafting onto polypropylene. 2. Solvent effect on graft-copolymerization of acrylonitrile by preirradiation method. *J. Polym. Sci., Part A: Polym. Chem.*, **27**, 53–62.
172. Misra, B.N., Sood, D.S., and Mehta, I.K. (1985) Grafting onto polypropylene. 1. Effect of solvents in gamma-radiation-induced graft-copolymerization of poly(acrylonitrile). *J. Polym. Sci., Part A: Polym. Chem.*, **23**, 1749–1757.
173. Gupta, B., Muzyyan, N., Saxena, S., Grover, N., and Alam, S. (2008) Preparation of ion exchange membranes by radiation grafting of acrylic acid on FEP films. *Radiat. Phys. Chem.*, **77**, 42–48.
174. Sportelli, L., Rosi, A., Bonincontro, A., and Cametti, C. (1987) Effect of gamma-irradiation on membranes of normal and pathological erythrocytes (beta-thalassemia). *Radiat. Environ. Biophys.*, **26**, 81–84.
175. Vijayavergiya, V. and Mookerjee, A. (1989) Effect of gamma-radiation on cation-transport across an artificial membrane. *J. Radioanal. Nucl. Chem. Art.*, **134**, 39–44.
176. Brandrup, J., Immergut, E.H., and Grulke, E.A. (1999) *Polymer Handbook*, 4th edn (John Wiley & Sons, Inc., New York).
177. Abu Seman, M.N., Khayet, M., and Hilal, N. (2010) Comparison of two different UV-grafted nanofiltration membranes prepared for reduction of humic acid fouling using acrylic acid and N-vinylpyrrolidone. *Desalination*, **287**, 19–29.
178. Bilongo, T.G., Remigy, J.C., and Clifton, M.J. (2010) Modification of hollow fibers by UV surface grafting. *J. Membr. Sci.*, **364**, 304–308.
179. Kaeselev, B., Pieracci, J., and Belfort, G. (2001) Photoinduced grafting of ultrafiltration membranes: comparison of poly(ether sulfone) and poly(sulfone). *J. Membr. Sci.*, **194**, 245–261.
180. Rahimpour, A. (2011) Preparation and modification of nano-porous polyimide (PI) membranes by UV photo-grafting process: Ultrafiltration and nanofiltration performance. *Korean J. Chem. Eng.*, **28**, 261–266.
181. Yu, H.Y., Zhou, J., Gu, J.S., and Yang, S. (2010) Manipulating membrane permeability and protein rejection of UV-modified polypropylene macroporous membrane. *J. Membr. Sci.*, **364**, 203–210.
182. Abu Seman, M.N., Khayet, M., Bin Ali, Z.I., and Hilal, N. (2010) Reduction of nanofiltration membrane fouling by UV-initiated graft polymerization technique. *J. Membr. Sci.*, **355**, 133–141.
183. Crivello, J.V., Belfort, G., and Yamagishi, H. (1995) Low fouling ultrafiltration and microfiltration aryl polysulfone. US Patent 5,468,390.
184. Khayet, M., Abu Seman, M.N., and Hilal, N. (2010) Response surface modeling and optimization of composite nanofiltration modified membranes. *J. Membr. Sci.*, **349**, 113–122.
185. Yamagishi, H., Crivello, J.V., and Belfort, G. (1995) Development of a novel photochemical technique for modifying poly(arylsulfone) ultrafiltration, membranes. *J. Membr. Sci.*, **105**, 237–247.

186. Kochkodan, V.M., Hilal, N., Goncharuk, V.V., Al-Khatib, L., and Levadna, T.I. (2006) Effect of the surface modification of polymer membranes on their microbiological fouling. *Colloid J.*, **68**, 267–273.
187. Yang, Q., Hu, M.X., Dai, Z.W., Tian, J., and Xu, Z.K. (2006) Fabrication of glycosylated surface on polymer membrane by UV-induced graft polymerization for lectin recognition. *Langmuir*, **22**, 9345–9349.
188. Yu, H.Y., He, J.M., Liu, L.Q., He, X.C., Gu, J.S., and Wei, X.W. (2007) Photoinduced graft polymerization to improve antifouling characteristics of an SMBR. *J. Membr. Sci.*, **302**, 235–242.
189. Yu, H.Y., Xu, Z.K., Lei, H., Hu, M.X., and Yang, Q. (2007) Photoinduced graft polymerization of acrylamide on polypropylene microporous membranes for the improvement of antifouling characteristics in a submerged membrane-bioreactor. *Sep. Purif. Technol.*, **53**, 119–125.
190. Hong, K.H., Liu, N., and Sun, G. (2009) UV-induced graft polymerization of acrylamide on cellulose by using immobilized benzophenone as a photo-initiator. *Eur. Polym. J.*, **45**, 2443–2449.
191. Malaisamy, R., Berry, D., Holder, D., Raskin, L., Lepak, L., and Jones, K.L. (2010) Development of reactive thin film polymer brush membranes to prevent biofouling. *J. Membr. Sci.*, **350**, 361–370.
192. Rana, D. and Matsuura, T. (2010) Surface modifications for antifouling membranes. *Chem. Rev.*, **110**, 2448–2471.
193. Van der Bruggen, B. (2009) Chemical modification of polyethersulfone nanofiltration membranes: a review. *J. Appl. Polym. Sci.*, **114**, 630–642.
194. Suzuki, S., Ogawa, T., and Hitotsuyanagi, N. (1985) Plasma-polymerized membranes of 4-vinylpyridine in reverse osmosis. *ACS Symp. Ser.*, **281**, 69–82.
195. Zhang, Y., Kang, E.T., Neoh, K.G., Huang, W., Huan, A.C.H., Zhang, H., and Lamb, R.N. (2002) Surface modification of polyimide films via plasma polymerization and deposition of ally[pentafluorobenzene. *Polymer*, **43**, 7279–7288.
196. Gancarz, I., Pozniak, G., Bryjak, M., and Frankiewicz, A. (1999) Modification of polysulfone membranes. 2. Plasma grafting and plasma polymerization of acrylic acid. *Acta Polym.*, **50**, 317–326.
197. Hoffman, A.S. (1996) Surface modification of polymers: physical, chemical, mechanical and biological methods. *Macromol. Symp.*, **101**, 443–454.
198. Kuzuya, M., Matsuno, Y., and Kondo, S.-I. (1993) The Nature of Plasmainduced Free Radicals of Polycondensed Polymers Studied by Electron Spin Resonance. International Symposium on Plasma Chemistry, Vol. 11, pp. 1386–1391.
199. Kuzuya, M., Ito, H., Kondo, S., Noda, N., and Noguchi, A. (1991) Electron spin resonance study of the special features of plasma-induced radicals and their corresponding peroxy radicals in polytetrafluoroethylene. *Macromolecules*, **24**, 6612–6617.
200. Kuzuya, M., Kondo, S., Sugito, M., and Yamashiro, T. (1998) Peroxy radical formation from plasma-induced surface radicals of polyethylene as studied by electron spin resonance. *Macromolecules*, **31**, 3230–3234.
201. Chiang, M.-H., Liao, K.-C., Lin, I.-M., Lu, C.-C., Huang, H.-Y., Kuo, C.-L., and Wu, J.-S. (2010) Modification of hydrophilic property of polypropylene films by a parallel-plate nitrogen-based dielectric barrier discharge jet. *IEEE Trans. Plasma Sci.*, **38**, 1489–1498.
202. Lai, J., Sunderland, B., Xue, J., Yan, S., Zhao, W., Folkard, M., Michael, B.D., and Wang, Y. (2006) Study on hydrophilicity of polymer surfaces improved by plasma treatment. *Appl. Surf. Sci.*, **252**, 3375–3379.
203. Roy, S., Yue, C.Y., Lam, Y.C., Wang, Z.Y., and Hu, H. (2010) Surface analysis, hydrophilic enhancement, ageing behavior and flow in plasma modified cylic olefin copolymer (COC)-based microfluidic devices. *Sens. Actuators B*, **150**, 537–549.

204. Slepička, P., Vasina, A., Kolskáb, Z., Luxbacherc, T., Malinskýd, P., Mackovád, A., and Švorčíka, V. (2010) Argon plasma irradiation of polypropylene. *Nucl. Instrum. Methods Phys. Res., Sect. B*, **268**, 2111–2114.
205. Wang, Y., Yin, S., Ren, L., and Zhao, L. (2009) Surface characterization of the chitosan membrane after oxygen plasma treatment and its aging effect. *Biomed. Mater.*, **4**, 35003–35009.
206. Teng, M.-Y., Lee, K.-R., Liaw, D.-J., Lin, Y.-S., and Lai, J.-Y. (2000) Plasma deposition of acrylamide onto novel asromatic polyamide membrane for pervaporation. *Eur. Polym. J.*, **36**, 663–672.
207. Lee, K.-R., Yu, S.-J., Huang, S.-L., Wang, D.-M., and Lai, J.-Y. (1998) Pervaporation of water-ethanol mixtures through plasma graft polymerization of polar monomer onto crosslinked polyurethane membrane. *J. Appl. Polym. Sci.*, **67**, 1789–1797.
208. Wang, Y.-C., Li, C.-L., Chang, P.-F., Fan, S.-C., Lee, K.-R., and Lai, J.-Y. (2002) Separation of water-acetic acid mixture by pervaporation through plasma-treated asymmetric poly(4-methyl-1-pentene) membrane and dip-coated with polyacrylic acid. *J. Membr. Sci.*, **208**, 3–12.
209. Zhao, Z.-P., Li, J., Wang, D., and Chen, C.-X. (2005) Nanofiltration membrane prepared from polyacrylonitrile ultrafiltration membrane by low-temperature plasma: 4. grafting of N-vinylpyrrolidone in aqueous solution. *Desalination*, **184**, 37–44.
210. Lee, K.-R., Chen, R.-Y., and Lai, J.-Y. (1992) Plasma deposition of vinyl acetate onto Nylon-4 membrane for pervaporation and evaporation separation of aqueous alcohol mixtures. *J. Membr. Sci.*, **75**, 171–180.
211. Ulbricht, M. and Belfort, G. (1996) Surface modification of ultrafiltration membranes by low temperature plasma II. Graft polymerization onto polyacrylonitrile and polysulfone. *J. Membr. Sci.*, **111**, 193–215.
212. Uyama, Y., Kato, K., and Ikada, Y. (1998) Grafting/characterization techniques/kinetic modeling. *Adv. Polym. Sci.*, **137**, 1–39.
213. Zhan, J., Liu, Z., Wang, B., and Ding, F. (2004) Modification of membrane surface charge by a low temperature plasma induced grafting reaction and its application to reduce membrane fouling. *Sep. Sci. Technol.*, **39**, 2977–2995.
214. Khulbe, K.C., Feng, C., and Matsuura, T. (2010) The Art of surface modification of synthetic polymeric membranes. *J. Appl. Polym. Sci.*, **115**, 855–895.
215. Chen, J., Li, J., Zhao, Z.-P., Wang, D., and Chen, C.-X. (2007) Nanofiltration membrane prepared from polyacrylonitrile ultrafiltration membrane by low-temperature plasma: 5. Grafting of styrene in vapor phase and its application. *Surf. Coat.Technol.*, **201**, 6789–6792.
216. Moravej, M., Yang, X., Nowling, G.R., Change, J.P., Hicks, R.F., and Babyan, S.E. (2004) Physics of high-pressure helium and argon radio-frequency plasma. *J. Appl. Phys.*, **96**, 7011–7017.
217. Park, J., Henins, T., Herrmann, H.W., Selwyn, G.S., Jeong, J.Y., Hicks, R.F., Shim, D., and Chang, S.C. (2000) An atmospheric pressure plasma source. *Appl. Phys. Lett.*, **76**, 288–290.
218. Schutze, A., Jeong, J.Y., Babayan, S.E., Park, J., Selwyn, G.S., and Hicks, R.F. (1998) The atmospheric-pressure plasma Jet: a review and comparison to other plasma sources. *IEEE Trans Plasma Science*, **26**, 1685–1694.
219. Zhao, Y., Taylor, J., and Hong, S. (2005) Combined influence of membrane surface properties and feed water qualities on RO/NF mass transfer, a pilot study. *Water Res.*, **39**, 1233–1244.

Appendix

Table A.1 Summary of chemically initiated graft polymerization methods.

Reference	Type of initiator	Monomer	Membrane support	Graft polymerization time[b]	Membrane performance	Grafted polymer characterization methods
[152]	2,2′-Azobis (isobutyramidine) dihydrochloride	3-Allyl-5,5-dimethylhydantoin	Aromatic polyamide (RO)	10–100 min (multistep)	$L_p \uparrow$, $R_s \downarrow$, BFR $\uparrow$	ATR-FTIR, XPS, contact angle, and surface charge
[86]	Redox initiators $K_2S_2O_8$ and $K_2S_2O_5$	Methacrylic acid, polyethyleneglycol methacrylate, and sulfopropylmethacrylate	PES (UF)	20–120 min (one step)	NR	ATR-FTIR
[104]	Redox system: potassium persulfate-sodium metabisulfite, Merck. Redox initiators $K_2S_2O_8$ and $Na_2S_2O_5$	Methacrylic acid, polyethyleneglycol methacrylate, vinylsulfonic acid, 3-sulfopropyl-methacrylate K-salt, and 2-acrylamido-2-methylpropane-sulfonic acid	Polyamide composite (RO)	NR (one step)	NR	ATR-FTIR
[66]	Redox initiators $K_2S_2O_8$ and $Na_2S_2O_5$	Methacrylic acid, polyethyleneglycol methacrylate	Polyamide composite (RO)	15–120 min (one step)	$L_p \downarrow$, R_s ($\approx$)	ATR-FTIR, XPS, and streaming potential
[132]	Redox initiators $K_2S_2O_8$ and $Na_2S_2O_5/K_2S_2O_5$	Acrylic acid, methacrylic acid, dimethylamino ethyl methacrylate, polyethyleneglycol ester of methacrylic acid, sulfopropyl methacrylate, hydroxyl-ethyl ester of methacrylic acid	Polyamide and polypiperazine-based (NF)	15–60 min (one step)	$L_p \downarrow$, $R_s \uparrow$, SFR, and CPFR $\uparrow$ (NR individually)	ATR-FTIR, EDX

Ref	Initiator	Monomer	Membrane	Time	Results	Characterization
[150]	4,4′-Azo-bis(4-cyanovaleryl chloride)	Styrene, acrylic acid, 4-vinylpyridine, and 4-vinylpyridine/1% divinylbenzene	Polyethylene (PE) and polypropylene (PP) (MF)	6–18 h (two steps)	$L_p \sim$	SEM, EDX
[167]	Redox initiators $K_2S_2O_8$ and Na_2SO_3	Methylacrylic acid	Phenolphthalein PES (UF/MF)	6 h (one step)	NR	ATR-FTIR, SEM, and graft yield
[166]	Redox system $K_2S_2O_8$-$K_2S_2O_3$	Methacrylic acid, acrylamide, and 3-sulfopropyl methacrylate	PES (NF)	30 min (one step)	$R_s \uparrow$, SFR $\uparrow$	IR
[165]	Redox system, potassium persulfate, and potassium metabisulfite. Redox initiators $K_2S_2O_8$ and $K_2S_2O_5$	Uncharged glycol ester of methacrylic acid, anionic sulfopropylmethacrylate, anionic 2-acrylamido-2-methyl propane sulfonate	Aromatic polyamide (RO)	20 min (one step)	$L_p \downarrow$, $R_s \uparrow$, BFR, SFR, and CP=R $\uparrow$ (NR individually)	ATR-FTIR, contact angle
[151]	Benzoyl peroxide	Methacrylic acid	PES (UF)	2–8 h (two steps)	$L_p \downarrow$	Graft yield, FTIR, XPS, contact angle, and SEM

Legend: $\uparrow$, increases; $\downarrow$, decreases; ($\approx$), less than 5% change; NR, not reported; L_p, permeability; R_s, R_c, R_p, and R_b are the rejection of salt(s), colloidal particle(s), protein(s), and biofoulant(s), respectively; SFR, CPFR, PFR, and BFR are the fouling resistance of salt(s), colloidal particle(s), protein(s), and biofoulant(s), respectively; and DG, degree of grafting.

[a]Graft polymerization time only refers to the reaction time and does not include the total processing time; the number of steps indicated does not include substrate cleaning.

Table A.2 Summary of UV-initiated graft polymerization methods.

Reference	UV-sensitive membrane or initiator	Monomer	Membrane support	Graft polymerization time[b]	Membrane performance	Grafted polymer characterization methods
[177]	Membrane: PES t_{exp} = 1–5 min λ = 365 nm Power = 21.7 mW/cm²	Acrylic acid, n-vinylpyrrolidone	PES (NF)	1–5 min (one step)	$L_p \downarrow$, $R_c \uparrow$, CPFR $\uparrow$	ATR-FTIR, AFM
[178]	Membrane: PSf Energy = 16.3–22.8 J cm⁻²	Diallyl dimethyl ammonium chloride	Hollow fiber PSf (UF)	NR (one step)	$L_p \downarrow$, $R_s \uparrow$	NR
[179]	Membrane: PES, PSf λ = 300 nm Energy = 20–1946 mJ m⁻²	N-2-Vinylpyrrolidone, 2-acrylamidoglycolic acid, and 2-acrylamido-2-methyl-1-propanesulfonic acid	PES (UF) and PSf (UF)	3–60 s (multistep)	$L_p \downarrow$, R_p (≈), PFR $\uparrow$	ATR-FTIR, contact angle, and DG
[184]	Membrane: PES t_{exp} = 37.9 s λ = 365 nm Power = 16.57 × 10⁴ mW m⁻²	n-Vinylpyrrolidone	PES (NF)	36.70 s to 5 min (two steps)	$L_p \downarrow$, $R_c \uparrow$ (L_p and R_c NR individually), CPFR $\uparrow$	NR
[186]	Photoinitiator: Benzophenone (for PES) or ether of benzoin (for PVF) in methanol t_{exp} = 1–5 min	2-Acrylamido-2-methyl-1-propanesulfonic acid, 2-hydroxyethyl methacrylate, and 2-(dimethylamino)ethyl methacrylate quaternized with methyl chloride	PVF (MF) and PES (MF)	1–5 min (multistep)	$L_p \downarrow$, BFR $\uparrow$	DG, AFM
[191]	Membrane: PES t_{exp} = 5–15 min λ = 300 nm Power = 25 ± 1 mW cm⁻²	[2-(Acryloyloxy)ethyl] trimethyl ammonium chloride, acrylic acid	PES (NF)	5–15 min (one step)	$L_p \downarrow$, BFR $\uparrow$	DG, ATR-FTIR, SEM, contact angle, and zeta potential
[144]	Membrane: FES t_{exp} = 5–15 min λ > 300 nm Power = 60 ± 10 mW cm⁻²	Poly(ethylene glycol) methacrylate	PES (UF)	1.5–3 min (one step)	$L_p \downarrow$, $R_p \uparrow$, PFR $\uparrow$	DG, contact angle, ATR-FTIR, and zeta potential

Appendix

Ref	Conditions	Monomer(s)	Time	Results	Characterization
[145]	Membrane: PES t_{exp} = up to 60 s λ = 300 nm	N-2-Vinylpyrrlidone, 2-hydroxyethyl methacrylate, acrylic acid, 2-acrylamidoglycolic acid, 3-sulfopropyl methacrylate, and 2-acrylamido-2-methyl-1-propanesulfonic acid	PES (UF)	Up to 60 s (two steps)	$L_p \downarrow$, $R_p \uparrow$, PFR $\approx$	DG, ATR-FTIR
[185]	Membrane: poly(arylsulfone) (PAS) t_{exp} = 2.5–10 min λ = 253.7 nm	2-Hydroxyethyl methacrylate, glycidyl methacrylate, and methacrylic acid	PAS (UF)	2.5–10 min (one step)	NR	ATR-FTIR, DG
[187]	Photoinitiator: benzophenone t_{exp} = 5–40 min λ = 300 nm	A-D-Allyl glucoside	Polypropylene (PP)	5–40 min (multistep)	NR	XPS, ATR-FTIR, contact angle, and grafting density
[181]	Photoinitiator: benzophenone t_{exp} = 10–60 min λ = 350–450 nm	2-Hydroxylethyl acrylate	PP (UF)	10–80 min (multistep)	$L_p \uparrow$, $R_p \downarrow$, PFR $\uparrow$	XPS, SEM, and DG
[188]	Photoinitiator: benzophenone t_{exp} = 5–30 min λ = 232–500 nm	Acrylamide	PP (UF)	5–30 min (multistep)	$L_p \uparrow$, BFR $\uparrow$	ATR-FTIR, DG, contact angle, and SEM
[189]	Photoinitiator: benzophenone t_{exp} = 5–30 min λ = 232–500 nm	2-Aminoethyl methacrylate	PP (UF)	5–30 min (multistep)	L_p varies, $R_b \downarrow$, BFR $\uparrow$	ATR-FTIR, XPS, DG, and SEM

Legend: $\uparrow$, increases; $\downarrow$, decreases; ($\approx$), less than 5% change; NR, not reported; L_p, permeability; R_s, R_c, R_p, and R_b are the rejection of salt(s), colloidal particle(s), protein(s), and biofoulant(s), respectively; SFR, CPFR, PFR, and BFR are the fouling resistance of salt(s), colloidal particle(s), protein(s), and biofoulant(s), respectively; and DG, degree of grafting.

[a] t_{exp} (exposure time), λ (UV source wavelength), and power and energy of UV source are listed as reported.

[b] Graft polymerization time only refers to the reaction time and does not include the total processing time; the number of steps indicated does not include substrate cleaning.

Table A.3 Summary of gamma-initiated graft polymerization methods.

Reference	Gamma treatment	Monomer	Membrane support	Graft polymerization time	Membrane performance	Grafted polymer characterization methods
[170]	Radio nuclide Co^{60} Total Dose = 5–30 kGy Dose rate = 4.2 kGy/h	Styrene	PES films	1.2–7.1 h (one step)	NR	FTIR, SEM, and TGA
[173]	Co^{60} gamma Total dose = 46.7 kGy Dose rate = 0.18 kGy h^{-1}	Acrylic acid	Fluorinated ethylene propylene (FEP) films	0.5–8 h (two steps)	NR	% swelling, ion exchange capacity, FTIR, X-ray diffraction, TGA, DSC, and DG
[138]	2100 Ci ^{60}Co Total dose = 4–20 Mrad Dose rate = 0.27 Mrad h^{-1}	Acrylic acid, acrylamide	Isotactic polypropylene (IPP) fibers	3–4 h (two steps)	NR	IR, TGA, DSC, and DG
[171]	2100 Ci ^{60}Co Total dose = 4–11 Mrad Dose rate = 0.27 Mrad h^{-1}	Acrylonitrile	IPP fibers	3–4 h (two steps)	NR	IR, TGA, DTG, DTA, and DG
[172]	2100 Ci ^{60}Co Total dose = 0.5–12 Mrad Dose rate = 0.35 Mrad h^{-1}	Acrylonitrile	IPP fibers	3–4 h (two steps)	NR	IR, TGA, and DG
[108]	^{60}Co Total dose = 10–40 kGy Dose rate = 4.51 kGy h^{-1}	2-Hydroxyethyl methacrylate	Polypropylene (PP) (MF/UF)	1–5 h (two steps)	$L_p \downarrow$, $R_p \uparrow$ PFR $\uparrow$	FTIR, elemental analysis, contact angle, SEM, UV–vis, and DG

Legend: ↑, increases; ↓ decreases; (≈), less than 5% change; NR, not reported; L_p, permeability; R_s, R_c, R_p, and R_b are the rejection of salt(s), colloidal particle(s), protein(s), and biofoulant(s), respectively; SFR, CPFR, PFR, and BFR are the fouling resistance of salt(s), colloidal particle(s), protein(s), and biofoulant(s), respectively; and DG, degree of grafting.

[a] Graft polymerization time only refers to the reaction time and does not include the total processing time; the number of steps indicated does not include substrate cleaning.

Table A.4 Summary of plasma-initiated graft polymerization methods.

Reference	Plasma treatment	Monomer	Membrane support	Graft polymerization time[b]	Membrane performance	Grafted polymer characterization methods
[77,131]	Ar $t_{exp} = 1\text{--}10$ min Power = 40 W Pressure = 160 mTorr	Acrylamide, acrylic acid	PES (MF/UF)	1 h (two steps)	$L_p \uparrow$, PFR $\uparrow$	ATR-FTIR, XPS, contact angle, and graft yield
[211]	He $t_{exp} = 30$ s Power = 25–50 W Pressure = 100–200 mTorr	2-Hydroxy-ethyl-methacrylate, acrylic acid, and methacrylic acid	PAN and PSf (UF)	1–4 h (three steps)	$L_p \uparrow$, PFR $\uparrow$, $R_p \uparrow$	ATR-FTIR, contact angle, XPS, and graft yield
[112]	He $t_{exp} = 10\text{--}90$ s Power = 25 W Pressure = 200 mTorr	n-Vinyl-2-pyrrolidone	PES (UF)	1 h (three steps)	$L_p \uparrow$, PFR $\uparrow$	Contact angle, ATR-FTIR, XPS, and graft density
[196]	Ar $t_{exp} = 15\text{--}600$ s Power = 30–120 W Pressure = 0.45–1.13 Torr	Acrylic acid	PS (UF)	1–15 h (three steps)	$L_p \downarrow$, $R_p \downarrow$, PFR $\uparrow$	Contact angle, ATR-FTIR, pore size, graft yield, and AFM
[139]	Ar $t_{exp} = 20\text{--}120$ s Power = 10 W Pressure = 5 mTorr	Acrylic acid	PAN (UF)	1–5 min (one step)	$L_p \uparrow$, $R_s \downarrow$	Contact angle, ATR-FTIR
[215]	Ar $t_{exp} = 60$ s Power = 60 W Pressure = 113 mTorr	Styrene	PAN (UF)	1–2 h (two steps)	$L_p \downarrow$, $R_{oil} \uparrow$	ATR-FTIR, contact angle, and pore size

Table A.4 (Continued).

Reference	Plasma treatment	Monomer	Membrane support	Graft polymerization time[b]	Membrane performance	Grafted polymer characterization methods
[209]	Ar t_{exp} = 60 s Power = 30–40 W Pressure = 113–150 mTorr	n-Vinylpyrrolidone	PAN (UF)	1–15.5 h (two steps)	$L_p \downarrow$, $R_s \uparrow$	ATR-FTIR, XPS
[120]	H_2/He t_{exp} = 10–40 s Power = 40 W Pressure = 760 Torr	Methacrylic acid, acrylamide	Polyamide TFC (NF)	0.5–2 h (three steps)	$L_p \uparrow$, $R_s \uparrow$, SFR $\uparrow$, PFR ($\approx$)	Contact angle, ATR-FTIR, and AFM

Legend: ↑, increases; ↓, decreases; (≈), less than 5% change; NR, not reported; L_p, permeability; R_s, R_c, R_p, and R_b are the rejection of salt(s), colloidal particle(s), protein(s), and biofoulant(s), respectively; SFR, CPFR, PFR, and BFR are the fouling resistance of salt(s), colloidal particle(s), protein(s), and biofoulant(s), respectively; and DG, degree of grafting.

[a] The first line refers to the plasma gas; t_{exp}, plasma exposure time.

[b] Graft polymerization time only refers to the reaction time and does not include the total processing time; the number of steps indicated does not include substrate cleaning.

6
Recent Advances in Ion Exchange Membranes for Desalination Applications
Chalida Klaysom, Bradley P. Ladewig, Gao Qing Max Lu, and Lianzhou Wang

6.1
Introduction

Ion exchange membranes (IEMs) have gained enormous attention over the past several decades because of their versatile applications in many important areas including energy and alternative clean energy, potable water production, wastewater treatment, food industry, and chemical industry [1–10]. Therefore, a solid understanding of IEMs in both material synthesis and principle theories of transport phenomena is of crucial importance for the rational development of the desired IEMs for various applications.

In this chapter, a comprehensive overview of IEMs covering the fundamentals, as well as the recent development of IEMs and their applications is provided. Although the main focus will be on IEMs for desalination application by electro-driven processes, the general concepts for developing IEMs suitable for different applications are also briefly summarized. The outline of this chapter is as follows:

1) the fundamentals of IEMs and their transport phenomena
2) the material development
3) the future perspective of IEMs.

6.2
Fundamentals of IEMs and Their Transport Phenomena

IEMs are selective membranes carrying the anionic or cationic charged groups that allow specific charged species to pass through, while rejecting others of the same charges as the membranes. On the basis of the processes the IEMs are applied to, they can be classified into three major types [11]:

- Separation processes involving the separation of a component from an electrolyte solution. They can be further subdivided based on the driving force for the transport of ionic species, such as electro-driven processes in electrodialysis (ED) and concentration gradient in dialysis or Donnan dialysis [2, 4, 6, 12, 13].

Functional Nanostructured Materials and Membranes for Water Treatment, First Edition.
Edited by Mikel Duke, Dongyuan Zhao, and Raphael Semiat.
© 2013 Wiley-VCH Verlag GmbH & Co. KGaA. Published 2013 by Wiley-VCH Verlag GmbH & Co. KGaA.

- Electrochemical reaction, in which a certain chemical is generated such as in chloride-alkaline process [1, 2, 14].
- Energy conversion devices for applications, such as fuel cells and batteries [2, 8, 15, 16].

To date, ED and fuel cells are the most important processes using IEMs. Because IEMs have been widely used in many applications, the properties required for the IEMs highly depend on the specific application. However, generic property requirements are (i) low ionic resistance, (ii) high transport number of counterions, (iii) high selectivity, and (iv) good thermal, mechanical, and chemical stability [17].

According to the charged functional groups of membranes, the IEMs are generally subdivided into cation exchange membranes (CEMs), anion exchange membranes (AEMs), and bipolar membranes (BMs). The BMs are the membranes having layers of CEM and AEM which are adjacent and laminated together. The CEMs with negatively charged functional groups are selective to cationic species and reject anionic species and vice versa for the AEMs. The definition of the Donnan exclusion is well known as the prevention of ions with the same charges as that of the membrane to transport through the membrane by electric repulsion. On the basis of the electroneutrality, fixed charges of the membrane are neutralized and filled with its counterions, making co-ions unable to freely move through an IEM. However, in an IEM containing large pores that are charge balanced and fully filled with electrolyte solution, the Donnan exclusion can be suppressed and thus results in less selective membranes. In addition, Donnan exclusion depends on several parameters such as the concentration of fixed ions, the concentration of electrolytes, and the interaction among charge functional groups of IEMs and their counterions [18].

Ion selectivity of IEMs is quantitatively expressed in terms of membrane permselectivity, which measures the ease with which the counterion migration occurs through an IEM compared to the co-ions, and is defined as [19, 20]:

$$P_s = \frac{t_i^m - t_i}{1 - t_i} \tag{6.1}$$

where P_s is the permselectivity of IEMs, t_i^m is the transport number of the counterion in the membrane, and t_i is that of the same ion in free solution at the same concentration. The term *transport number* refers to the fraction of total current carried by counter ions through an IEM (normally more than 0.9). In addition, the selectivity of an IEM may be represented in terms of permeated equivalent of the particular ion relative to NaCl, which is the major component of nearly all natural saline water. For example, the transport number of a cation relative to sodium is defined as follows:

$$P_{Na}^M = \frac{t_M^m / t_{Na}^m}{C_M / C_{Na}} \tag{6.2}$$

where P_{Na}^M is the permeated equivalent of cation relative to sodium ion, t_M^m is the transport number of cations through membranes, t_{Na}^m is the transport number of sodium ion through membranes, and C_M and C_{Na} are the concentration of the cation and the sodium ion, respectively. A similar expression is also used for anions relative to the chloride ion.

6.2.1
Ion Transport through IEMs

The Nernst–Planck flux equation is one of the most widely used equations for describing transport phenomena in both solution and IEMs because of its relatively simple relation. The Nernst–Planck equation is based on the assumption of the independent migration of cations and anions in solution and membranes. Although the ions in solution are treated as an independent compartment, the electroneutrality has to be balanced in which there are no excess charges in the system. The transport of ions can be expressed in terms of diffusion, migration, and convection as defined by the following equations:

$$J_i = J_{i(d)} + J_{i(m)} + J_{i(conv)} \tag{6.3}$$

$$J_i = D_i \frac{dC_i}{dx} - \frac{z_i F C_i D_i}{RT} \frac{d\varphi}{dx} + vC_i \tag{6.4}$$

where J_i is the flux of component i, v the velocity, C the concentration, D the diffusion coefficient, x the direction coordinate, z the valence charge, F the Faraday constant, R the gas constant, T the temperature, and φ the electrical potential. Furthermore, these forms are defined as follows:

- *Diffusion* is the movement of molecular components caused by chemical potential gradient. In the IEM system, the dominant term for ion movement by diffusion is the concentration gradient.
- *Migration* is the movement of ions due to an electrical potential gradient, which is normally developed by the application of external power source.
- *Convection* is the movement of mass caused by mechanical force such as hydrodynamic flow or stirring. This term is normally of less importance in the solid membrane and can be neglected in the system when there is no flow or strong stirring condition.

The Nernst–Planck model neglects the factor of interaction among different ions and solvents in the real situation. Consequently, the irreversible thermodynamic approach, which takes the effects of fluxes of heat, electricity, momentum, and interaction among individual components and solvents into account, is considered more reasonable and practical to apply in the real system. In this section, the only basic properties and phenomena of ion exchange capacity (IEC) are presented. The detailed explanation for the transport phenomena in membranes can be acquired from the literature [17, 20–26].

6.2.2
Concentration Polarization and Limiting Current Density

Owing to the difference in ion mobility in selective membranes and in electrolyte solution, a decrease in electrolyte concentration takes place near one side of the membrane surface and the accumulation takes place on the other side of the surface. As a result, the concentration gradient develops near the membrane surface, referring to concentration polarization.

As a consequence of the concentration polarization development, if the concentration exceeds the solubility limits, precipitation will occur in the concentrate compartment. The scale on the membrane surface creates an extra resistance and can damage the membranes. In the dilute compartment, when the ion concentration near the membrane interface reaches zero there is a drastic increase in the resistance and potential drop across the membrane, resulting in high energy consumption and enhanced water dissociation. The water dissociation has dramatic effects on the electro-driven process, responsible for the loss of current utilization and undesirable consequences from the pH changes. The increased pH of solution near the CEMs' interface in the concentrated cell further enhances the precipitation and scaling phenomena. In contrast, the decreased pH of the solution near the AEM can damage the membranes. Therefore, the concentration polarization effects are required to be minimized especially to avoid water dissociation phenomena.

The concentration polarization can be reduced by reducing the current density and the thickness of its boundary layer near the membrane interface. The thickness of the boundary layer is determined by the hydrodynamic flow, which depends on the cell and spacer design and the feed flow velocity. At the provided constant velocity, the current density will reach the maximum value when salt concentration near the membrane interface in the depleting solution is reduced to nearly zero. This maximum current density is called limiting current density (LCD), and can be defined by the following equation based on the classical polarization theory:

$$i_{\lim} = \frac{FDC}{z\delta\left(t_i - t_i\right)} \tag{6.5}$$

where $i_{\lim}$ is the LCD and δ is the diffusion boundary layer. The definitions of F, D, and C can be found in Equation 6.4. In practical ED, the operating condition of the system is limited by the LCD. It was believed that operating the ED over the LCD leads to the reduction of current efficiency and other unpredictable phenomena such as the scaling on the membrane surface.

6.2.2.1 The Overlimiting Current Density
On the basis of classical polarization, there are no more ions to transport and thus no current that exceeds the limiting current. However, in practice, current density over the limiting current is normally observed. In the past, it was believed that the water dissociation is responsible for the overlimiting current density as the OH− and H+ that can migrate through the membrane and carry the current are

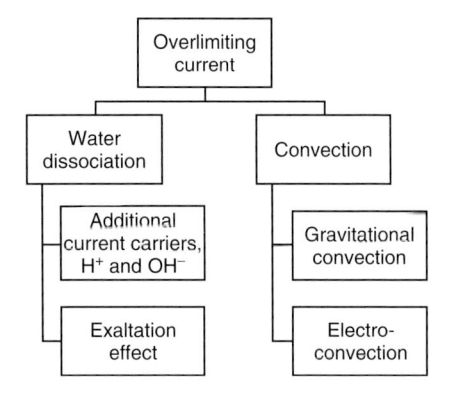

Figure 6.1 Scheme of the mechanism for overlimiting current transfer.

generated. However, it has been demonstrated and reported in several experimental studies that the water dissociation is not responsible for the origin of overlimiting current [20, 23, 27–39]. The transport of OH− and H+ only partly contributes to the current over the limiting current. The origin of the current over the limiting one is still unclear and is now a subject of intensive discussion [20, 23, 27–39].

To date, several models have been developed to describe the mechanism of overlimiting current for a better understanding of its phenomena and its role in an application, consequently leading to the design of membranes that can overcome or minimize the effects of concentration polarization phenomena. The state-of-the-art mechanism describing the origin of overlimiting current by coupled effects is summarized in Figure 6.1 [23, 25, 28].

6.2.2.2 Water Dissociation

The water dissociation is generally expressed by Equation 6.6.

$$H_2O \rightleftharpoons H^+ + OH^- \tag{6.6}$$

In IEMs, the water splitting reaction occurs within a thin layer near the membrane interface in the depleting side. It was found that the water splitting in electrodriving membrane processes is higher than that in normal water solution, as the fixed charge groups of the IEMs can also be involved in the reaction as catalysts [23, 25, 28]. Nikonenko et al. [23] reported the range of ionic functional groups of IEMs in the order of increasing water dissociation as shown below:

$$-N^+(CH_3)_3 < -SO_3^- < -PO_3H < =NH, -NH_2^+ < -COO^- < -PO_3^{2-}$$

In addition, water dissociation becomes more rapid with the presence of metallic hydroxides such as $Mg(OH)_2$, and $Fe(OH)_3$, which are normally precipitated on the CEM surface and act as a catalyst for the water dissociation reaction [23, 25, 29].

The counterion transfer can also be increased with the occurrence of water splitting; the generated H+ or OH− near the membrane surface attracts the ions from the adjacent bulk solution toward the membranes (an exaltation effect) [28]. For example, the OH− near CEM surface attracts cations from the bulk depleting solution toward the boundary layer near the membrane surface, resulting in the increase of ions to transport through the membranes.

6.2.2.3 Gravitational Convection

Gravitational convection provides additional counterion transfer by the enhancement of solution mixing. This type of convection is developed because of nonuniform distribution of temperature or solution density near the membrane interface [40]. The high resistance that develops on the dilute side because of concentration polarization, can lead to Joulean heat production and the formation of a temperature gradient. The concentration density and temperature gradient correspond to the mixing and movement of the counterion, contributing to the overlimiting current [40].

Many research studies have investigated the effect of natural convection and the gravitational convection by mounting membranes in two different configurations, vertical and horizontal position [23, 28, 30, 33, 35, 40–44]. It has been proved that when membrane is rearranged in the horizontal position with the dilute compartment underneath the membrane, the gravitational diffusion layer is stabilized and the free convection is minimized. In that case, the counterions that transfer under the influence of convection are neglected. The current over the limiting value was still observed in this case; implying that the transport of the ions may be contributed by the gravitational convection and may also be coupled by other phenomena, such as electroconvection.

6.2.2.4 Electroconvection

Electroconvection is due to the nonuniform electric field near the membrane surface, which causes the turbulent movement in the boundary layer and the transport of counterions. This phenomenon is highly influenced by the surface inhomogeneity of the membranes and more pronounced in the depleting solution. The mechanism governing the electroconvection in a membrane is believed to be electroosmotic slip of second kind or "electroosmosis II" referring to the occurence of electroosmosis due to an interaction between an electric field and the induced space charge near the membrane interface in the diffusion boundary layer [35]. The strength of the electric field is likely to be dependent on the Strokes radius of ions of the space charge layers. The larger the Stroke radius the stronger the space charge layer, contributing to the electroconvection and the enhancement of liquid motion [43].

6.2.3
Structure and Surface Heterogeneity of IEMs

Bulk and surface morphologies of IEMs play an important role on the membranes' properties, especially on the electrochemical properties and transport phenomena of the membranes.

The inner membrane morphology (bulk morphology) determines, to a large extent, the electrical conductivity, transport phenomena, permselectivity, diffusion, and hydraulic permeability of the IEMs. A microheterogeneous model is normally used for describing the transport characteristics of membranes in particular to measure the degree of membrane structural heterogeneity and correlate with membrane electrochemical behavior [32, 45, 46]. In the microheterogeneous model of IEMs, an IEM consists of a gel phase and intergel phase of filling solution. The gel phase is composed of fixed charged functional groups, which are considered to have a relatively uniform distribution on polymer chains and compensated solution of the fixed charged functional groups (Figure 6.2).

Many works also referred to a two-phase model that combines volume fraction of the inert phase into the gel phase and considered them as a quasi-homogeneous region, namely joint-gel phase. The use of two-phase model to explain the electrochemical behavior of various commercial IEMs was reported [32, 45]. They concluded that the selectivity of IEMs depends on the volume fraction of joint-gel phase and intergel phase. As the volume fraction of the intergel phase that is considered as the nonconducting region increases, the Donnan exclusion becomes less effective and thus results in a less selective membrane. When three different electrolyte solutions were used, it was found that the fraction of the intergel phase was considerably independent to the electrolyte solution. Therefore, the fraction of intergel phase was concluded to directly relate to the morphology of tested membranes and independent to the electrolyte solution [45].

The surface heterogeneity of the IEMs has been intensively investigated recently, mostly by chronopotentiometry and current voltage (i–v) characteristics [30–33, 40, 46]. Chronopotentiometry is a powerful characterization technique that measures the correspondent potential of a system imposed to constant current. It has been widely applied for investigating the transport phenomena of the IEM, especially in the diffusion-controlled boundary layers to better understand the overlimiting current behavior of a membrane [30, 33, 40, 43, 47, 48]. The shapes and characteristic values of chronopotentiograms depend on many parameters such as membrane resistance, surface and structure properties of membranes, the testing, and the hydrodynamic conditions [40]. The typical shape of a chronopotentiogram is presented in Figure 6.3.

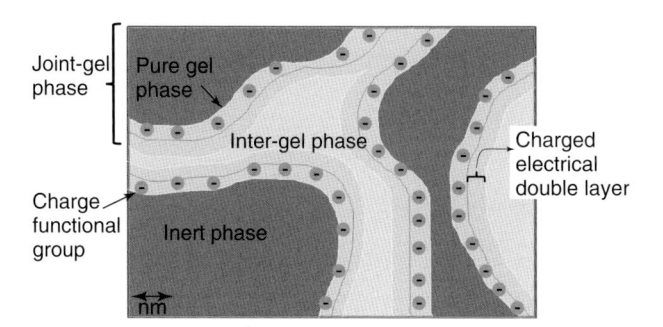

Figure 6.2 Scheme representative structure of IEM according to the micro-heterogeneous model. (Source: Adapted from C. Larchet et al. [43].)

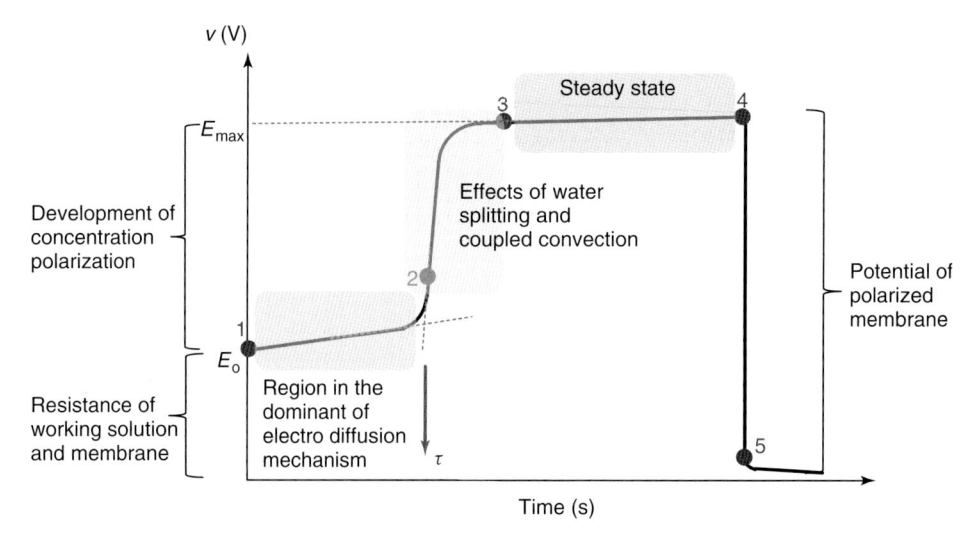

Figure 6.3 A representative chronopotentiogram of IEM.

When a constant current density is applied to the testing cell, the initial potential difference (E_o) is suddenly built up; its height is related to ohmic potentials of the resistance of the testing system (electrolyte and membrane). The potential difference gradually increases as a function of time because of the gradual increase in ohmic resistance of the system (region I from point 1 to 2). The slope of this part depends on the capacity of the charged electrical double layer at the membrane interface. The transport phenomenon of this region is governed mainly by electrodiffusion process. When the concentration of the depleting solution near the membrane interface reaches zero, suddenly the sharp developed potentials occur (at the infection point (2)). The transport mechanism of the IEMs changes mainly to a coupled convection. After that the curve reaches the steady state where the potential difference levels off (E_{max}). Note that the transition point can only be observed when the current higher than the LCD is applied to drive the change of transport phenomena on the membrane surface from electrodiffusion to other mechanisms. The difference between the initial potential and the potential at steady state (ΔE) indirectly relate to the thickness of concentration gradient [47]. Under the concentration polarization theory, the better the transport property the membrane is, the larger ΔE will be observed. As membranes with high transport property can transport ionic species from one side to the other better than the membranes with lower transport property, the ΔE between its two interfaces is thus bigger.

The i–v characteristic is another useful technique to investigate the concentration polarization of membrane interface and the LCD. Typically, the shape of i–v curve depends on the nature and surface properties of the membranes, the property of electrolyte solution, and the hydrodynamic condition in the testing process [30, 37, 38, 49, 50]. This technique records the corresponding current density with

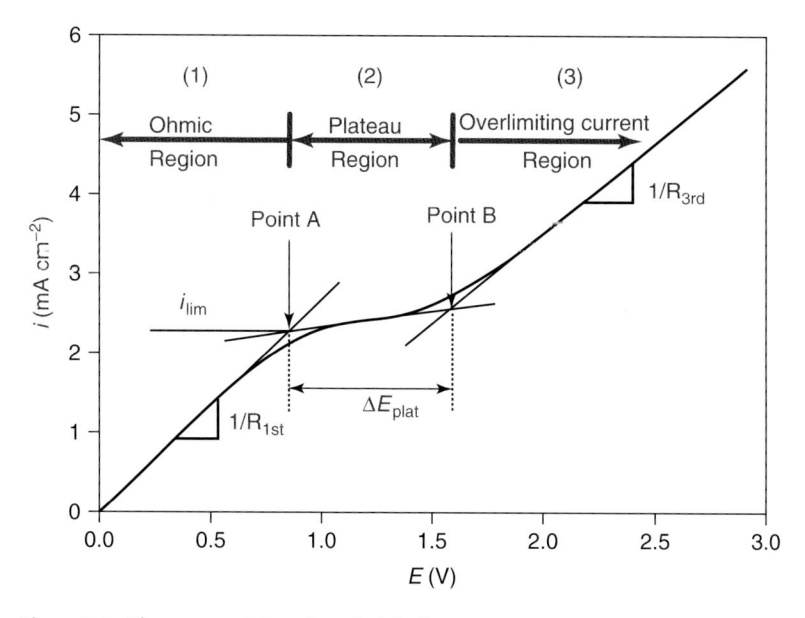

Figure 6.4 The representative characteristic $i-v$ curve.

the stepwise potential difference. Typically, an $i-v$ curve can be divided into three regions, as shown in Figure 6.4.

The region (1), the Ohmic region, is the region where the current density is linearly proportional to potential difference. When the ion concentration in the boundary layer near membrane interface decreases toward zero and there are a limited number of ions that can carry the current, the current reaches the plateau range in that region (2). The LCD occurs at the inflection point that can be estimated from intersection of the tangents between region (1) and (2). The corresponding current above the plateau region can be referred to a overlimiting current region (region (3)), which is believed to be influenced dominantly by the coupled convection effect.

It was found that surface heterogeneity of the membranes has a high impact on the membrane electrochemical behavior and the development of concentration polarization. Specifically, the presence of nonconducting regions reduces the active transfer areas, resulting in locally higher current density near the conducting regions compared to the overall current density across entire membrane area (Figure 6.5). This phenomenon has a strong influence on the characteristic curves of chronopotentiograms and $i-v$ curve, as shown in Figure 6.6a,b.

Membranes with homogeneous and heterogeneous surfaces obtained different characteristic curves in a chronopotentiogram (Figure 6.6a). While heterogeneous membranes gave a shorter transition time and more diffuse curve near the inflection region, the more homogeneous membranes possess longer transition times and almost vertical curve from inflection point to the steady state [51]. This is due to the nonuniform current line distribution of the heterogeneous membranes. As

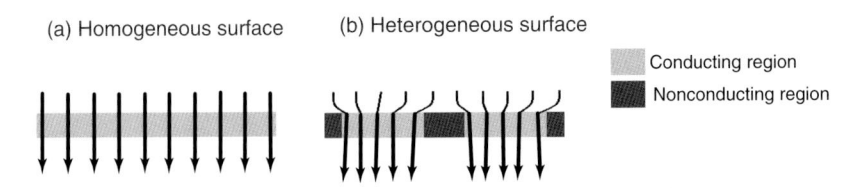

Figure 6.5 Current line distribution through (a) homogeneous surface and (b) heterogeneous surface of ion exchange membranes.

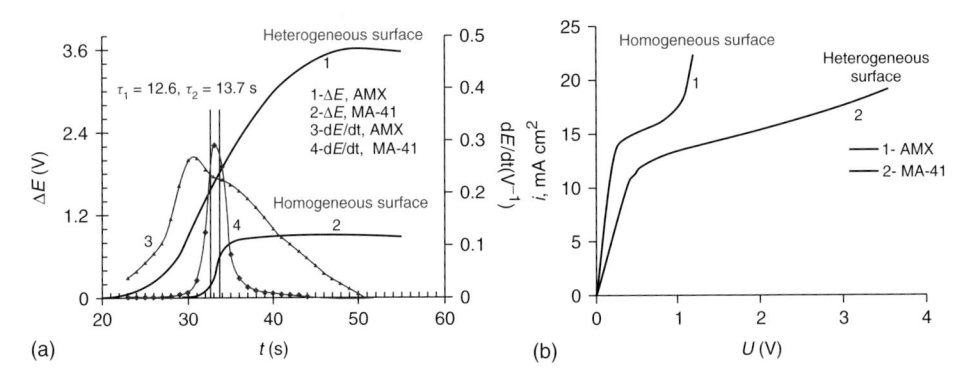

Figure 6.6 Different shapes and characteristic curves of homogeneous and heterogeneous commercial membranes (AMX from Tokuyama Soda and MA-41 from Schekino Production): (a) chronopotentiograms and their derivatives of a homogeneous AMX and a heterogeneous MA-41 and (b) i–v curve of AMX and MA-41 membrane. (Source: Adapted from N. Pismenkaia et al. [40]).

in the heterogeneous membrane, the local LCD near the conducting region is higher than the average current density of the whole surface, the ionic species are removed faster and thus the concentration decreases quickly and the potential difference develops rapidly at the beginning of the process, resulting in reaching the transition time sooner. Furthermore, the time the heterogeneous membranes required to achieve a quasi-steady state is slower compared to a more homogeneous one as the slow growth of the potential drop is observed from the inflection point.

The degree of surface heterogeneity also has a strong impact on the $i-v$ development of the concentration polarization visible in the $i-v$ curves (Figure 6.6b) [52]. It was found that the lower LCD was obtained in the case of heterogeneous membranes. Balster et al. [27] also studied the cause and phenomena of the overlimiting current using microheterogeneity model and microtopology technique. They concluded that surface heterogeneity of membranes has a great influence on the shape of the $i-v$ curve and its plateau length. In general, the more heterogeneous the membrane surface the longer the plateau length. If the space distance between the conducting and nonconducting regions is in the same magnitude with the boundary-layer thickness, the shortening of plateau length can normally be observed.

6.3 Material Development

This section provides an overview of the progress in the development of IEMs, including the various approaches made to the membrane modification. Various types of IEMs have been developed for different applications. Some commercial IEMs, manufacturers, and their properties are shown in Table 6.1 [18, 53, 54].

6.3.1 The Development of Polymer-Based IEMs

The search for new materials for IEMs has kept intensifying in order to supply robust membranes for the existing applications and to expand the opportunity to new potential applications. Polymer IEMs can be prepared via three approaches depending on the starting materials.

1) Introduction of charged moieties to polymer chains followed by the formation of membranes.
2) Polymerization of monomers containing charged moieties. The charged polymers then undergo film processing to form membranes.
3) Introduction of functional charged groups on the already film-formed membranes.

6.3.1.1 Direct Modification of Polymer Backbone

Polyarylene polymers containing aromatic pendant groups on polymer backbones such as poly aryl sulfone, poly aryl ketone, polybenzimidazole (PBI), polyphenylquinoxalines (PPQs), and polyphenylene oxides (PPOs) are attractive as new polymer matrix for IEMs due to several reasons: (i) their mechanical and thermal stability, (ii) processibility, (iii) low cost, and last but not the least, (iv) the ability to chemically modify the polymer backbone via the electrophilic substitution at their aromatic skeletons [55–59]. The introduction of charged moieties by direct polymer backbone modification via the electrophillic substitution is widely used owing to its simplicity and reproducibility under defined conditions [59].

Poly(aryl sulfone) Thermoplastic poly (aryl sulfone) such as polyether sulfone (PES) is one of the most widely studied polymers because of its excellent mechanical, thermal, and chemical stability. Moreover, PES possesses a high degree of process flexibility [60]. A number of research projects have focused on the development of PES as IEMs to replace expensive Nafion membranes. This type of polymer has been prepared with various structures, referring to slightly different pendent groups in the polymer back bones (as shown in Figure 6.7), available from different suppliers. Consequently, there are several ways to introduce charged functional groups (predominantly sulfonate groups ($-SO_3H$)) into the polymer backbones. Table 6.2 provides different path ways of the electrophillic substitution reactions applied to different PES structures.

Table 6.1 Properties of some commercial IEMs.

Membrane	Type	Thickness (mm)	Water content (%)	Area resistance (Ω cm^{-2})[a]	IEC (mequiv g^{-1})	Permselectivity (%)[b]
Tokuyama Soda Co. Ltd, Japan						
Neosepta CMX	CEM, PS/DVB	0.14–0.20	25–30	1.8–3.8	1.5–1.8	97
Neosepta AMX	AEM, PS/DVB	0.12–0.18	25–30	2.0–3.5	1.4–1.7	95
Neosepta CMS	CEM, PS/DVB	0.15	38	1.5–2.5	2.0	—
Neosepta ACM	AEM, PS/DVB	0.12	15	4.0–5.0	1.5	—
Asahi Glass Co. Ltd, Japan						
CMV	CEM, PS/DVB	0.15	25	2.9	2.4	95
AMV	AEM, PS/BTD	0.14	19	2.0–4.5	1.9	92
HJC	CEM, heterogeneous	0.83	51	—	1.8	—
Ionic Inc., USA						
61CZL386	CEM, heterogeneous	0.63	40	9	2.6	—
103PZL183	AEM, heterogeneous	0.60	38	4.9	1.2	—
Dupont Co., USA						
Nafion 117	CEM, fluorinated	0.20	16	1.5	0.9	97
Nafion 901	CEM, fluorinated	0.40	5	3.8	1.1	96
RAI Research Corp., USA						
R-5010-H	CEM, LDPE	0.24	20	8.0–12.0	0.9	95
R-5030-L	AEM, LDPE	0.24	30	4.0–7.0	1.0	83
R-1010	CEM, fluorinated	0.10	20	0.2–0.4	1.2	86
R-1030	AEM, fluorinated	0.10	10	0.7–1.5	1.0	81
CSMCRI, Bhavangar India						
IPC	CEM, LDPE/HDPE	0.14–0.16	25	1.5–2	1.4	97
IPA	AEM, LDPE/HDPE	0.16–0.18	15	2.0–4.0	0.8–0.9	92
HGC	CEM, PVC	0.22–0.25	14	4.0–6.0	0.7–0.8	87
HGA	AEM, PVC	0.22–0.25	12	5.0–7.0	0.4–0.5	82

Table 6.1 (continued).

Membrane	Type	Thickness (mm)	Water content (%)	Area resistance (Ω cm^{-2})^a	IEC (mequiv g^{-1})	Permselectivity (%)^b
FuMa-Tech GmbH, Germany						
FKE	CEM, —	0.05–0.07	—	<3	>1	>98
FTCM-A	CEM, PA	0.50–0.60	—	<10	>2.2	>95
FTCM-E	CEM, PET	0.50–0.60	—	<10	>2.2	>95
FAD	AEM, —	0.08–0.10	—	<0.8	>1.5	>91
FTAM-A	AFM, PA	0.50–0.60	—	<8	>1.7	>92
FTAM-E	AEM, PET	0.50–0.60	—	<8	>1.7	>92

^{a}Measured: 0.5 mol dm^{-3} NaCl and
^{b}0.1/0.01 mol dm^{-3} NaCl and 0.1/0.5 mol dm^{-3} KCl for membranes from FuMa-Tech GMbH at 25 °C.

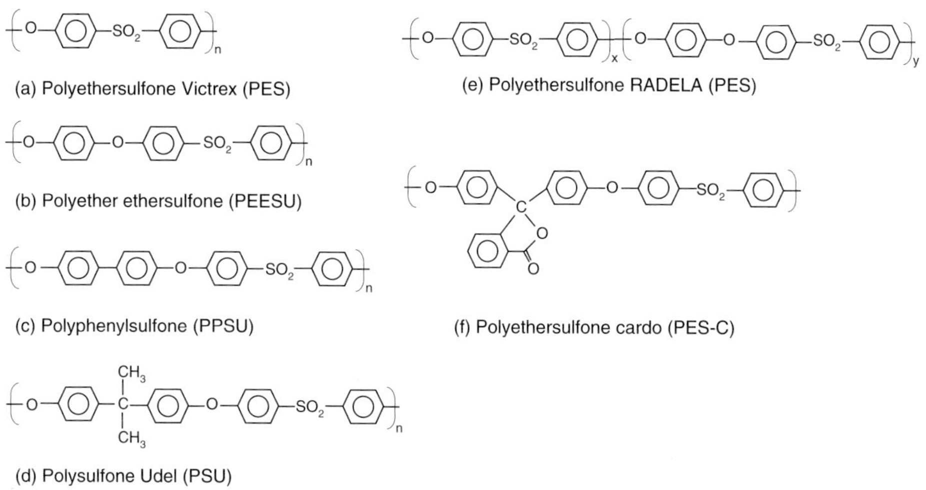

(a) Polyethersulfone Victrex (PES)

(b) Polyether ethersulfone (PEESU)

(c) Polyphenylsulfone (PPSU)

(d) Polysulfone Udel (PSU)

(e) Polyethersulfone RADELA (PES)

(f) Polyethersulfone cardo (PES-C)

Figure 6.7 Different chemical structures of poly (aryl sulfone) base polymers.

The AEM based on PES can also be prepared by conventional chloromethylation and quateramination [69, 70]. Although the resultant AEM showed high ionic conductivity and good IEC, which is an expression of the equivalent number of fixed charge groups per unit of dry membrane weight, the chloromethylation step requires the use of toxic solvent of chloromethylether that can be harmful to human health.

Poly(aryl ether ketone) The polymer family of poly (aryl ether ketone) has also been intensively studied recently [8, 59, 71–78]. The modification of the polymer backbone can be carried out in the same manner as for PES polymer family.

Table 6.2 Sulfonation reaction for PES polymers.

Sulfonation reagent/process	Polymer	Solvent	Note
Chlorosulfonic acid [58, 60–64]	PES	H_2SO_4, Chloroform, Dichloromethane (DCM), Dichloroethane (DCE)	• Simplicity • Difficulty for the chemical disposal of concentrated H_2SO_4 acid • Hard to control the undesirable side reaction
Methylation-sulfonation-reaction [65, 66]	PSU	N-methyl pyrolidone (NMP)	• Good quantitative and region-specific reaction control • Complicated procedure
Sulfurtrioxide triethyl phosphate (SO_3-TEP) [67, 68]	PES	DCM	• Minimize or eliminate the possible side reactions • Convenient and relatively inexpensive • Difficult to control the sulfonation reaction to a specific stable position

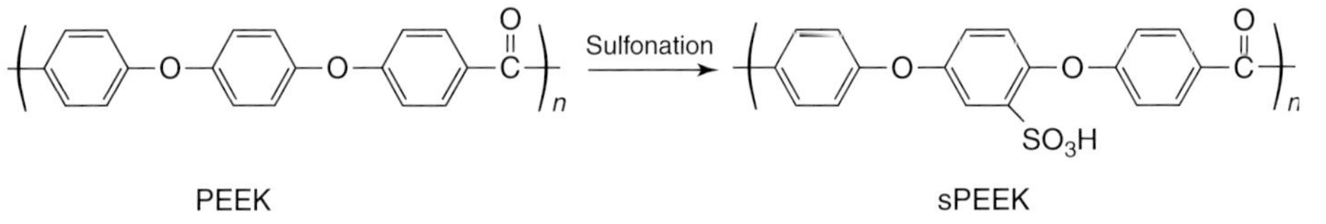

Figure 6.8 Chemical structures of polyether ether ketone (PEEK) and sulfonated polyether ether ketone (sPEEK).

Figure 6.8 depicts the chemical structure of poly (ether ether ketone) (PEEK, Victrex) and the sulfonated poly (ether ether ketone) (sPEEK).

The most commonly used method for preparation of CEM from PEEK is by dissolving the polymer in H_2SO_4 acid [71, 73, 74]. The sPEEK with the degree of sulfonation (DS) of around 90% showed a high IEC of 2.48 mequiv g^{-1} with similar proton conductivity of Nafion. However, it is worth noting that the resultant sPEEK with the high DS exhibited poor thermal stability that makes it difficult for the fabrication into membrane films in melt processes [72]. Moreover, sPEEK with the high DS can uptake water molecules up to four molecules per sulfonate group and increase the number of water uptake up to eight molecules per unit functional group when it was formed in the films, resulting in high degree of water swollen and poor dimensional stability. This poor melt processability and dimensional stability limited the application of the sPEEK. Therefore, much effort has been focused on the strategies to overcome the aforementioned problems of the sPEEK. Blending the sPEEK with other polymers or the introduction of cross-linking agents to the sPEEK are powerful strategies to improve the poor thermal and mechanical stability of the sPEEK membrane. The sPEEK can also be cross-linked either by reaction bonding or thermal treatment to improve the dimensional stability of membranes [79].

Polyphenylene Oxides (PPO) AEMs based on the PPO have been prepared either by chloroacetylation and quaternary amination or by bromination and amination processes [80–83]. Also, CEMs can be prepared by the same strategy of bromination and then sulfonation reaction as shown in Figure 6.9 [59, 84].

The new path way for preparing AEMs as proposed in the Figure 6.9 provides an advantage of avoiding the use of toxic chemical such as chloromethyl methylether, normally used in the conventional preparation procedure of AEMs. Via the approach illustrated in Figure 6.9a, the membranes with high IEC can be obtained. However, the thermal stability of the resultant membranes still requires more improvement. This problem can be solved by the introduction of bromination substitution (Figure 6.9b), which can occur on both aryl and benzyl position. As a consequence, the amination can occur in both positions and can also create cross-linking to some extend among the functional groups of the membranes. It has been reported that the position of the moieties groups has a high influence on the properties of the resultant membrane [80]. The balance of the functional groups on benzyl and aryl position allows the tuning of membrane properties, desirable for different application needs.

6.3.1.2 Direct Polymerization from Monomer Units

The direct synthesis of polymer from monomer units provides excellent opportunity in tailoring polymer composition that allows the control of amount and distribution of the ionogenic functional groups along the polymer backbones. This also advances the tuning of both microstructure and properties of the IEMs.

Polyethersulfone (PES) The direct polymerization from monomer units offers some advantages on the precise control of sulfonate groups on the aromatic rings of the polymer backbones, and the feasibility to tune the molecular weight of the polymer, thus leading to the polymer with higher mechanical stability and better processibility.

The sulfonated polyethersulfone (sPES) via direct polymerization of monomers containing charge moieties have been successfully reported [85–90]. In practice, sPES can be obtained by the direct aromatic nucleophilic substitution and polycondensation of dichlorodiphenylsulfone (DCDPS), 3,3′-disulfonate dichlorodiphenylsulfone (SDCDPS), and biphenol as shown in the Figure 6.10.

The resultant membrane based on sPES obtained from direct polymerization of the monomer units exhibited high IEC up to around 3 mequiv g^{-1}. However, at high DS, the membranes suffer from high swollen degree, resulting in low

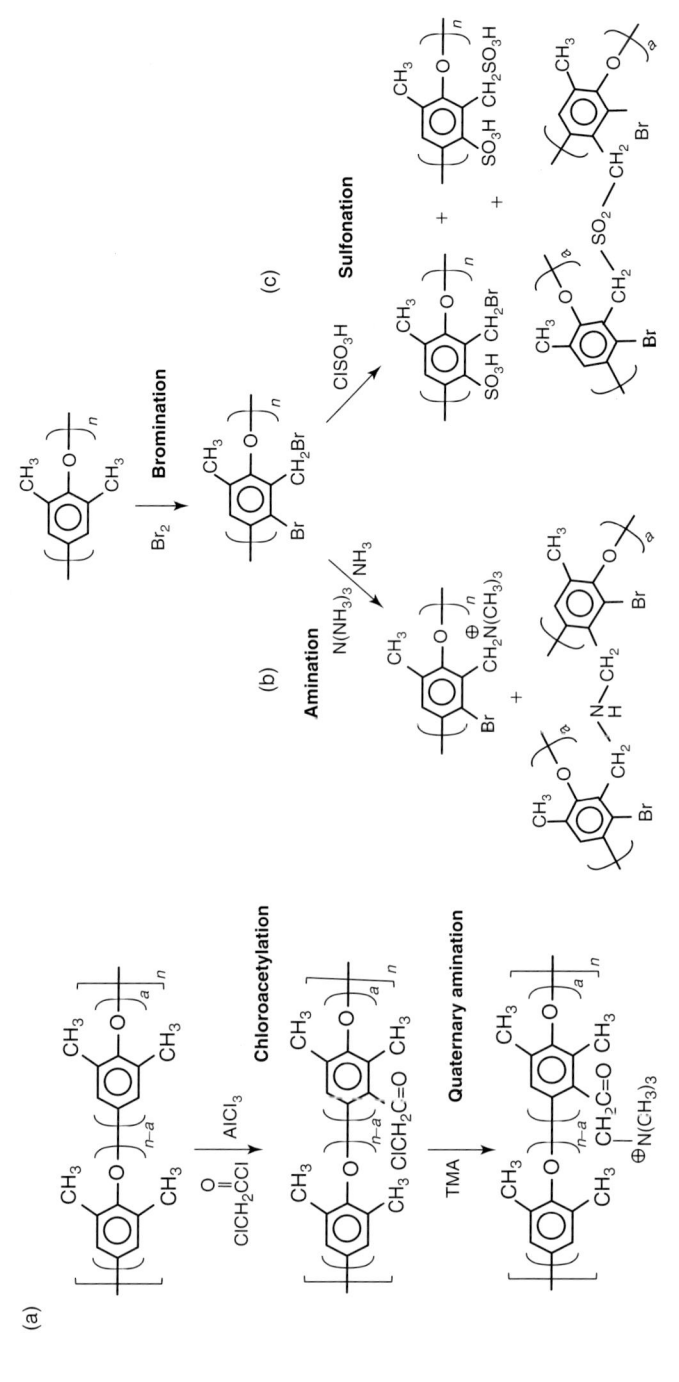

Figure 6.9 Main reactions and structures of IEMs from PPO: (a) anion exchange membranes employing Friedel-Crafts chloroacetylation, (b) anion exchange membrane prepared by bromination and amination, and (c) cation exchange membrane prepared by bromination and sulfonation reaction.

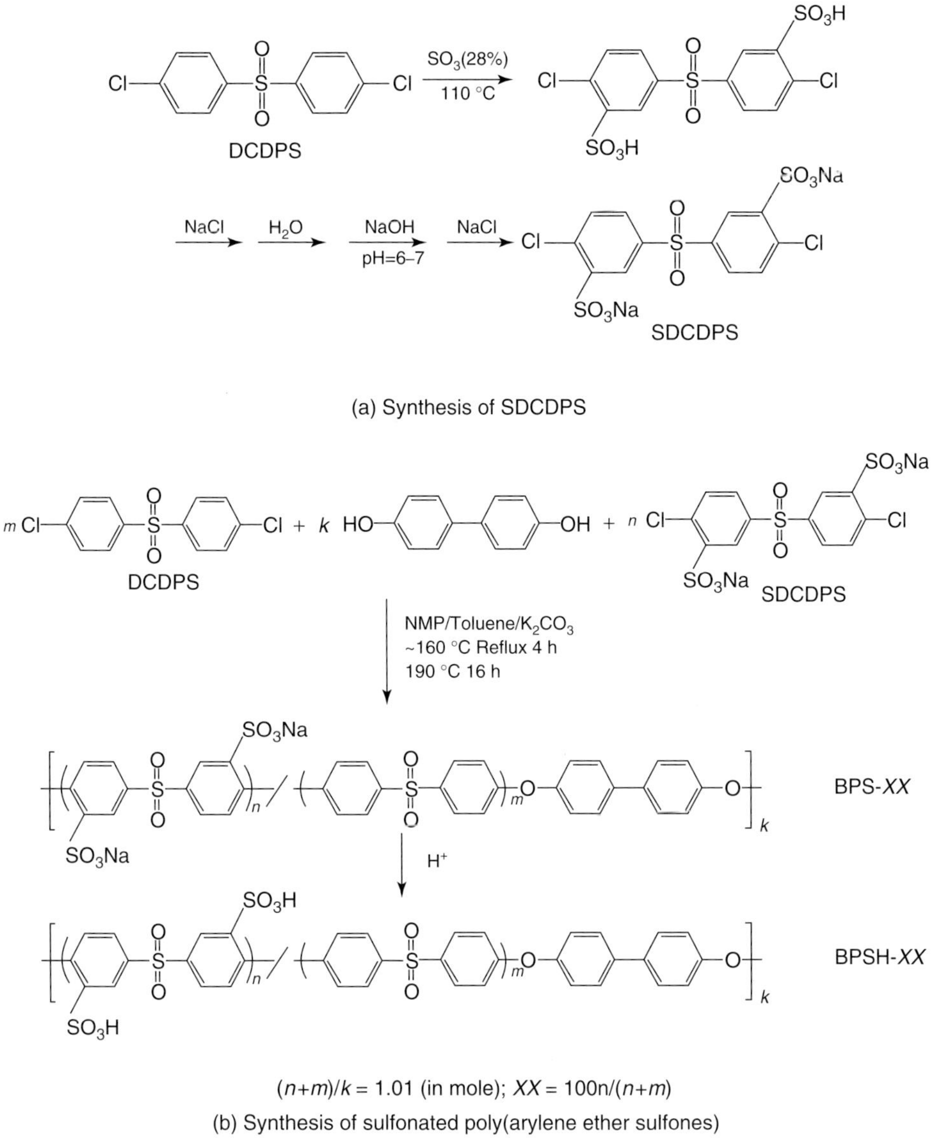

(a) Synthesis of SDCDPS

(n+m)/k = 1.01 (in mole); XX = 100n/(n+m)
(b) Synthesis of sulfonated poly(arylene ether sulfones)

Figure 6.10 (a) Synthesis of sDCDPS and (b) sulfonated poly (arylene ether sulfone).

mechanical stability. To improve the dimensional stability of the membrane, a certain degree of cross-linking was introduced [55]. The cross-linked membranes can have significantly improved mechanical stability, while still maintaining good conductivity.

6.3.1.3 Charge Induced on the Film Membranes

The IEM can also be prepared by forming the non-ionogenic polymer films first, and subsequently by the introduction of charged functional groups onto the formed polymer films.

The radiation-induced grafting is a versatile technique to introduce functional groups onto different membrane substrates [91, 92]. The grafting technique offers several advantages to the membrane development, including feasibility in processes, well-defined composition, and ability to tailor membrane properties with a variety material selection, including surface substrates, grafting materials, and induced functionalities [91, 93]. The film substrates can be porous or nonporous membranes. Typical examples include hydrocarbon polymer-based films of polyethylene (PE), polypropylene (PP), polyalkene (polyalkene nonwoven fabrics (PNF)), and fluorocarbon polymer-based films of polyvinylidene fluoride (PVDF), polytetrafluoroethylene (PTFE), poly(tetrafluoroethylene-co-hexafluoropropylene) (FEP), and poly(ethylene-co-tetrafluoroethylene) (ETFE) [15, 16, 94–97]. The chemically stable fluorinated carbon films are normally selected for applications that require robust membranes to withstand severe working conditions [98, 99]. For the grafting agents, there are two major types: (i) functional monomers such as acrylic acid and methacrylic acid that can be attached directly to the substrate as charged functional groups and (ii) nonfunctional monomers such as styrene, N-vinylpyridine, and vinylbenzylchloride that can be further chemically modified or conversed into ion exchangeable groups.

CEMs with carboxylic acid functional groups (–COOH) can be prepared by either direct grafting of acrylic monomers or by indirect grafting of epoxy acrylate monomers on the polymer films (PE, PNF), followed by postgrafting treatment with sodium iminodiacetate [95, 98]. The membranes with sulfonated groups (–SO_3H) are normally prepared by grafting polymer films (PE, FEP, PNF, or PTFE) with styrene and subsequently sulfonating the grafted films with chlorosulfonic acid, or sulfuric acid in dichloromethane or tetrachloroethane for carbon tetrachloride solution [100–102]. It is worth noting that the nature and properties of grafted film substrate has to be taken into account when selecting the suitable sulfonating agents.

The AEMs can also be prepared in the same manner as the CEMs with sulfonate groups as depicted in Figure 6.11 [95, 103, 104]. First, vinylbenzylchloride or glycidyl methacrylate (GMA) is grafted on the film substrates (PNF), followed by amination reaction to convert the functional groups to amine derivatives.

The IEMs prepared form grafting technique showed high conductivity and good chemical, thermal, and mechanical stability which make them interesting for industrial applications. The grafting method was proved powerful as the properties of the IEMs can be conveniently controlled. However, most IEMs prepared by the grafting technique are still at a laboratory scale; the commercial success is still limited [105]. This may be due to the constrained access to the radiation source and the difficulty in the reproduction of uniformly grafted membranes

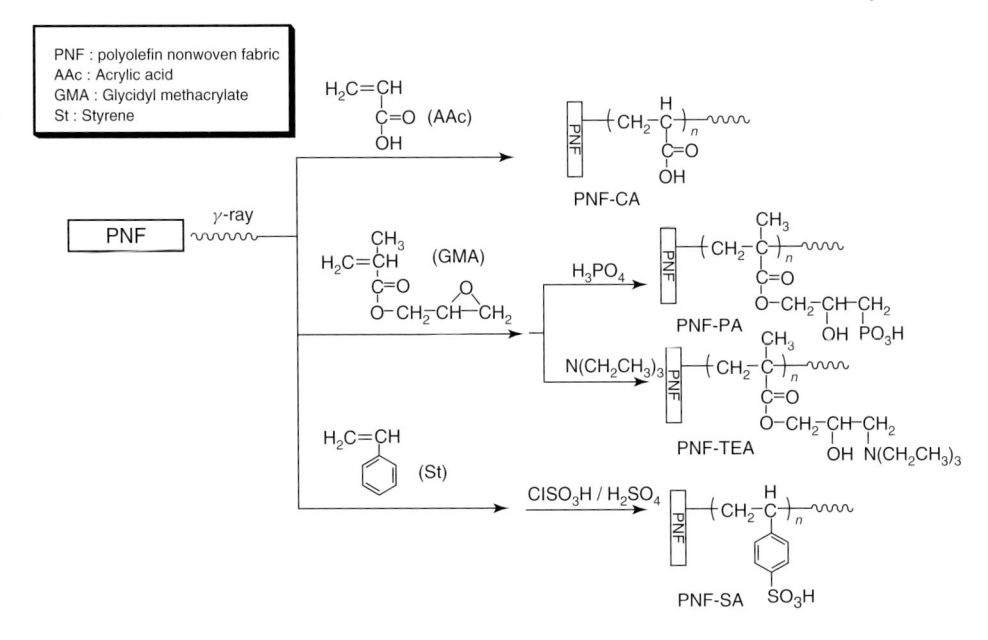

Figure 6.11 Schematic diagram for the preparation the IEM by grafting method [95].

on a large scale. This leaves big challenges for feasibility study, scale up, and commercialization.

6.3.2
Composite Ion Exchange Membranes

At this point, one can obviously notice that it is difficult to acquire all the targeted properties in one IEM to satisfy the requirements of industrial applications. For instance, to obtain high conducting membranes, high chemical modifications should be applied to the membrane, which will subsequently reduce the mechanical stability. Hence, innovative composite concept that enables the structural and functional tailoring of different materials may offer an alternative approach for the development of new IEMs with excellent electrochemical properties and good mechanical stability.

Inorganic–organic composite materials have gained increased attention because of its specific properties arising from synergistic effects among the components in the composite. The inorganic compartment usually offers vehicles for carrying extra charge functional groups, electrical properties, and enhancement of chemical, thermal, and mechanical stabilities. In contrast, the organic counterpart provides opportunities for chemical modifications, structure flexibility, and processibility on large scale.

The composite IEMs can be prepared by several routes such as sol–gel process, blending, *in situ* polymerization, molecular self-assembly [106–111]. Table 6.3 provides examples of composite IEMs prepared from different routes and their

Table 6.3 Preparation routes of composite IEMs and the resultant membrane properties.

Composite system	Preparation route	Property	Application
Nafion-silica [112–114]	Sol–gel	The composites showed significant reduction of methanol crossover, however, their proton conductivity was diminished	DMFCs
Nafion-sulfonated mesoporous silica [115]	Blending	The composite exhibited the enhanced proton conductivity and reduced methanol crossover up to 30% reduction compared to the pristine Nafion	DMFCs
PEO-[Si(OEt)$_3$]$_2$SO$_3$H [81]	Sol–gel	• Estimated pore diameter of 1–4 nm • Thermal stability up to 265 °C • IEC of 0.4–1.0 mequiv g^{-1}	NF
sPES-silica [116]	Sol–gel	High proton conductivity of 63.6 mS cm^{-1} and good IEC comparable to Nafion, the commercial membrane	DMFCs
sPES-sulfonated mesoporous silica [117–121]	Blending	The composite showed good IEC, ionic conductivity, transport properties while maintain decent mechanical and thermal stability	ED Water splitting
PVA-zirconium phosphate [122]	Blending	• Proton conductivity of 1–10 ms cm^{-1} at 50%RH • Reduced methanol crossover	DMFCs
PVA-silica [123–125]	Sol–gel	• IEC of 0.84–1.43 mequiv g^{-1} • Proton conductivity of 20–110 mS cm^{-1} at room temperature with fully hydrated condition	Electrodriving process, DMFC
sPEK/sPEEK-SiO$_2$, TiO$_2$, ZrO$_2$ [126]	Sol–gel	Though the composites showed up to 60-fold reduction of methanol flux, their proton conductivity was affected by 10–30% reduction	DMFCs

Note: DMFCs, direct methanol fuel cells; NF, nano filtration; ED, electrodialysis; PVA, polyvinyl alcohol; and PEO, polyethylene oxide.

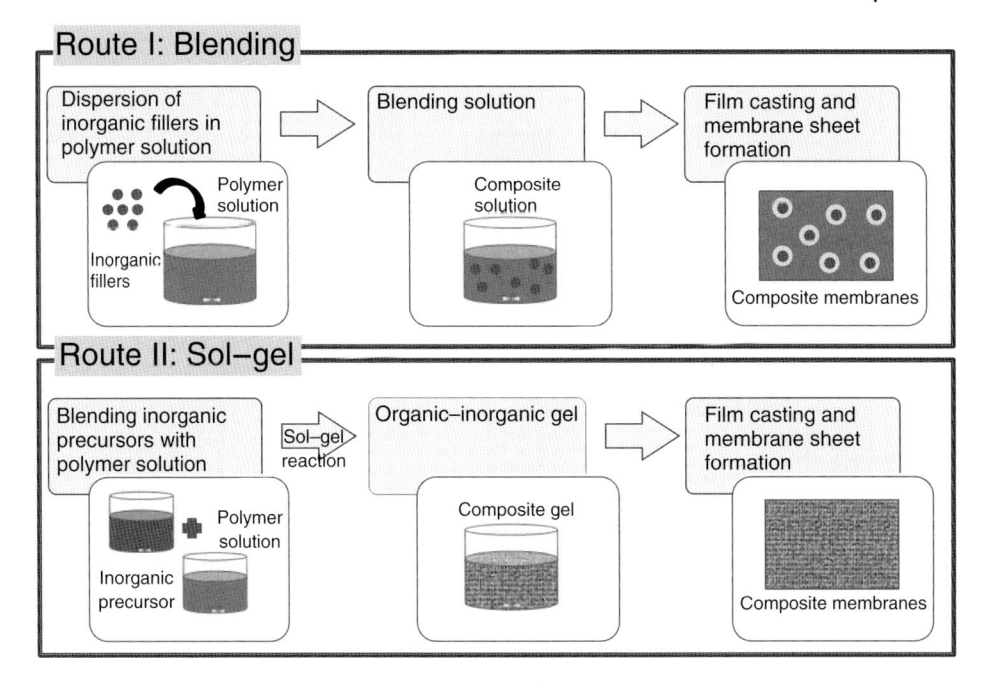

Figure 6.12 The most commonly used preparation methods for composite membranes: Route I blending and route II sol–gel method.

properties for specific applications. It is worth noting that the most frequently used routes are physical blending and sol–gel technique as depicted in Figure 6.12. For the first approach, the resultant membranes normally show phase separation from aggregated fillers, causing mechanical instability of the membranes. In contrast to the first approach, sol–gel method offers better interconnection between two domains.

The great challenges in developing composite materials are the incompatibility between the distinct compartments, aggregation of fillers, and their phase separation. However, this problem can be minimized by enhancing the interaction among them via covalent bond, hydrogen bond, and electrostatic interaction. Frequently used strategies for improving the interaction between inorganic and organic matrix are; (i) functionalization of inorganic fillers or/and polymer matrix and (ii) introduction of the inorganic filler on the polymer chains [127]. In the later case, the inorganic fillers can be fixed to the polymer by several approaches such as attaching the initiator or functional group that can further be polymerized with the polymer matrix.

As it can be noticed from Table 6.3, though the development of composite IEMs for fuel cell applications has been well established, there are very rare works applying the same composite concept to develop other types of electro-driven membrane processes [117–121].

Recently, our research group has developed a series of composite IEMs based on sPES and sulfonated mesoporous silica. By synergistic approaches of membrane preparation technique, the so-called, two-step phase inversion and the presence of surface functionalized silica nanofillers, membranes with controllable porosities and desired properties were successfully prepared. The synthesis route of these composites is depicted in Figure 6.13 [117].

The two-step phase inversion technique allows the control of membrane porosities by tuning the aging time before precipitating the partly dried film in water bath. As the viscosity and solvent/nonsolvent exchange rate altered accordingly, the pore sizes and the porosity of membrane were able to be controlled. The surface functionalized silica acted as the vehicle carrying extra charged functional groups for ion exchange, resulting in the improvement of water uptake, conductivity, transport properties, and electrochemical behavior of the resultant composites. However, at a high percent loading, the reduction of such properties was observed

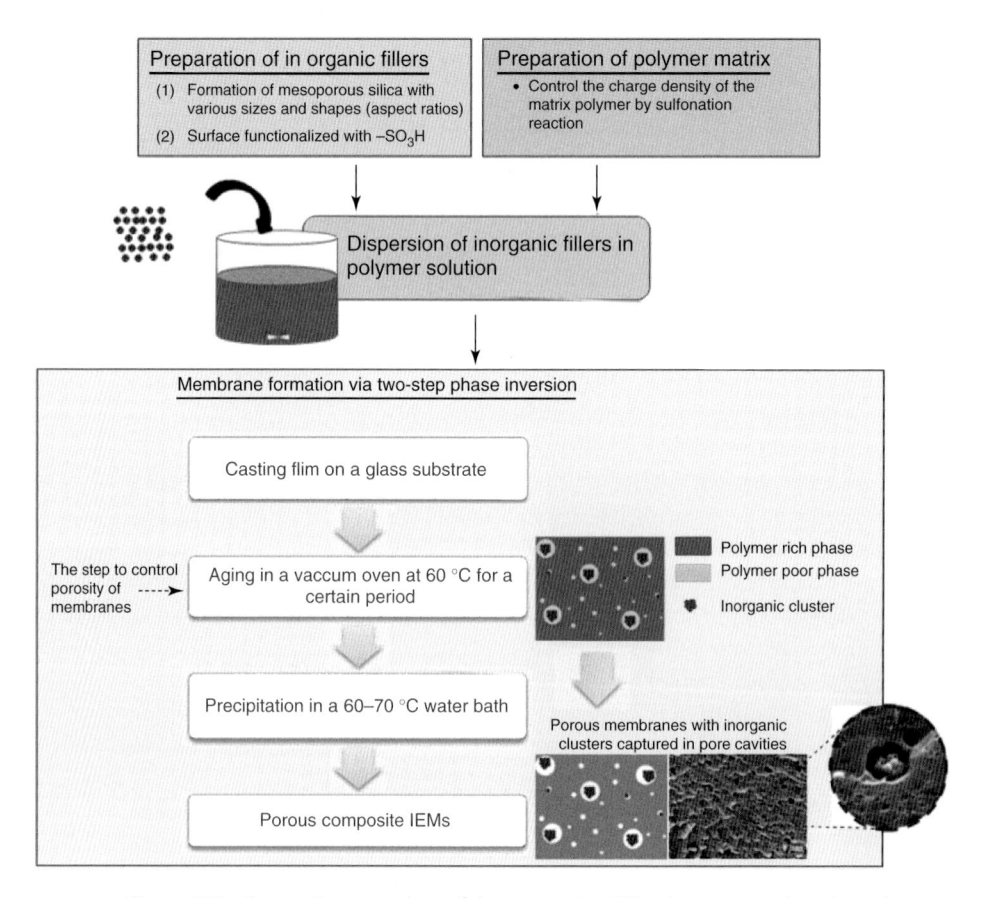

Figure 6.13 Preparation procedure of the composite IEMs via two-step phase inversion.

as the agglomerates interfere with the charged functional groups of the polymer matrix. Therefore, it is crucially important to optimize the loading to prepare membranes with overall good properties.

It is important to mention that the choices of preparation routes have to be wisely selected for a designed composite. Although good interaction between fillers and matrix give advantages for mechanical and thermal stability, it is normally at the cost of the reduction of functional groups for ion exchange and suppression of the membrane conductivity [106, 112, 113, 123, 126]. In many cases, the IEC and conductivities of the membranes decreased because of the incorporation of inorganic fillers, interfering with the functional groups of the polymer matrix [112, 113, 123, 126]. Therefore, optimization is of crucial importance and the percolation studies are recommended.

6.3.3
Membranes with Specific Properties

Although the properties and performance of the IEMs have been significantly improved over the last five decades, they are still insufficient for separating specific ionic species in some niche applications. Several strategies, including modifying membrane surface and their hydrophilicity, changing the structure and porosity of membranes, and incorporating special functional groups in polymer chains or fillers have been proposed .

In general, permselectivity among ions in a mixed solution through selective membranes depends on the affinity of the components with the membrane and the migration rate of each component in the membrane phase [128, 129].

Sieving monovalent ions such as NaCl from other multivalent ions has been studied mainly by modification of IEMs, making the membrane matrix dense or forming a thin dense layer on the surface. This is the simplest concept to change permselectivity of IEMs. One simple method to make membranes dense is to increase cross-linkage of the membranes, resulting in tunable pore sizes of the membranes and then adjustable selectivity between bulkier molecules and smaller molecules [134, 135]. However, the result appears to be not that simple. The transport number of ions depends not only on the size of ions but also on the Gibbs hydration energies of the ions. Although increasing cross-linkage enhances the permselectivity of the membrane, it also increases electrical resistance of the membrane. Therefore, conducting polymers were applied and a dense layer on the membrane surface is more preferable to avoid the increase of resistance. Figure 6.14 shows some chemical structures of conducting polymers.

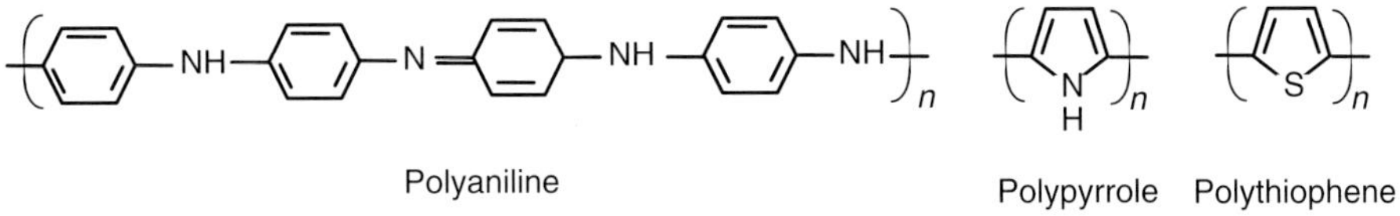

Figure 6.14 Structures of conducting polymers.

The conductive nature of conducting polymers is associated with unique conjugate bonds along the polymer backbones that allows electrons to freely transport through when an electrical potential is applied [138]. Polypyrrole is extremely rigid and has good affinity to IEMs [79, 136, 138]. It can easily form highly dense structure or tightly bonded layer on the membrane surface by impregnation of conducting polymers into the membrane matrix or ion exchange with the cation exchangeable groups on the membrane surface [137]. Moreover, polypyrrole possesses secondary amino groups, which affect the permselectivity between charged ions. It has been reported that the permselectivity of divalent ions compared to NaCl decreased because of the synergistic sieving effect of dense polypyrrole layer and the difference of repulsion forces among ions with the membranes [139]. For more examples of membrane surface modification by conducting polymer, polyaniline (PANI) has also been reported to affect the selectivity of monovalent ions from multivalent ions [138, 140–142]. It has been proposed that a reduction in transport number of divalent ions compared with monovalent ions through both composite PANI/AEMs and CEMs may be a result of three factors; the electrostatic repulsions increasing the difficulty for multivalent ions to pass through the membranes, the decrease of hydrophilicity of membranes resulting in the reduction of interaction with more hydrated ions, and the partial reduction or increment in the surface charge density of AEM/PANI and CEM/PANI.

The formation of a thin cationic charged layer on the surface of CEMs has been reported to be another effective method to separate monovalent ions from multivalent ions [128–130, 143, 144]. Cationic polyelectrolytes such as the amino groups of polyethyleneimine which adsorbed on IEMs without ion exchanging with sulfonic groups of the membranes are thought to be due to highly branched structure of polyethyleneimine. Therefore, most positive charges from the modifier and negative charges from the CEMs remain and the presence of these positive charges results in difficulties for multivalent ions to pass through the membranes owing to higher repulsion force.

Sata et al. has systematically investigated the influence of membrane hydrophilicity on the selectivity of anion species compared to chloride ions [17, 131, 132, 134, 135, 137]. It is apparent that the transport number of anions through AEMs was predominated by the hydration energy of anions rather than their hydrated size [128]. For instance, AEMs with increased hydrophilicity showed the significant decrease of transport number of fluoride ions, which are strongly hydrated anions compared to chloride ions. A similar trend was also observed in the case of sulfate ions. In contrast, for nitrate ions with similar hydrated ion size but less hydration energy compared to the chloride ions, the mobility ratio of the nitrate ions to chloride ions was increased with the decreased hydrophilicity of the AEM. Therefore, the understanding of the relationship between hydrophilicity of AEMs and hydration energy of anions is of great importance. It is expected that by decreasing hydrophilicity of the AEMs, less hydrated anions can permeate through the membranes more easily than strongly hydrated anions [17] (Table 6.4).

Ether compounds such as ethylene glycol, dipropylene glycol, and dimethyl ether have been used to change the hydrophilicity of IEMs in order to change

Table 6.4 Ionic radius, hydrate ionic radius, and hydration energy of ion species [158, 159].

Ion	Ionic radius (nm)	Hydrated ionic radius (nm)	Hydration energy (kJ mol^{-1})
Na$^+$	0.095	0.365	407
Mg^{2+}	0.074	0.429	1921
Ca^{2+}	0.099	0.349	1584
Cl$^-$	0.181	0.347	376
NO$_3^-$	0.189	0.340	270
SO$_4^{2-}$	0.230	0.380	1138

permselectivity of membranes. Owing to the existence of ether groups and alcoholic groups, these compounds possess a hydrophilic nature, which is expected to make stronger hydrated ions permeate through the membranes more easily than less hydrated ions [131–133]. However, because these compounds are soluble in water, it was found that glycol compounds were dissolved from the membranes during ED, which is undesirable.

6.3.3.1 Improving Antifouling Property

Fouling is one of the most serious problems for membrane separation processes including IEM ED. It decreases the performance efficiency of the membranes. It is well known that AEMs often suffer from fouling more than CEMs do, because many organic foulants in solution are in anionic forms and can therefore stick to the surface of AEMs. There are many techniques that have been proposed to prevent fouling of IEMs. These may be grouped into two main categories, by promoting foulants to permeate easily through the membranes and by prohibiting foulants to pass through IEMs.

The structures of membranes were loosened to allow organic macroions to easily permeate through as a method to facilitate the transmission of foulants [145]. However, there was an undesirable effect of the decrease of ion selectivity of IEMs by loosening the structure of the membranes.

On the other hand, to prevent foulants penetrating through membranes a dense structure or thin layer of positively charged group on the membrane surface was introduced. Unfortunately, as a consequence, the resistance of membranes increased dramatically.

Recently, new techniques to improve resistance against organic fouling without significant increase of electrical resistance of the membranes have been patented [146].

It has been proved that the fixation of polyether compounds on the surface or inside the membranes prevents ion exchange groups of the membranes from the direct contact with organic macroions, which prohibits the absorption of them. To avoid the ether compounds dissolving during ED, the compounds were fixed by decomposing the ether bond of polyether compounds with an ether bound cleavage reagent such as a mixture of acetic anhydride-p-toluene sulfonic acid, phosphoric acid, and hydrobromic acid. The results demonstrated excellent fouling resistance and low electrical resistance (1.5–3.4 Ω cm^{-2}) in the membranes.

The surface property of membranes is one of the most important factors affecting organic fouling in membrane separation processes. It has been reported that the colloid and microbial foulants favor the adsorption on hydrophobic surface [147, 148]. Thus, the hydrophilic modification of the membrane surface becomes a potential method to prevent fouling. TiO$_2$/polymer self-assembly method has been used to modify hydrophilicity of polymeric membranes to decrease the fouling rate [147, 148]. This method is based on the ionic attraction between the oppositely charged species of TiO$_2$ particles and membranes, which promotes the adhesion between them [147–149]. The result showed that organic fouling in membranes modified with TiO$_2$ nanoparticles decreased compared with the unmodified one because of the increase of hydrophilicity of the membrane. Moreover, the adsorbed foulants on the tested membranes were more easily removed by shear force than those on unmodified membranes.

6.4
Future Perspectives of IEMs

The first industrial application of IEMs was in the field of desalination by ED. The recent development of fundamental theory has led to expanded applications to various fields such as electrodialysis reversal (EDR), electrodeionization (EDI), bipolar membrane electrodialysis (BMED), and fuel cell [2]. In this section, the recently developed IEMs and their perspectives particularly relevant to ED process, water purification, and desalination fields are highlighted.

Although desalination is an important technology that offers a solution for the world water crisis, the energy consumption, and production cost of this technology remains a great challenge.

Tremendous efforts have been dedicated to reduce the energy consumption of desalination technology and to bring down the water production cost to affordable levels. Besides the development of novel membranes, new process designs for operating systems that can work more effective are increasingly concerned [150, 151].

6.4.1
Hybrid System

Another strategy for reducing the overall cost of desalination is the use of hybrid system, a system that combines two or more desalination processes [84, 150, 152, 153]. The hybrid desalination system has recently gained intensive attention as it can offer a better separation performance for specific industrial separation and a process with optimized utilization of material and energy [84, 150, 152]. Electrodriven processes such as ED has been integrated with pressure-driven membrane operation such as microfiltration (MF), ultrafiltration (UF), nanofiltration (NF), and reverse osmosis (RO). Such hybrid systems showed better performance in water treatment and desalination compared to their conventional process [5–7, 154, 155]. Recently, a novel RO–ED hybrid system was proposed by Pellegrino et al. [5] that aimed to reduce the osmotic pressure at the interface of RO membrane, to reduce the energy consumption, and to improve the water recovery of the system. Their design is shown in Figure 6.15; the hollow-fiber RO membranes were packed in ED compartments between CEM and AEM. The stack of the hollow-fiber RO

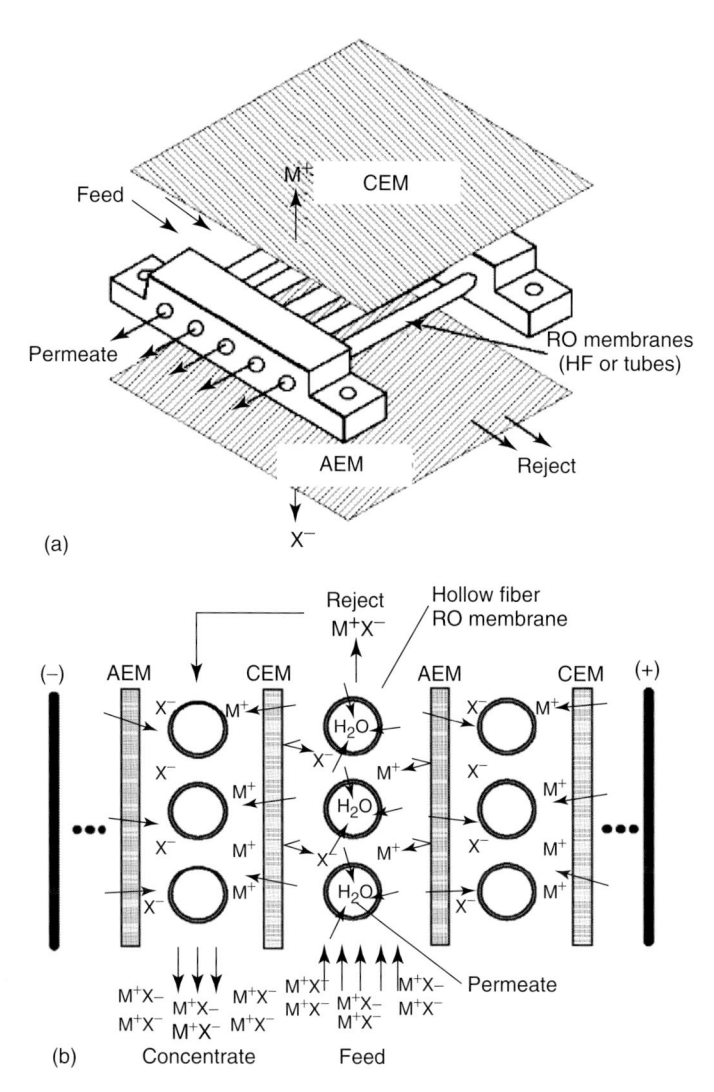

Figure 6.15 Schematic (a) cell pair in the hybrid RO-ED system and (b) stack configuration for the hybrid RO–ED. The M^+ and X^- represents cations and anions, respectively [5].

membranes may be considered to act as the spacer in the conventional ED system that helps the well mixing of the feed solution and promotes the turbulent flow of the system. This in turn suppresses the concentration polarization effects near the membrane interface. Their initial analysis of this new hybrid system design under different operating conditions indicated that the energy saving of 10–20% can be obtained.

However, the modeling analysis from this work also indicated the limiting factor of the conductivity of the IEMs on the energy consumption of the system. Therefore, the robust membranes are vital to be further developed for supporting the use of such a novel conceptual system design.

6.4.2
Small-Scale Seawater Desalination

Capital- and energy-intensive desalination plants that require seawater conveyance and reliable power supply for the system are still impractical and uneconomical at small scale for personal use or small villages in a remote area. Recently, research groups in Massachusetts Institute of Technology and Pohang University of Science and Technology have designed a novel small seawater desalination unit, the so-called ion-concentration polarization (ICP) desalination system (Figure 6.16) with the possibility of battery-powered operation [157, 156]. New system for small-scale desalination by ICP, is considered as the latest state-of-the-art system capable of continuous seawater desalination with high salt rejection of 99% at low power consumption of less than 3.5 Wh l^{-1} [156].

This work represents another example of novel conceptual nanoscale design based on theoretical study that have been proved by experimental implementation to be an excellent tool for providing fresh water for personal use in remote areas. Again, porous IEMs suitable to be used in such an innovative concept have to be further developed.

6.5
Conclusions

The present overview on the development of IEMs has addressed the vital role of IEMs involved in desalination and water purification applications. The properties and separation capabilities determine the performance of membranes. In the past decades, interdisciplinary approaches have been applied to the development of new IEMs. IEMs can be designed and prepared by a number of strategies varying from basic polymer reactions to advanced nanotechnology via molecular design and architectural tailoring of composite materials. However, to satisfy the membrane requirements for specific applications, the targeted properties of IEMs have to be set up, and appropriate synthesis routes for IEMs toward the targets should be specified accordingly. To accomplish this, fundamental understanding of the relationship among preparation conditions, structures, and functions of the IEMs

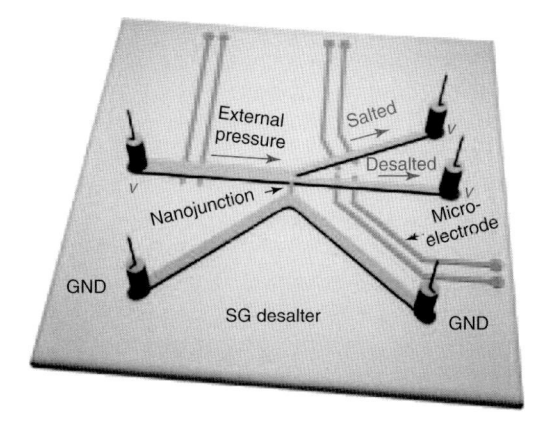

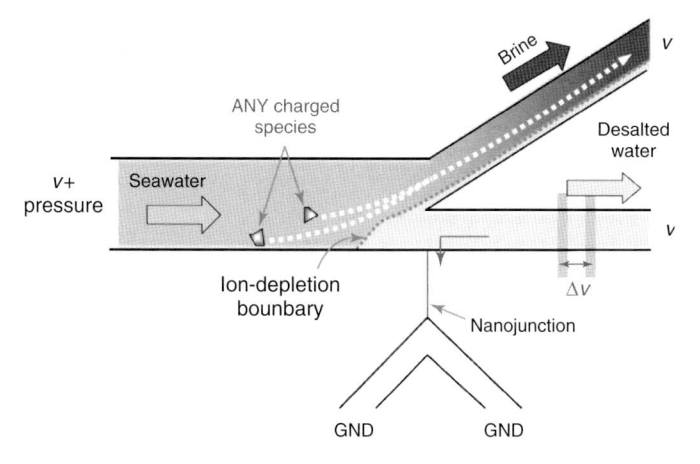

Figure 6.16 ICP desalination scheme: (a) micro/nanofluidic desalination system with embedded microelectrode for the measuring potential drop and (b) electrokinetic desalination operation associated with the external pressure [156].

have to be established. Furthermore, more efforts and attention should be paid to the following studies:

- percolation study of the membrane preparation, structure, and properties;
- elucidation of transport phenomena of newly prepared membranes, as well as the search for their new application;
- evaluation of new membranes in an application;
- the study of surface chemistry and architecture of membranes to reduce the effects of concentration polarization; and
- the modification of surface chemistry toward high fouling resistance properties of IEMs.

From the viewpoint of material engineering, nanotechnology offers powerful tools to tailor and develop novel membranes. Among several approaches presented in this chapter, the concept of composite design stands out as an interesting strategy for developing IEMs. The properties of membranes can be designed via the selection of component materials and the tailoring of desired functionalities to each component. Encouragingly, the preparation procedure of composite membranes is relatively simple, and thus, has a high potential for large-scale commercialization. Although this concept has drawn much attention in fuel cell application, the utilization of this strategy in the development of IEMs for desalination is still very limited, leaving us much room for development and comprehensive study of composite design and their applications in desalination.

Apart from the material development, the system design and their operation optimization should also be further developed. In the desalination application which is one of the major practical areas for the IEMs, new systems to bring down the energy consumption and production cost require more progress in comparison to more widely used processes such as RO. Such system designs that can combine with renewable energy sources or can work more effectively should also be emphasized for the sustainable development of energy-saving and environment friendly systems.

References

1. Schoeman, J.J., Steyn, A., and Makgae, M. (2005) Evaluation of electrodialysis for the treatment of an industrial solid waste leachate. *Desalination*, **186**, 273–289.
2. Koter, S. and Warszawski, A. (2000) Electromembrane processes in environment protection. *Pol. J. Environ. Stud.*, **9**, 45–56.
3. Ortiz, J.M. et al., (2006) Photovoltaic electrodialysis system for brackish water desalination: Modeling of global process. *J. Membr. Sci.*, **274**, 138–149.
4. Ortiz, J.M. et al., (2005) Brackish water desalination by electrodialysis; batch recirculation operation modeling. *J. Membr. Sci.*, **252**, 65–75.
5. Pellegrino, J., Gorman, C., and Richards, L. (2007) A speculative hybrid reverse osmosis/electrodialysis unit operation. *Desalination*, **214**, 11–30.
6. Turek, M. and Dydo, P. (2008) Comprehensive utilization of brackish water in ED-thermal system. *Desalination*, **221**, 455–461.
7. Xu, T. and Huang, C. (2008) Electrodialysis-based separation technologies: a critical review. *Am. Inst. Chem. Eng.*, **54**, 3147–3159.
8. Luo, Q. et al., (2008) Preparation and characterization of Nafion/SPEEK layered composite membrane and its application in vandium redox flow battery. *J. Membr. Sci.*, **325**, p. 553–558.
9. Hansen, H.K., O.L.M., and Villumsen, A. (1999) Electrical resistance and transport numbers of ion-exchange membranes used in electrodialytic soil remediation. *Sep. Sci. Technol.*, **34**, p. 2223–2233.
10. Smalley, R.E. (2005) Future global energy prosperity: The terawat challenge. *MRS Bull.*, **30**, p. 412–417.
11. Strathmann, H., (2004) *Ion-Exchange Membrane Separation Processes*, Membrane Science and Technology Series, Vol. 9, Elsevier, Amsterdam and Boston, xi, 348 p.
12. Dzyazko, Y.S. and Belyakov, V.N. (2004) Purification of a diluted nickel solution

containing nickel by a process combining ion exchange and electrodialysis. *Desalination*, **162**, 179–189.
13. Sadrzadeh, M. et al., (2008) Separation of lead ions from wastewater using electrodialysis: Comparing mathematical and neural network modeling. *Chem. Eng. J.*, **144**, 431–441.
14. Balster, J., Stamatialis, D.F., and Wessling, M. (2004) Electro-catalytic membrane reactors and the development of bipolar membrane technology. *Chem. Eng. Process.*, **43**, 1115–1127.
15. Tsang, E.M.W. et al., (2009) Nanostructure, morphology and properties of fluorous copolymers bearing ionic grafts. *Macromolecules*, **42**, 9467–9480.
16. Qiu, J. et al., (2009) Performance of vanadium redox flow battery with a novel amphoteric ion exchange membrane synthesized by two-step grafting method. *J. Membr. Sci.*, **324**, 215–220.
17. Sata, T. (2002) *Ion Exchange Membranes. Preparation, Characterization, Modification and Application*, The Royal Society of Chemistry, Cambridge.
18. Nagarale, R.K., Gohil, G.S., and Shahi, V.K. (2006) Recent developments on ion-exchange membranes and electro-membrane processes. *Adv. Colloid Interface Sci.*, **119** (2–3), 97–130.
19. Division, National Chemical Research Laboratory Process Development (1960) *Demineralization by Electrodialysis*, Butterworths Scientific, London.
20. Lakshminarayanaiah, N. (1969) *Transport Phenomena in Membranes*, Academic Press, Inc., New York.
21. Wilson, J.R. (1960) *Demineralization by Electrodialysis*, The University Press, Glasgow, London.
22. Mulder, M. (1991) *Basic Principles of Membrane Technology*, Kluwer Academic Publishers, Dordrecht.
23. Nikonenko, V.V. et al., (2010) Intensive current transfer in membrane systems; Modelling, mechanisms and application in electrodialysis. *Adv. Colloid. Interface Sci.*, **160**, 101–123.
24. Baker, R.W. (2004) *Membrane Technology and Applications*, John Wiley & Sons, Ltd.
25. Tanaka, Y. (2007) Ion exchange membrane-Fundamentals and applications. *J. Membr. Sci.*, **12**, 139–186.
26. Lakshminarayanaiah, N. (1965) Transport phenomena in artificial membranes. *Chem. Rev.*, **65**, 491–565.
27. Balster, J. et al., (2007) Morphology and microtopology of cation-exchange polymers and the origin of the overlimiting current. *J. Phys. Chem.*, **111**, 2152–2165.
28. Zabolotsky, V.I. et al., (1998) Coupled transport phenomena in overlimiting current electrodialysis. *Sep. Purif. Technol.*, **14**, 255–267.
29. Tanaka, Y. and Seno, M. (1986) Concentration polarization and water dissociation in ion-exchange membrane electrodialysis. *J. Chem. Soc., Faraday Trans.*, **82** (1), 2065–2077.
30. Krol, J.J., Wessling, H., and Strathmann, H. (1999) Chronopotentiometry and overlimiting ion transport through monopolar ion exchange membranes. *J. Membr. Sci.*, **162**, 155–164.
31. Krol, J.J., Wessling, H., and Strathmann, H. (1999) Concentration polarization with monopolar ion exchange membranes: current–voltage curves and water dissociation. *J. Membr. Sci.*, **162**, 145–154.
32. Lee, H.-J. et al., (2008) Influence of the heterogeneous structure on the electrochemical properties of anion exchange membranes. *J. Membr. Sci.*, **320**, 549–555.
33. Block, M. and Kitchener, J.A. (1966) Polarization phenomena in commercial ion-exchange membranes. *J. Electrochem. Soc.*, **113** (9), 947–953.
34. Tanaka, Y. (2004) Concentration polarization in ion-exchange membrane electrodialysis: The events arising in an unforced flowing solution in a desalting cell. *J. Membr. Sci.*, **244** (1–2), 1–16.
35. Pismenskaia, N.D. et al., (2007) Coupled convection of solution near the surface of ion-exchange membranes in intensive current regimes. *Russ. J. Electrochem.*, **43**, 325–345.
36. Patel, R.D. and Lang, K.-C. (1977) Polarization in ion-exchange membrane electrodialysis. *Ind. Eng. Chem. Fundam.*, **16**, 340–348.

37. Aguilella, V.M. et al., (1991) Current–voltage curves for ion-exchange membranes. Contributions to the total potential drop. *J. Membr. Sci.*, **61**, 177–190.
38. Ray, P. et al., (1999) Transport phenomenon as a function of counter and co-ions in solution: chronopotentiometric behavior of anion exchange membrane in different aqueous electrolyte solutions. *J. Membr. Sci.*, **160**, 243–254.
39. Choi, J.-H., Park, J.-S., and Moon, S.-H. (2002) Direct measurement of concentration distribution within the boundary layer of an ion-exchange membrane. *J. Colloid Interface Sci.*, **251**, 311–317.
40. Pismenskaya, N. et al., (2004) Chronopotentiometry applied to the study of ion transfer through anion exchange membranes. *J. Membr. Sci.*, **228**, 65–76.
41. Rubinshtein, I. et al., (2002) Experimental verification of the electroosmotic mechanism of overlimiting conductance through a cation exchange electrodialysis membrane. *Russ. J. Electrochem.*, **38**, 853–863.
42. Metayer, M., Bourdillon, C., and Selegny, E. (1973) Concentration polarization on ion-exchange membranes in electrodialysis with natural convection. *Desalination*, **13**, 129–146.
43. Larchet, C. et al., (2008) Application of chronopoteniometry to determine the thickness of diffusion layer adjacent to an ion-exchange membrane under natural convection. *Adv. Colloid. Interface Sci.*, **139**, 45–61.
44. Loza, N.V. et al., (2006) Effect of modification of ion-exchange membrane MF-4SK on its polarization characteristics. *Russ. J. Electrochem.*, **42** (8), 815 822.
45. Elattar, A. et al., (1998) Comparison of transport properties of monovalent anions through anion-exchange membranes. *J. Membr. Sci.*, **143** (1–2), 249–261.
46. Volodin, E. et al., (2005) Ion transfer across ion-exchange membranes with homogeneous and heterogeneous surfaces. *J. Colloid Interface Sci.*, **285**, 247–258.
47. Vyas, P.V. et al., (2003) Studies of the effect of variation of blend ratio on permselectivity and heterogeneity of ion-exchange membranes. *J. Colloid Interface Sci.*, **257**, 127–134.
48. Wilhelm, F.G. et al., (2001) Chronopotentiometry for the advanced current–voltage characterisation of bipolar membranes. *J. Electroanal. Chem.*, **502**, 152–166.
49. Choi, J.-H., Lee, H.-J., and Moon, S.-H. (2001) Effects of electrolytes on the transport phenomena in a cation-exchange membrane. *J. Colloid Interface Sci.*, **238**, 188–195.
50. Meng, H. et al., (2005) A new method to determine the optimal operating current (I_{lim}') in the electrodialysis process. *Desalination*, **181**, 101–108.
51. Pismenskaia, N. et al., (2004) Chronopotentiometry applied to the study of ion transfer through anion exchange membranes. *J. Membr. Sci.*, **228**, 65–76.
52. Ibanez, R., Stamatialis, D.F., and Wessling, M. (2004) Role of membrane surface in concentration polarization at cation-exchange membranes. *J. Membr. Sci.*, **239**, 119–128.
53. http://www.fumatech.com/Startseite/Produkte/fumasep/Ionenaustauscher-membranen/ (accessed 1 March 2008).
54. Lee, H.-J. et al., (2002) Designing of an electrodialysis desalination plant. *Desalination*, **142**, 267–286.
55. Feng, S. et al., (2009) Synthesis and characterization of crosslinked sulfonated poly(arylene ether sulfone) membranes for DMFC applications. *J. Membr. Sci.*, **335**, 13–20.
56. Iojoiu, C. et al., (2005) Mastering sulfonation of aromatic polysulfones: Crucial for membranes for fuel cell application. *Fuel Cells*, **5** (3), 344–354.
57. Johnson, B.C. et al., (1984) Synthesis and characterization of sulfonated poly(arylene ether sulfones). *J. Polym. Sci. Polym. Chem. Ed.*, **22**, 721–737.
58. Klaysom, C. et al., (2010) Synthesis and characterization of sulfonated polyethersulfone for cation-exchange membranes. *J. Membr. Sci.*, **368**, 48–53.
59. Roziere, J. and Jones, D.J. (2003) Non-fluorinated polymer materials for

proton exchange membrane fuel cells. *Annu. Rev. Mater. Res.*, **33**, 503–555.
60. Dai, H. *et al.*, (2007) Development and characterization of sulfonated poly(ether sulfone) for proton exchange membrane materials. *Solid State Ion.*, **178**, 339–345.
61. Nagarale, R.K. *et al.*, (2005) Preparation and electrochemical characterization of sulfonated polysulfone cation-exchange membranes: Effects of the solvents on the degree of sulfonation. *J. Appl. Polym. Sci.*, **96**, 2344–2345.
62. Li, Y. and Chung, T.-S. (2008) Highly selective sulfonated polyethersulfone (sPES)-based membranes with transition metal counterions for hydrogen recovery and natural gas separation. *J. Membr. Sci.*, **308**, 128–135.
63. Guan, R. *et al.*, (2005) Polyethersulfone sulfonated by chlorosulfonic acid and its membrane characteristics. *Eur. Polym. J.*, **41**, 1554–1560.
64. Klaysom, C. *et al.*, (2011) Preparation of porous ion-exchange membranes (IEMs) and their applications. *J. Membr. Sci.*, **371**, 37–44.
65. Kerres, J., W.C., and Reichle, S. (1996) New sulfonated engineering polymers via the metalation route. I. Sulfonated poly(ethersulfone) PSU Udel via metalation-sulfination-oxidation. *J. Polym. Sci., Part A: Polym. Chem.*, **34**, 2421–2438.
66. Kerres, J. *et al.*, (1998) Development and characterization of crosslinked ionomer membranes based upon sulfinated and sulfonated PSU crosslinked PSU blend membranes by disproportionation of sulfinic acid groups. *J. Membr. Sci.*, **139**, 211–225.
67. Byun, I.S., Kim, I.C., and Seo, J.W. (2000) Pervaporation behavior of asymmetric sulfonated polysulfones and sulfonated poly(ether sulfone) membranes. *J. Appl. Polym. Sci.*, **76**, 787–798.
68. Noshay, A. and Robeson, L.M. (1976) Sulfonated polysulfone. *J. Appl. Polym. Sci.*, **20**, 1885–1903.
69. Li, L. and Wang, Y. (2005) Quaternized polyethersulfone Cardo anion exchange membrane for direct methanol alkaline fuel cells. *J. Membr. Sci.*, **262**, 1–4.
70. Evram, E. *et al.*, (1997) Polymers with pendant functional group. III. Polysulfones containing viologen group. *J. Macromol. Sci., A Pure Appl. Chem.*, **34**, 1701–1714.
71. Bailly, C. *et al.*, (1987) The sodium salts of sulphonated poly(aryl-ether-ether-ketone)(PEEK): preparation and characterization. *Polymer*, **28**, 1009–1016.
72. Jin, X. *et al.*, (1985) A sulphonated poly(aryly ether ketone). *Br. Polym. J.*, **17**, 4–10.
73. Kobayashi, T. *et al.*, (1998) Proton-conducting polymers derived from poly(ether-etherketone) and poly(4-phenoxybenzoyl-1,4-phenylene). *Solid State Ion.*, **106**, 219–225.
74. Zaidi, S.M.J. *et al.*, (2000) Proton conducting composite membranes from polyether ether ketone and heteropolyacids for fuel cell applications. *J. Membr. Sci.*, **173**, 17–34.
75. Do, K.N.T. and Kim, D. (2008) Synthesis and characterization of homogeneously sulfonated poly(ether ether ketone) membranes: Effect of casting solvent. *J. Appl. Polym. Sci.*, **40**, 1763–1770.
76. Fang, J. and Shen, P.K. (2006) Quaternized poly(phthalazinon ether sulfone ketone) membrane for anion exchange membrane fuel cells. *J. Membr. Sci.*, **285**, 317–322.
77. Trotta, F. *et al.*, (1998) Sulfonation of polyetheretehrketone by chlorosulfuric acid. *J. Appl. Polym. Sci.*, **70**, 478–483.
78. Drioli, E. *et al.*, (2004) Sulfonated PEEK-WC membranes for possible fuel cell applications. *J. Membr. Sci.*, **228**, 139–148.
79. Cui, W., Kerres, J., and Eigenberger, G. (1998) Development and characterization of ion-exchange polymer blend membranes. *Sep. Purif. Technol.*, **14**, 145–154.
80. Tongwen, X. and Weihua, Y. (2001) Fundamental studies of a new series of anion exchange membranes: membrane preparation and characterization. *J. Membr. Sci.*, **190**, 159–166.
81. Wu, C. *et al.*, (2005) Synthesis and characterizations of new negatively charged organic–inorganic hybrid materials,

Part II. Membrane preparation and characterizations. *J. Membr. Sci.*, **247**, 111–118.

82. Xu, C., Xu, T., and Yang, W. (2003) A new inorganic–organic negatively charged membrane: membrane preparation and characterization. *J. Membr. Sci.*, **224**, 117–125.

83. Tongwen, X. and Zha, F.F. (2002) Fundamental studies on a new series of anion exchange membrane: Effect of simultaneous amination-crosslinking processes on membranes ion-exchange capacity and dimensional stability. *J. Membr. Sci.*, **199**, 203–210.

84. Xu, T. (2005) Ion exchange membranes: state of their development and perspective. *J. Membr. Sci.*, **263**, 1–29.

85. Kang, M.-S. et al., (2003) Electrochemical characterization of sulfonated poly(arylene ether sulfone) (S-PES) cation-exchange membranes. *J. Membr. Sci.*, **216** (1–2), 39–53.

86. Krishnan, N.N. et al., (2006) Synthesis and characterization of sulfonated poly(ether sulfone) copolymer membranes for fuel cell applications. *J. Power Sources*, **158**, 1246–1250.

87. Sankir, M. et al., (2006) Synthesis and characterization of 3,3′-disulfonated-4,4′-dichlorodiphenyl sulfone (SDCDPS) monomer for proton exchange membranes (PEM) in fuel cell application. *J. Appl. Polym. Sci.*, **100**, 4595–4602.

88. Ueda, M. et al., (1993) Synthesis and characterization of aromatic poly(ether sulfone)s containing pendant sodium sulfonate groups. *J. Polym. Sci., A Polym. Chem.*, **31**, 853–858.

89. Kim, Y.S. et al., (2004) Sulfonated poly(arylene ether sulfone) copolymer proton exchange membranes: composition and morphology effects on the methanol permeability. *J. Membr. Sci.*, **243**, 317–326.

90. Wang, F. et al., (2001) Synthesis of highly sulfonated poly(arylene ether sulfone) random (statistical) copolymers via direct polymerization. *Marcomol. Symp.*, **175**, 387–395.

91. Kato, K. et al., (2003) Polymer surface with graft chains. *Prog. Polym. Sci.*, **28**, 209–259.

92. Choi, S.-H., Lee, K.-P., and Nho, Y.C. (1999) Electrochemical properties of polyethylene membrane modified with sulfonic acid group. *Korea Polym. J.*, **7**, 297–303.

93. Nasef, M.M. and Saidi, H. (2005) Structure–property relationships in radiation grafted poly(tetrafluoroethylene)-graft-polystyrene sulfonic acid membranes. *J. Polym. Res.*, **12**, 305–312.

94. Sunaga, K. et al., (1999) Characteristics of porous anion-exchange membranes prepared by cografting of glycidyl methacrylate with divinylbenzene. *Chem. Mater.*, **11**, 1986–1989.

95. Lee, K.-P., Choi, S.-H., and Kang, H.-D. (2002) Preparation and characterization of polyalkene membranes modified with four different ion-exchange groups by radiation-induced graft polymerization. *J. Chromatogr. A*, **948**, 129–138.

96. Gupta, B. and Chapiro, A. (1989) Preparation of ion-exchange membranes by grafting acrylic acid into pre-irradiated polymer films-1. grafting into polyethylene. *Eur. Polym. J.*, **25**, 1137–1143.

97. Yamee, B. et al., (2010) Hybrid polymer-silicon proton conducting membranes via a pore-filling surface-initiated polymerization approach. *Appl. Mater. Interfaces*, **2**, 279–287.

98. Gupta, B. and Chapiro, A. (1989) Preparation of ion-exchange membranes by grafting acrylic acid into pre-irradiated polymer films – 2.Grafting into teflon-FEP. *Eur. Polym. J.*, **25**, 1145–1148.

99. Gubler, L. et al., (2009) Radiation grafted fuel cell membranes based on co-grafting of alpha-methylstyrene and methacrylonitrile into a fluoropolymer base film. *J. Membr. Sci.*, **339**, 68–77.

100. Gupta, B., Buchi, F.N., and Scherer, G.G. (1994) Cation exchange membranes by pre-irradiation grafting of styrene into FEP films. I. Influence of synthesis conditions. *J. Polym. Sci., A Polym. Chem.*, **32**, 1931–1938.

101. Gupta, B. et al., (1996) Crosslinked ion exchange membranes by radiation grafting of styrene/divinylbenzene into FEP films. *J. Membr. Sci.*, **118**, 213–238.

102. Nasef, M.M., Saidi, H., and Nor, H.M. (2000) Cation exchange membranes

by radiation-induced graft copolymerization of styrene onto PFA copolymer films. III. Thermal stability of the membranes. *J. Appl. Polym. Sci.*, **77**, 1877–1885.
103. Danks, T.N., Slade, R.C.T., and Varcoe, J.R. (2003) Alkaline anion-exchange radiation-grafted membranes for possible electrochemical application in fuel cell. *J. Mater. Chem.*, **13**, 712–721.
104. Herman, H., Slade, R.C.T., and Varcoe, J.R. (2003) The radiation-grafting of vinylbenzyl chloride onto poly(hexafluoropropylene-co-tetrafluoroethylene) films with subsequent conversion to alkaline anion-exchange membranes: optimisation of the experimental conditions and characterisation. *J. Membr. Sci.*, **218**, 147–163.
105. Nasef, M.M. and Hegazy, E.-S.A. (2004) Preparation and applications of ion exchange membranes by radiation-induced graft copolymerization of polarmonomers onto non-polar films. *Prog. Polym. Sci.*, **29**, 449–561.
106. Lavorgna, M. *et al.*, (2007) Hybridization of Nafion membranes by the infusion of functionalized siloxane precursors. *J. Membr. Sci.*, **294**, 159–168.
107. Oh, S.-Y. *et al.*, (2010) Inorganic–organic composite electrolytes consisting of polybenzimidazole and Cs-substituted heteropoly acids and their application for medium temperature fuel cells. *J. Mater. Chem.*, **20**, 6359–6366.
108. Mauritz, K.A. *et al.*, (2004) Self-assembled organic/inorganic hybrids as membrane materials. *Electrochim. Acta*, **50**, 565–569.
109. Zou, J., Zhao, Y., and Shi, W. (2004) Preparation and properties of proton conducting organic–inorganic hybrid membranes based on hyperbranched aliphatic polyester and phosphoric acid. *J. Membr. Sci.*, **245**, 35–40.
110. Duvdevani, T. *et al.*, (2006) Novel composite proton-exchange membrane based on silica-anchored sulfonic acid (SASA). *J. Power Sources*, **161**, 1069–1075.
111. Wu, C., Xu, T., and Yang, W. (2005) Synthesis and characterizations of novel, positively charged poly(methyl acrylate)–SiO2 nanocomposites. *J. Membr. Sci.*, **41**, 1901–1908.
112. Li, C. *et al.*, (2006) Casting Nafion-sulfonated organosilica nano-composite membranes used in direct methanol fuel cells. *J. Membr. Sci.*, **272**, 50–57.
113. Chen, W.-F. and Kuo, P.-L. (2007) Covalently cross-linked perfluorosulfonated membranes with polysiloxane framework. *Macromolecules*, **40**, 1987–1994.
114. Ladewig, B.P. *et al.*, (2007) Nafion-MPMDMS nanocomposite membranes with low methanol permeability. *Electrochem. Commun.*, **9**, 781–786.
115. Lin, Y.-F. *et al.*, (2007) High proton-conducting Nafion/-SiO$_3$H functionalized mesoporous silica composite membranes. *J. Power. Sources*, **171**, 388–395.
116. Shahi, V.K. (2007) Highly charged proton-exchange membrane: Sulfonated poly(ether sulfone)-silica polyelectrolyte composite membranes for fuel cells. *Solid State Ion.*, **117**, 3395–3404.
117. Klaysom, C. *et al.*, (2011) Preparation of porous composite ion-exchange membranes for desalination application. *J. Mater. Chem.*, **21**, 7401–7409.
118. Klaysom, C. *et al.*, (2011) The influence of the particle size of surface functionalized additives on composite ion-exchange membranes. *J. Phys. Chem. C*, **115**, 15124–15132.
119. Klaysom, C. *et al.*, (2011) The effects of aspect ratio of inorganic fillers on structure, property, and performance of composite ion-exchange membranes. *J. Colloid Interface Sci.*, **363**, 431–439.
120. Marschall, R. *et al.*, (2012) Composite proton conducting polymer membranes for clean hydrogen production with solar light in a simple photoelectrochemical compartment cell. *Int. J. Hydrogen Energy*, **37**, 4012–4017.
121. Klaysom, C. *et al.*, (2010) Synthesis of composite ion-exchange membranes and their electrochemical properties for desalination applications. *J. Mater. Chem.*, **20**, 4669–4674.
122. Helen, M., Viswanathan, B., and Srinivasa Murthy, S. (2006) Fabrication and properties of hybrid membranes

based on salts of heteropolyacid, zirconium phosphate and polyvinyl alcohol. *J. Power Sources*, **163**, 433–439.
123. Fu, R.-Q., Hong, L., and Lee, J.Y. (2007) Membrane design for direct ethanol fuel cells: a hybrid proton-conducting interpenetrating polymer network. *Fuel Cells*, **1**, 52–61.
124. Nagarale, R.K. et al., (2005) Preparation of organic–inorganic composite anion-exchange membranes via aqueous dispersion polymerization and their characterization. *J. Colloid Interface Sci.*, **287**, 198–206.
125. Binsu, V.V., Nagarale, R.K., and Shahi, V.K. (2005) Phosphonic acid functionalized aminopropyl triethoxysilane-PVA composite material: organic–inorganic hybrid proton-exchange membranes in aqueous media. *J. Mater. Chem.*, **15**, 4823–4831.
126. Nunes, S.P. et al., (2002) Inorganic modification of proton conductive polymer membranes for direct methanol fuel cells. *J. Membr. Sci.*, **203**, 215–225.
127. Kickelbrick, G. (2003) Concepts for the incorporation of inorganic building blocks into organic polymers on a nanoscale. *Prog. Polym. Sci.*, **28**, 83–114.
128. Sata, T. (2000) Studies on anion exchange membranes having permselectivity for specific anions in electrodialysis-effect of hydrophilicity of anion exchange membranes on permselectivity of anions. *J. Membr. Sci.*, **167**, 1–31.
129. Sata, T., Sata, T., and Yang, W. (2002) Studies on cation-exchange membranes having permselectivity between cations in electrodialysis. *J. Membr. Sci.*, **206**, 31–60.
130. Sata, T. and Izuo, R. (1978) Modification of properties of ion exchange membranes V. Structure of cationic polyelectrolyte on the surface of cation exchange membranes. *Colloid Polym. Sci.*, **256**, 757–769.
131. Sata, T., Mine, K., and Matsusaki, K. (1998) Change in transport properties of anion-exchange membranes in the presence of ethylene glycols in electrodialysis. *J. Colloid Interface Sci.*, **202**, 348–358.
132. Sata, T. et al., (1998) Changing permselectivity between halogen ions through anion exchange membranes in electrodialysis by controlling hydrophilicity of the membranes. *Faraday Trans.*, **94** (1), 147–153.
133. Sata, T. et al., (1999) Transport properties of cation exchange membranes in the presence of ether compounds in electrodialysis. *J. Colloid Interface Sci.*, **219**, 310–319.
134. Sata, T., Teshima, K., and Yamaguchi, T. (1996) Permselectivity between two anions in anion exchange membranes crosslinked with various diamines in electrodialysis. *J. Polym. Sci., A Polym. Chem.*, **34** (8), 1475–1482.
135. Sata, T. and Nojima, S. (1999) Transport properties of anion exchange membranes prepared by the reaction of crosslinked membranes having chloromethyl groups with 4-vinylpyridine and trimethylamine. *J. Polym. Sci., B Polym. Phys.*, **37**, 1773–1785.
136. Sata, T. (1991) Properties of ion-exchange membranes combined anisotropically with conducting polymers. 2. Relationship of electrical potential generation to preparation conditions of composite membranes. *Chem. Mater.*, **3** (5), 838–843.
137. Sata, T., Yamaguchi, T., and Matsusaki, K. (1996) Preparation and properties of composite membranes composed of anion-exchange membranes and polypyrrole. *J. Phys. Chem.*, **100** (41), 16633–16640.
138. Pellegrino, J. (2003) The use of conducting polymers in membrane-based separations. *Ann. N. Y. Acad. Sci.*, **984**, 289–305.
139. Gohil, G.S., Binsu, V.V., and Shahi, V.K. (2006) Preparation and characterization of mono-valent ion selective polypyrrole composite ion-exchange membranes. *J. Membr. Sci.*, **280**, 210–218.
140. Nagarale, R.K. et al., (2004) Preparation and electrochemical characterization of cation- and anion-exchange/polyaniline composite membrane. *J. Colloid Interface Sci.*, **277**, 162–171.

141. Amado, F.D.R. et al., (2004) Synthesis and characterisation of high impact polystyrene/polyaniline composite membranes for electrodialysis. *J. Membr. Sci.*, **234**, 139–145.
142. Sata, T. et al., (1999) Composite membranes prepared from cation exchange membranes and polyaniline and their transport properties in electrodialysis. *J. Electrochem. Soc.*, **146** (2), 585–591.
143. Stair, J.L., Harries, J.J., and Bruening, M.L. (2001) Enhancement of the ion-transport selectivity of layered polyelectrolyte membranes through cross-linking and hybridization. *Chem. Mater.*, **13**, 2641–2648.
144. Sata, T. and Mizutani, Y. (1979) Modification of properties of ion exchange membranes. VI Electrodialytic transport properties of cation exchange membranes with a electrodeposition layer of cationic polyelectrolytes. *J. Polym. Sci., Polym. Chem. Ed.*, **17**, 1199–1213.
145. Hodgdon, R.B., E.W., and Alexander, S.S. (1973) Macroreticular anion exchange membranes for electrodialysis in the presence of surface water foulants. *Desalination*, **13** (2), 105–220.
146. Aritomi, T. and M. Kawashima, Ion-exchange membrane. US Patent 6830671 B2, Editor 2004, Tokuyama Corporation, Japan. pp. 1–14.
147. Bae, T.-H. and Tak, T.-M. (2005) Preparation of TiO_2 self-assembled polymeric nanocomposite membranes and examination of their fouling mitigation effects in a membrane bioreactor system. *J. Membr. Sci.*, **266** (1–2), 1–5.
148. Bae, T.-H., Kim, I.-C., and Tak, T.-M. (2006) Preparation and characterization of fouling-resistant TiO_2 self-assembled nanocomposite membranes. *J. Membr. Sci.*, **275**, 1–5.
149. Liu, Y., Wang, A., and Claus, R. (1997) Molecular self-assembly of TiO_2/polymer nanocomposite films. *J. Phys. Chem. B*, **101**, 1385–1388.
150. Younos, T. and Tulou, K.E. (2005) Energy needs, consumption and sources. *J. Contemp. Water Res. Educ.*, **132**, 27–38.
151. Semiat, R. (2000) Desalination: present and future. *Water International*, **25**, 54–65.
152. Xu, T. and Huang, C. (2008) Electrodialysis-based separation technologies: a critical review. *Am. Inst. Chem. Eng.*, **54**, 3146–3159.
153. Koprivnjak, J.F., Perdue, E.M., and Pfromm, P.H. (2006) Coupling reverse osmosis with electrodialysis to isolate natural organic matter from fresh waters. *Water Res.*, **40**, 3385–3392.
154. Van der Bruggen, B. et al., (2003) Electrodialysis and nanofiltration of surface water for subsequent use as infiltration water. *Water Res.*, **37**, 3867–3874.
155. Kim, J.-O. et al., (2006) Development of novel wastewater reclamation system using microfiltration with advanced new membrane material and electrodialysis. *Mater. Sci. Forum*, **510–511**, 586–589.
156. Kim, S.J. et al., (2010) Direct seawater desalination by ion concentration polarization. *Nat. Nanotechnol.*, **5**, 297–301.
157. Shannon, M.A. (2010) Water desalination: fresh for less. *Nat. Nanotechnol.*, **5**, 248–250.
158. Walha, K. et al., (2007) Brackish groundwater treatment by nanofiltration, reverse osmosis and electrodialysis in Tunisia: performance and cost comparison. *Desalination*, **207**, 95–106.
159. Tansel, B. et al., (2006) Significance of hydrated radius and hydration shells on ionic permeability during nanofiltration in dead end and cross flow modes. *Sep. Purif. Technol.*, **51**, 40–47.

7
Thin Film Nanocomposite Membranes for Water Desalination
Dan Li and Huanting Wang

7.1
Introduction

Membrane-based separation processes have attracted remarkable attention because of their potential advantages over other existing processes for many industrial applications such as gas separation, liquid separation, and pervaporation [1–6]. Desalination is a separation process for removing dissolved salts and other minerals from various water sources, such as seawater and brackish water, to produce fresh water for human and animal consumption, irrigation, and other industrial uses [6]. Among various desalination technologies developed so far, including distillation processes (e.g., vapor compression, multistage flash, and multieffect distillation) and membrane processes (e.g., electrodialysis, nanofiltration, and reverse osmosis), reverse osmosis (RO) process has become one of the major desalination processes over the past 40 years, and shared over 44% of desalting production capacity and around 80% of desalination plants installed worldwide [2]. Because of its relatively lower energy cost and simplicity, the RO process is believed to continue to be a leading desalination technology [6, 7].

Commercially available RO membranes are mainly derived from two basic types of polymers: cellulose acetate (CA) and aromatic polyamides (PAs) [8]. The CA membranes, which were the industry standard through the 1960s to the mid-1970s, are usually made from CA, triacetate (CTA), cellulose diacetate (CDA), or a blend of them in the form of asymmetric configuration [9]. Because of their neutral surface and tolerance to a low level of free chlorine, CA membranes exhibit a relatively stable performance in the applications where the feed water has a high fouling potential (e.g., municipal effluent and surface water sources) [10, 11]. Furthermore, CA membranes are relatively of low cost because they are derived from abundant naturally occurring polymers [9, 11]. However, CA membranes have some drawbacks, such as a narrow operating pH range (4.5–7.5), susceptibility to biological attack, structural compaction under high pressure and low upper temperature limit [9]. Therefore, the current RO membrane market is dominated by PA thin film composite (TFC) membranes. A typical structure of commercial TFC membrane is illustrated in Figure 7.1a. The PA TFC membrane

Functional Nanostructured Materials and Membranes for Water Treatment, First Edition.
Edited by Mikel Duke, Dongyuan Zhao, and Raphael Semiat.
© 2013 Wiley-VCH Verlag GmbH & Co. KGaA. Published 2013 by Wiley-VCH Verlag GmbH & Co. KGaA.

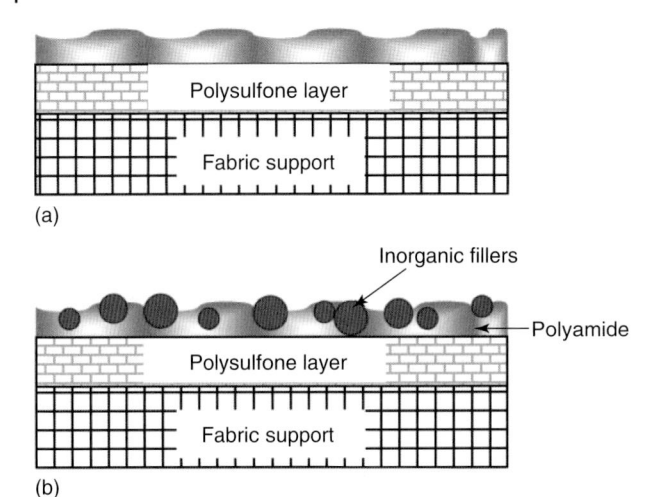

Figure 7.1 Conceptual illustration of PA (a) TFC and (b) TFN membrane structures [23].

consists of three layers: a polyester web acting as the structural support (120–150 μm thick), a polysulfone (PS) microporous interlayer (~40 μm), and an ultrathin polyamide barrier layer on the upper surface (~0.2 μm) that is fabricated by interfacial polymerization process [12]. When compared with CA membranes, the interfacially polymerized PA membranes exhibit superior water flux, salt and organic rejection; higher pressure compaction resistance; wider operating temperature range (0–45 °C) and pH range (from 1 to 11); and higher stability to biological attack [8, 13]. Therefore, this class of membranes is widely used in commercial single-pass seawater desalination plants around the world, and they offer a combination of high flux and high selectivity, which is unmatched by the CA membranes [14, 15]. However, the PA membranes exhibit some drawbacks in desalination, such as chlorine and fouling susceptibility, which may worsen the membrane performance by shortening membrane life, and reducing flux or salt rejection [2, 16–22]. A great deal of effort has been made to modify the properties of PA TFC membranes by chemical or physical modification, but the improvement is still unsatisfactory [6]. Therefore, the development of the new membrane materials with improved desalination performance remains as one of the major research activities in the field of membrane science.

Inorganic membranes possess attractive characteristics such as excellent chemical and microstructural stabilities for desalination application [24–26]. For instance, modeling studies showed that the small-pore zeolites, such as zeolite ZK-4 with a pore size of 4.2 Å, were promising promise for use in membrane-based desalination [27]; zeolites allow small molecules (e.g., water) to pass through their microporous channels and block large molecules (e.g., hydrated ions). Despite the potential advantages of the inorganic membranes, their practical application to desalination is still hindered by the high cost of membrane fabrication, the complication of handling, and the difficulty in scaling up of thin and defect-free membranes, as compared with polymer membranes [28, 29]. The deficiencies of organic polymeric materials and the comparative disadvantages of inorganic membranes have prompted the development of high-performance membrane materials to overcome these limitations.

Emerging mixed matrix membranes (MMMs) have shown great potential in terms of the membrane development. The MMMs consist of an organic polymer as the continuous phase and fillers as the dispersed phase [4]. MMMs are also known as composite membranes or nanocomposite membranes where nanosized fillers are incorporated. The investigation into MMMs started from gas separation application and pervaporation, which was first reported in the 1970s following the discovery of a delayed diffusion time lag effect for CO_2 and CH_4 in a rubbery polymer with the addition of 5A zeolite [30]. There are numerous examples for the use of a wide variety of filler materials in the fabrication of MMMs, including porous fillers (e.g., zeolites, carbon molecular sieves, carbon nanotubes, metal organic framework – MOF, polyhedral oligomeric silsesquioxane – POSS, and mesoporous silica) [23, 30–39] and nonporous fillers (e.g., silica and metal oxides) [40–49]. The role of the fillers in the membrane materials is dependent on the interaction between the polymer matrix and the filler phase and on the structure of fillers as well. In general, the addition of inorganic particles into a polymer matrix leads to the structural change of polymer, that is, disrupting polymer chain packing and increasing the void/free volumes of polymers. Hence, the presence of inorganic particles in a polymer matrix may result in an enhancement of permeability and sometimes permselectivity [4, 45, 50–54]. Furthermore, some microporous fillers are capable of discriminating different types of molecules on the basis of their sizes and shapes, thus resulting in increased selectivity of the MMMs. However, the selectivity of the MMMs may be compromised by interfacial voids or defects because of poor interactions between the organic polymer and the filler particles. When mesoporous fillers (pore sizes of 2–50 nm) are used, polymer chains may be able to penetrate into the mesopores of the fillers in the membrane fabrication process [31]. In this case, the application of mesoporous materials may improve the contact of filler/polymer and result in a selective film. Organic-functionalized filler particles are shown to have better interfacial compatibility with the polymer matrix, and the resultant nanocomposite membranes exhibit enhanced gas selectivities [55, 56].

In the twenty-first century, the concept of MMMs was further developed for water treatment [57] and, more recently, extended to the fabrication of RO (including nanofiltration) membranes for water desalination [6]. Zeolite A-PA TFN RO membranes were first reported by Hoek and coworkers [23, 58]. Figure 7.1b shows a typical structure of thin film nanocomposite (TFN) membranes: inorganic nanoparticles (e.g., zeolites) embedded throughout the thin film layer (e.g., PA) on the top of nonwoven polyester fabric supported PS layer [23]. Until now, a number of TFN membranes have been investigated for their potential application for water desalination, as listed in Table 7.1.

In terms of the polymer matrix, current studies are centered on the application of aromatic PA as the thin selective layer of TFN membranes that is formed via interfacial polymerization [23, 32, 33, 35, 41–44, 60], despite the fact that layer-by-layer assembled polyelectrolytes (PEs) or sulfonated poly (arylene ether sulfone) have

Table 7.1 Summary of the progress in the fabrication and membrane properties of PA TFN RO membranes over the past decade.

Polymer matrix	Inorganic filler	Preparation method	TFN membrane properties (compared with PA TFC membranes)	References
Polyamide (MPD-TMC)[a]	Synthesized zeolite LTA (NaA)[b] nanocrystals (~50–150 nm)	Interfacial polymerization with adding zeolite LTA (NaA)[b] in TMC-hexane	Smoother, more hydrophilic, and negatively charged PA surfaces; higher pure water permeability with equivalent solute rejections	[23]
Polyamide (MPD-TMC)	Synthesized zeolite A (NaA)[b] crystals (~100, ~200, ~300 nm)	Interfacial polymerization with adding zeolite LTA (NaA)[b] in TMC-isoparaffin	More permeable, negatively charged, and thicker than PA TFC films; greater enhancement in permeability, with the addition of smaller nanocrystals	[33]
Polyamide (MPD-TMC)	Synthesized zeolite A (NaA or AgA)[b] nanocrystals (~140 nm)	Interfacial polymerization with adding zeolite A (NaA and AgA)[b] in TMC-isoparaffin	Higher water permeability with similar salt rejection; more hydrophilic, smooth interfaces, and limited bactericidal activity for AgA-TFN membranes	[32]
Polyamide (MPD-TMC)	Zeolite A (NaA)[b] crystals (~250 nm)	Interfacial polymerization with adding zeolite A (NaA)[b] in TMC-isoparaffin	Different post-treatment changing the molecular structure of membranes; first demonstration of TFN membranes with commercially relevant seawater RO separation performance	[59]
Polyamide (MPD-TMC)	Zeolite Y crystals (~250 nm)	Interfacial polymerization with dispersing zeolite Y in TMC-hexane-ethanol pre-seeding solution	Compact and flat surface morphology; higher permeability and comparable salt rejection with increased zeolite loading (less than 0.4 wt%)	[60]
Polyamide (MPD-TMC)	Commercial silica nanoparticles LUDOX® HS-40 (~16 nm) and TEOS hydrolyzed silica (~3 nm)	Interfacial polymerization with adding silica in MPD aqueous solution	Tunable pore radius, increasing number of pores, and higher thermal stability; high water flux and lower salt rejection, in particular at higher loading of silica	[44]
Polyamide (MPD-BTC)[a]	Commercial silver nanoparticles (~50–100 nm)	Interfacial polymerization with adding silica in BTC-HCFC[a]	Slightly lower flux and higher rejection; higher antibiofouling effect	[43]
Polyamide (MPD-TMC)	Synthesized TiO$_2$ nanoparticles (~10 nm or less)	Self-assembly of TiO$_2$ on the neat MPD-TMC TFC membrane surface	Higher salt rejection and lower flux; higher photocatalytic bactericidal efficiency under UV light illumination	[41]

Membrane	Nanomaterial	Method	Results	Ref.
Polyamide (MPD-TMC)	Synthesized TiO$_2$ nanoparticles (~10 nm or less)	Self-assembly of TiO$_2$ on the neat MPD-TMC TFC membrane surface	Slightly higher flux and unchanged salt rejection; higher photocatalytic bactericidal efficiency under UV light illumination	[40]
Polyamide (MPD-TMC)	Commercial TiO$_2$ nanoparticles (~30 nm)	Interfacial polymerization with adding TiO$_2$ in TMC-HCFC^a	Enhanced surface hydrophilicity; comparable water flux and higher salt rejection with limited amount of TiO$_2$	[42]
Polyamide	Single-walled carbon nanotubes (SWCNTs) (diameter: 0.8–1.4 nm)	Interfacial polymerization with adding SWCNTs in non-polar solvents, for example, hexane or the mixture of hexane and chloroform	Higher salt rejection and significantly improved flux	[61]
Polyamide (MPD-TMC)	Commercial multiwalled carbon nanotubes (MWCNTs) (diameter: 9–12 nm; length: 10–15 μm)	Interfacial polymerization with adding MWCNTs in MPD aqueous solution	Slightly lower salt rejection and flux; improved chlorine resistance by increasing MWCNTs loading	[35]
Polyelectrolytes (PAA-PAH)^c	Commercial multiwalled carbon nanotubes (MWCNTs) (diameter: 9–12 nm; length: 10–15 μm)	Layer-by-layer (LBL) assembly with adding MWCNTs in PAA^c solution	Improved thermal stability; improved chlorine resistance	[34]
Sulfonated poly (arylene ether sulfone)-polyamide (MPD-TMC) copolymer	Synthesized mesoporous silica nanoparticles (~100 nm)	Interfacial polymerization with adding SiO$_2$ in TMC-cyclohexane	Higher water flux and similar salt rejection	[36]

^{a}MPD, *m*-phenylenediamine; TMC, trimesoyl chloride; BTC, 1,3,5-benzene tricarbonyl chloride; and HCFC, 1,1-dichloro-1-fluoroethane.
^{b}LTA, Linde Type A zeolite; AgA, zeolite LTA in the silver form; NaA, zeolite LTA in the sodium form.
^{c}PAA, poly (acrylic acid); PAH, poly (allylamine hydrochloride).

also been investigated in the development of TFN RO membranes [34, 36]. Similar to gas separation and pervaporation MMMs, the inorganic fillers incorporated into TFN RO membranes can be classified into two main groups, that is, porous and nonporous materials. As porous inorganic fillers, zeolite molecular sieves have attracted great attention because their extremely narrow pore size distribution ensures superior size and shape selectivities [23, 32, 33, 59, 60]. Apart from zeolites, other porous materials, such as mesoporous silica and carbon nanotubes, have also been reported for their use in TFN RO membranes [34–36]. Nonporous materials, including titanium dioxide, silica, and silver particles, have been often applied to tailor the structure of polymer matrix and add more functionality [40–44]. Studies have revealed that the addition of inorganic fillers (e.g., zeolite, carbon nanotubes, silica, silver, and titanium dioxide) could improve the properties of aromatic PA membranes, including salt rejection, water permeability, fouling, and chlorine resistance as well as thermal or mechanical stability [23, 32–36, 40–44, 59, 60]. Importantly, the synthesis of filler materials and the membrane formation processes need to be carefully controlled to fabricate high-performance TFN membranes. The methods for the preparation of fillers and nanocomposite membranes are exemplified in the following sections.

7.2
Fabrication and Characterization of Inorganic Fillers

As described earlier, in the fabrication of TFN RO membranes, the membrane configuration that the inorganic particles are added into the selective polymer layer is directly adopted from gas separation and pervaporation MMMs [4, 57]. In particular, nanosized fillers (e.g., ~100 nm) are required to achieve good dispersion in a polymer matrix of the thin selective layer in TFN membranes as they best match the characteristic thin film thickness and can make greater enhancement of permeability than large particles (e.g., ~300 nm) [33]. The good dispersibility of synthesized nanosized zeolites is a prerequisite for the formation of homogeneous and uniform TFN membranes. Excellent compatibility and interaction between the dispersed nanoparticles and the polymer matrix are crucial for the improvement of membrane desalination performance, such as flux enhancement or fouling mitigation [57].

Many wet chemistry methods, such as the sol–gel technique and hydrothermal synthesis, have been developed to synthesize inorganic filler nanoparticles. The first example of such sol–gel dates back to the work of Ebelmen about the synthesis of silica in 1846 [62]; however, the sol–gel science only really boomed in the 1980s [63]. Since then, there has been a large increase in the available literature [63–71], suggesting that sol–gel processing is a very popular and reliable method to produce a wide variety of high-quality materials. Sol–gel processing has been applied in the production of oxide powders, such as titanium oxide and silica, with uniform and small particle sizes and varied morphologies. It involves the transition from a liquid "sol" into a solid "gel," which can be ordinarily divided into a series of steps, including the formation of solution, gelation, aging,

Hydrolysis:
$$Si(OR)_4 + 4H_2O \longrightarrow Si(OH)_4 + 4ROH$$
$$\text{Silicic acid,}$$

Where R = Vinyl, Alkyl, or Aryl groups.

Condensation:

(a) Water condensation:
$$\equiv Si(OH) + (OH)Si \equiv \longrightarrow \equiv Si\text{–}O\text{–}Si \equiv + H_2O$$

(b) Alcohol condensation:
$$\equiv Si(OH) + (OR)Si \equiv \longrightarrow \equiv Si\text{–}O\text{–}Si \equiv + ROH$$

The overall reaction can be written as:
$$Si(OR)_4 + 2H_2O \longrightarrow SiO_2 + 4ROH$$

Figure 7.2 Sol–gel reaction occurring during silica network formation: hydrolysis and condensation [63, 72].

drying, dehydration, and densification [71]. Figure 7.2 shows a sol–gel reaction occurring during the formation of silica networks starting from silicon alkoxide [63, 72]. Typically, the reaction mechanism of the sol–gel process is based on the hydrolysis and condensation of molecular precursors in a solution, originating from a liquid "sol" of nanometric particles. The subsequent condensation and inorganic polymerization lead to a three-dimensional metal oxide network denominated wet "gel" [65]. The further drying and heat treatments are sometimes needed to acquire the final crystalline state or lead a wet "gel" converted into dense particles [65, 72]. The main processing parameters that influence the structures and properties of the "gel" include the nature and concentration of precursors employed, solution pH and temperature, agitation, and additives used [65, 72, 73]. In particular, the porosity of silica particles may differ greatly depending on the variation of preparation methods, which is also the case in the sol–gel synthesis of other inorganic fillers, such as TiO_2.

In addition to silica nanoparticles, photocatalytic TiO_2 nanoparticles have also been added into TFN RO membranes [40–42]. The TFN desalination membranes containing TiO_2 nanoparticles showed improved fouling resistance and antibacterial property, and the formation of the biofilm on the membrane surface was reduced during desalination processes [40, 41]. In the preparation of TiO_2 nanoparticles via the sol–gel method, titanium (IV) alkoxide is hydrolyzed under acidic conditions, and the resulting titanium hydroxide (Ti-OH) species are then condensed into titanium oxide networks [74]. In particular, the hydrolysis reaction occurs on the addition of water and a small amount of acid as the catalyst, resulting in the formation of Ti-OH species through the removal of alkyl group (OR). The condensation process follows the hydrolysis reaction, leading to the formation of Ti-O-Ti bonds. The reported TiO_2-PA TFN membranes are all with the incorporation of ∼10 nm or smaller metal oxide particles with surface hydroxyl groups, which are believed to be the active sites to form hydrogen bonds with carbonyl groups in PA polymer [40]. The potential bonding between TiO_2 and polymer can help enhance the compatibility between inorganic and organic phases, and such bonding is also expected to exist in other TFN membranes with the addition of oxide particles, such as silica. However, until now, there is no direct experimental evidence confirming this hypothesis.

Hydrothermal synthesis is another important method for nanoparticle preparation, and normally referred to as the reaction conducted in the aqueous solutions within sealed vessels, such as autoclaves with or without Teflon liners, under controlled temperature and autogenous pressure [74]. The temperature can be elevated well above the boiling point of water, resulting in an increase in pressure. Many researchers have used the hydrothermal method to prepare various types of particles or nanoparticles, in particular zeolites. Some review articles devoted specifically to this method can be found in the literature [75–79].

Typically, in the hydrothermal synthesis of zeolites, reactants, such as those containing silica and alumina, are mixed together with a cation source to form a homogeneous solution or gel. The resultant aqueous reaction mixture, usually in a basic (high pH) medium, is heated in a sealed plastic bottle or autoclave for zeolite crystallization. Similar to any other types of crystals, zeolite crystals are produced via nucleation and grown in the synthesis solution/gel. *Nucleation* is defined as a process where the small aggregates of precursors give rise to nuclei (called *embryos*). As the synthesis proceeds, the nuclei grow into zeolite crystals [80], which can be recovered by repeated filtration, washing, and drying [77].

Various types of zeolite nanocrystals, such as zeolite X (faujasite - FAU), can be obtained at the early crystallization of initial gel systems without involving organic structure-directing agents (SDAs), and their primary crystal sizes are in the order of nanometers [81, 82]. However, the nanocrystals from such syntheses typically form aggregates of larger sizes and broad particle size distributions [76]. Furthermore, in order to collect nanocrystals with a good yield, the synthesis process needs to be controlled by adjusting crystallization temperature or time. At low temperatures, zeolite crystallization rate is often slow, and a long synthesis time of about several days or even weeks is required. It is noted that the synthesis of highly dispersible zeolite nanocrystals from SDA-free synthesis systems is still a difficult task. In contrast, in the zeolite synthesis solutions/gels with SDAs, SDA molecules not only play the role of a pore-filling agent and structurally direct the crystallization toward specific zeolitic structures, but also play an important role in controlling the crystallization rate in the growth of zeolite nanocrystals [76, 83]. The colloidal zeolite nanocrystals so obtained contain SDAs in their voids, and the removal of the SDAs leads to the opening of zeolitic channels. In most cases, the removal of SDAs from zeolite structures requires calcination in air or oxygen, which usually leads to irreversible aggregation [84–86]. Wang and coworkers reported the use of an organic polyacrylamide polymer network as a temporary barrier to retain the dispersibility of nanocrystals during the removal of SDAs (as illustrated in Figure 7.3) [86]. This method has been applied to synthesize redispersible zeolite A crystals for the fabrication of zeolite-PA TFN membranes. Specially, the synthesis of zeolite A nanocrystals can be easily controlled in a Na_2O-SiO_2-Al_2O_3-H_2O gel

7.2 Fabrication and Characterization of Inorganic Fillers | 171

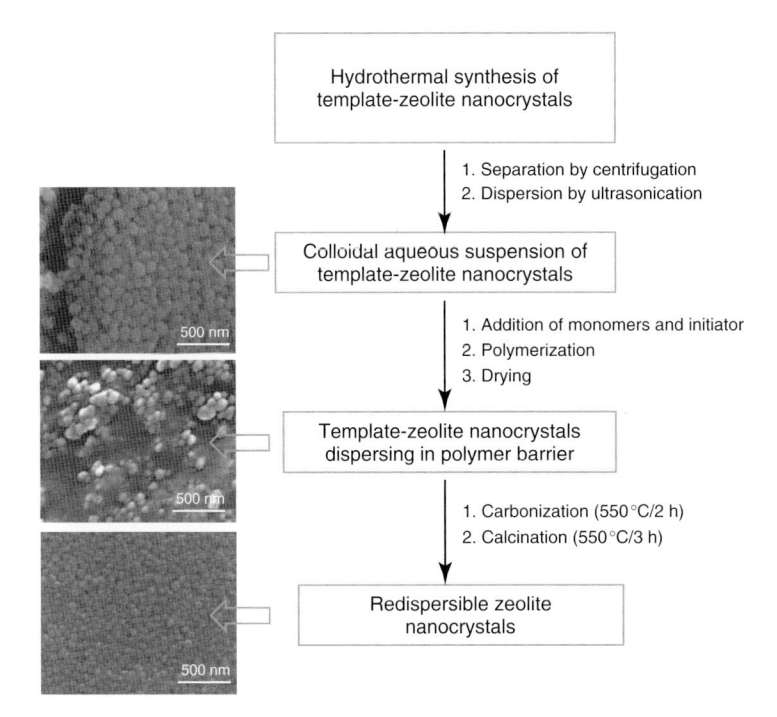

Figure 7.3 Preparation protocol and SEM images for colloidal suspensions of template-removed silicalite nanocrystals [86]. (Source: Reproduced by permission of The Royal Society of Chemistry.)

system with the addition of tetramethylammonium hydroxide (TMAOH) as SDA [23, 32, 33]. TMAOH assists the formation of uniform zeolite A nanocrystals without aggregation [76]. After the synthesis, it is necessary to remove TMA$^+$ and open up the zeolite micropores without nanocrystal aggregation for the subsequent preparation of TFN membranes. SDA-free zeolite A nanocrystals with good redispersiblity can be obtained using polyacrylamide hydrogels-assisted calcination [23, 32, 33]. By using this method, zeolite A nanocrystals with sizes ~100, 200, and 300 nm have been studied in the fabrication of TFN RO membranes [23, 32, 33].

The properties, such as morphologies, particle sizes, and surface functionality of inorganic filler particles, are the key to successful fabrication of TFN RO membranes. As shown in Figure 7.4, zeolite nanoparticles are characterized by a number of techniques such as X-ray diffraction (XRD), scanning electron microscopy (SEM), transmission electron microscopy (TEM), dynamic light scattering (DLS), and gas sorption. Uniform particle sizes and good dispersibility are desired for TFN membrane fabrication; the particle sizes and size distribution can be determined by measuring a large number of particles imaged via SEM or TEM. DLS is commonly used to determine the particle size distribution of particles and their dispersion in various solvents. The crystal structure of particles can be analyzed by powder XRD. In the case of zeolite nanoparticles, XRD provides the information about the

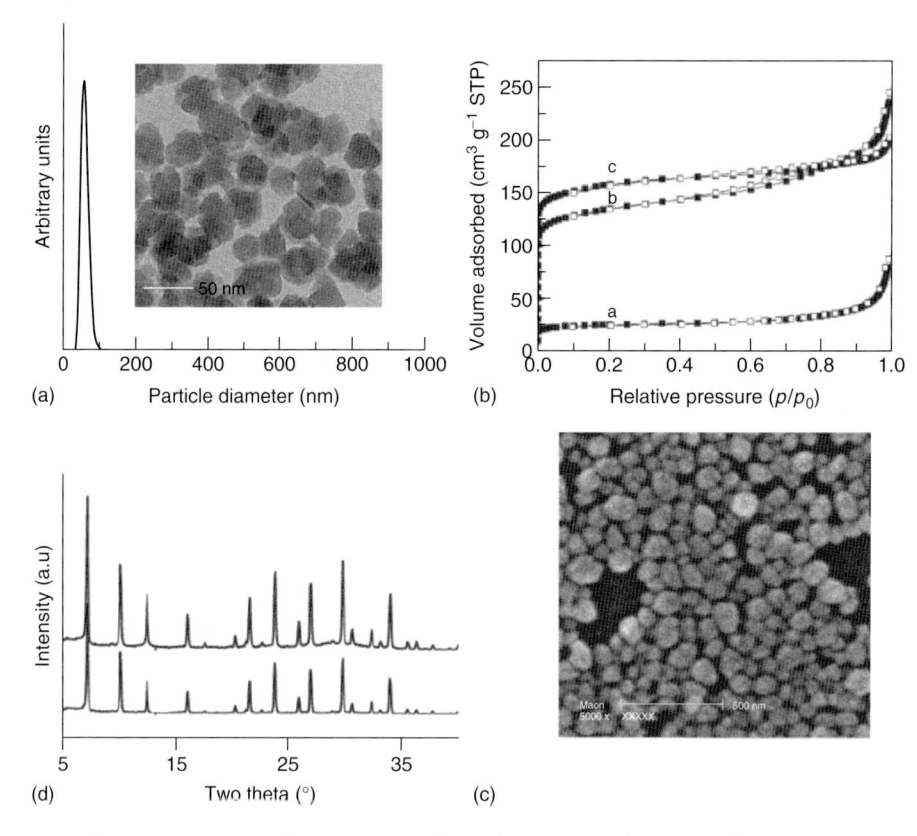

Figure 7.4 Common characterizations for zeolite nanocrystals: (a) particle size distribution of silicalite-1 by dynamic light scattering (DLS). The inset in (a) shows the corresponding TEM image [76], (b) nitrogen adsorption–desorption isotherm, (c) SEM image, and (d) XRD patterns of LTA-type zeolite [87].

specific crystal structures, determining the types of zeolites and their crystallinity. For porous particles, gas (nitrogen) sorption can provide strong evidence that pores of particles are still accessible after various treatment and functionalization. This information would be useful to understand the transport behavior of the resultant TFN membranes and determine whether water molecules can pass through the pores of fillers while hydrated ions are rejected.

7.3
Fabrication and Characterization of TFC/TFN Membranes

7.3.1
Interfacial Polymerization

A great deal of the recent studies has been conducted on the selection of suitable substrate and optimization of the thin film layer toward the enhanced separation performance of membranes, including flux and salt rejection, fouling and chlorine resistances, and thermal stability [6].

Polysulfone (PS) has been commonly used as the substrate in the laboratory scale or industrial fabrication of PA TFC membranes, owing to its stable chemical resistance in a wide range of pH. Moreover, PS is relatively hydrophilic, and especially suitable for the interfacial polymerization of PA in aqueous solution [88]. Studies revealed that further improvement of PA substrate, such as via plasma treatment, assisted in the preparation of high-performance PA membranes [88–90]. The use of inorganic nanoparticles in PS led to the formation of nanocomposite substrate materials with enhanced mechanical stability, which resisted physical compaction of its supported PA membranes [91]. To develop TFC membranes with high operating temperature and chemical stability, considerable efforts have been made to explore other polymers as the support layers, including poly (pathalazinone ether sulfone ketone) (PPESK) [92, 93], poly (phthalazinone ether amide) (PPEA) [94], polyvinylidene fluoride (PVDF) [88], and polypropylene (PP) [89].

The nature of the thin layer significantly affects the performance of TFC membranes. Many routes have been described to form an ultrathin barrier layer in the TFC membranes [8]. Among those methods, the interfacial polymerization is the most widely used technique for the fabrication of commercial PA TFC membranes [8, 95]. As shown in Figure 7.5a, the typical interfacial polymerization involves a reaction between two monomers – a diamine (e.g., 1,3-phenylenediamine (MPD)) and a diacid chloride (e.g., trimesoyl chloride (TMC)) [8]. Typically, the aqueous amine solution initially impregnates the nonwoven polyester fabric supported PS membrane. An ultrathin film (skin), well under half a micrometer thick, is quickly formed at the interface and remains attached to the PS support after the polymerization reaction takes place at the interface between two immiscible solvents [8, 96]. The resultant thin film active layer of aromatic PA TFC RO membranes is composed of the cross-linked form of three amide linkages and the linear form with pendant free carbonxylic acid (Figure 7.5a). The properties of PA TFC membranes can be controlled by carefully varying the interfacial polymerization process, which is dependent on several variables, such as the monomer concentrations and types, selection of organic solvents, reaction and curing temperature/time, and the use of additives [6, 8].

Some recent studies have reported the introduction of functional bonds into PA networks by using different monomers for the modification of membrane performance. As shown in Figure 7.5b, the novel "PA-urea" RO membranes comprised aromatic PA with functional bonds such as urea (−NHCONH-) showed both superior water flux and salt rejection to the traditional commercial TMC-MPD membrane [97]. The TFC RO membranes with PA-urethane as the active layer included the amide functional group -CONH-, urethane functional group -OCONH-, and the hydroxyl functional group -OH. The optimized PA-urethane TFC membrane had better desalination performance than that of the PA TFC membrane fabricated under the same

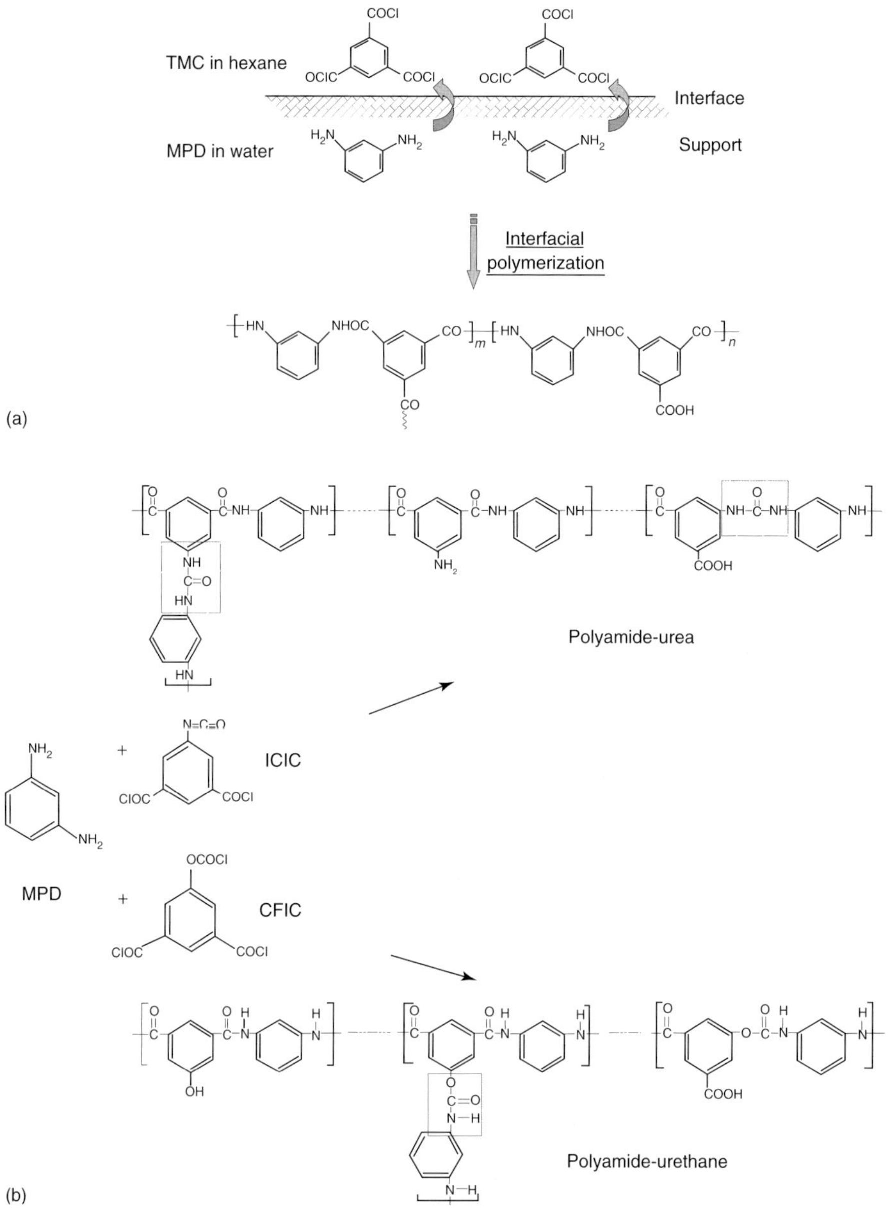

Figure 7.5 (a) The polyamide derived from *m*-phenylenediamine (MPD) and trimesoyl chloride (TMC) via the interracial polymerization [6, 89]. (Source: Reproduced by permission of The Royal Society of Chemistry). (b) Interfacial polymerization of polyamide-urea and urethane [14, 97].

conditions [14, 98, 99]. However, traditional MPD-TMC PA materials have been used in most cases of TFN membrane fabrication.

7.3.2
Interfacial Polymerization with Inorganic Fillers

A number of methods, including the addition of inorganic nanoparticle precursors and preformed inorganic nanoparticles, have been reported for the preparation of inorganic–organic nanocomposite membranes for various separation applications. In Figure 7.6a, the *in situ* method entails the use of nanoparticle precursors. For *in situ* preparation of composite membranes, the precursors of the fillers are added to the polymer solution and the particles are then formed during the process of membrane formation. In this approach, the inorganic particles are generally covalently linked with the polymer, thereby enabling good dispersion of fillers inside the polymer matrix [100, 101].

In the fabrication of TFN membranes, the widely reported preparation method is the addition of the preformed or commercial inorganic fillers into a monomer solution (Figure 7.6b). For instance, in the fabrication of TFN membranes, silica and multiwalled carbon nanotubes (MWCNTs) were dispersed in the MPD aqueous phase [35, 44], whereas zeolites, TiO_2, or silver nanoparticles were suspended in the

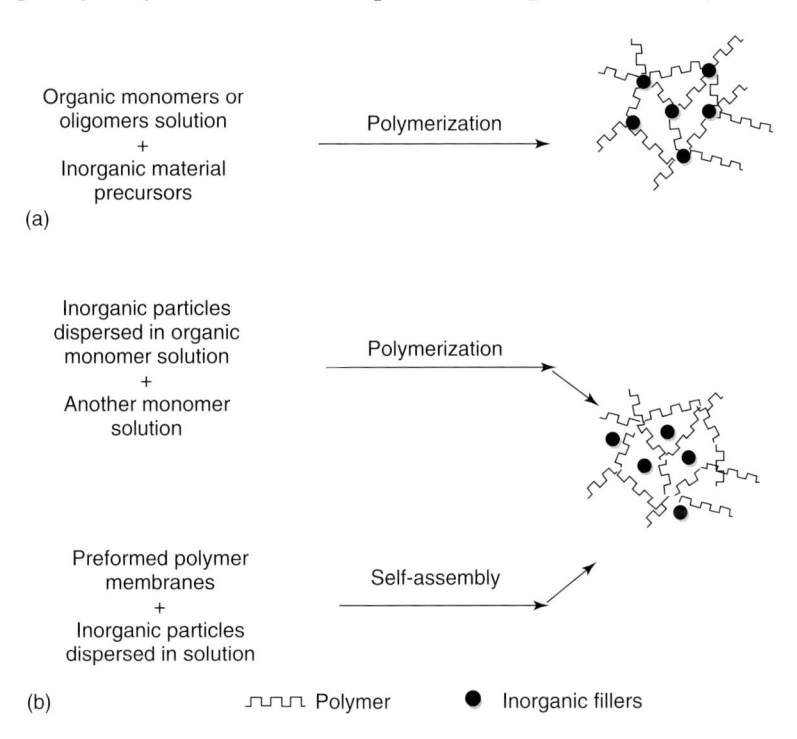

Figure 7.6 Preparation of nanocomposite membranes using inorganic nanoparticle precursors and preformed inorganic nanoparticles. (Source: Modified from Ref. [100].)

TMC organic phase (e.g., isoparaffin, hexane, and 1,1-dichloro-1-flueroethane) [23, 32, 33, 42, 43, 59, 60]. In this method, the dispersion of particles in the monomer solution is accomplished by means of mechanical stirring or ultrasonication to reduce particle agglomeration. The prevention of particle agglomeration is always critically important to produce a uniform membrane and avoid phase separation between the inorganic filler and the organic matrix. For instance, MWCNTs tend to aggregate because of the van der Waals interaction. To achieve uniform dispersion of MWCNTs in a polymer matrix, a surfactant Triton X-100 was added in the MPD aqueous solution [35]. Therefore, by direct addition of the inorganic fillers to monomer solutions, the PA TFN membranes can be fabricated after the interfacial polymerization of diamine and diacid chloride monomers.

In the fabrication of zeolite-PA TFN membranes, zeolite nanocrystals can be dispersed in either TMC-hexane solution as illustrated in Figure 7.7a or MPD aqueous solution. The early studies were mainly focused on the interfacial polymerization with the addition of zeolite A nanocrystals in TMC-hexane solution (Figure 7.7a) [23, 32, 33]; however, the dispersion of zeolite nanocrystals in a nonpolar organic solvent, such as hexane, needs to be carefully controlled; otherwise solute permeable defects would be produced in the resultant TFN membranes, resulting in lower solute rejection [60]. However, when zeolite nanocrystals are impregnated and covered with an excessive amount of MPD aqueous solution, the PA layer may be formed beyond the sizes of zeolite nanocrystals, causing a slight decrease of membrane permeability. More recently, Tsura and coworkers reported a novel "pre-seeding"-assisted synthesis of high-performance zeolite – PA TFN membranes for water desalination [60]. As shown in Figure 7.7b, zeolite Y nanoparticles dispersed in hexane with ethanol as the cosolvent were deposited on an MPD-impregnated PS support to form a pre-seeded substrate. The pre-seeding process led to a better zeolite-PA contact after the secondary interfacial polymerization [60]. Therefore, this process is promising way to produce high-performance inorganic-organic TFN desalination membranes without nonselective defects or voids.

Differing from the aforementioned two methods, the self-assembly method may be considered as a post-treatment to integrate inorganic fillers, such as TiO_2, with PA membranes after the interfacial polymerization (Figure 7.6c) [40, 41]. The dipping of pre-formed PA TFC membranes in TiO_2 aqueous solution would result in the monolayer adsorption of TiO_2 on PA surfaces with pendant free –COOH. This self-assembly method does not require controlling the complex interfacial polymerization arising from the direct addition of inorganic fillers into a monomer aqueous or solvent solution. However, this method is vulnerable to easy detachment of TiO_2 from PA surface, leading to the loss of TiO_2 nanoparticles [42]. In contrast, when TiO_2 nanoparticles are dispersed in a monomer solution, the interfacial polymerization would achieve a higher loading of TiO_2 nanoparticles throughout the PA layer due to the confinement of nanoparticles, leading to more robust structure of TiO_2 – PA TFN membranes [40–42].

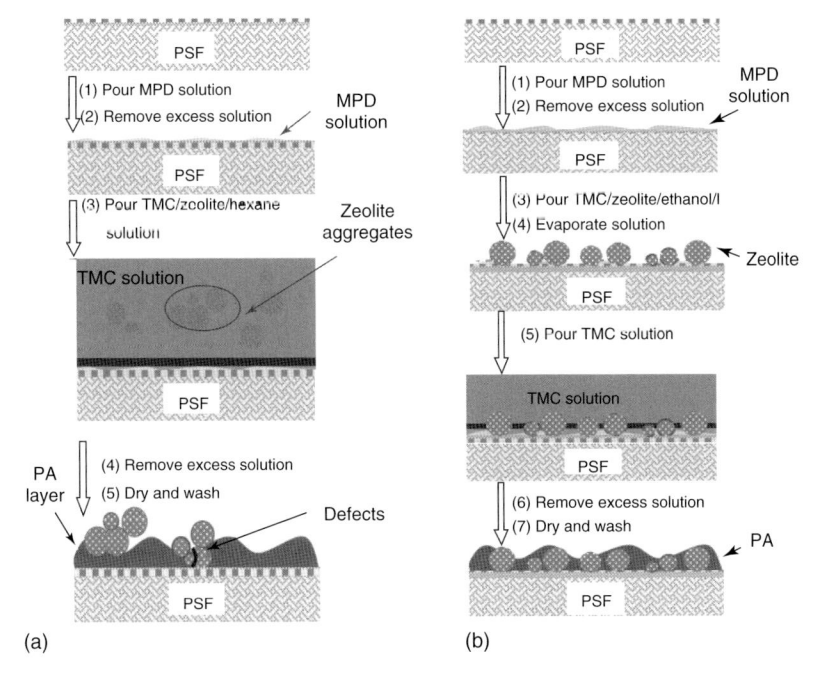

Figure 7.7 A schematic representation of the zeolite-polyamide TFN membranes fabricated on the PS supports via (a) zeolite added in the TMC-hexane solution and (b) pre-seeding-assisted synthesis [60]. (Source: Reproduced by permission of The Royal Society of Chemistry (RSC) for the Centre National de la Recherche Scientifique (CNRS) and the RSC.)

7.3.3
Characterization of TFN or TFC Membranes

Understanding the relationships among the chemistry, structure, and transport properties of desalination membranes is of particular importance in membrane research. Some important characteristics of membranes, including chemical composition, hydrophilicity/hydrophobicity, charge density, and surface morphology, may greatly affect the separation performance in a specific application.

Electron microscopy, such as SEM and TEM, is commonly used to examine the microstructure and morphologies of the polymer TFC or TFN membranes and to determine the thin barrier layer thickness (Figure 7.8). The surface of the conventional PA TFC membrane appears uniform and "leaflike," exhibiting a hill and valley microstructure (Figure 7.8). The addition of inorganic fillers, such as zeolites, may result in a compact and flat surface morphology for TFN membranes (Figure 7.8) [23, 60]. In an electron microscope, a focused electron beam interacts with the atoms in a sample and element-specific X-rays are generated. These X-rays can be detected with an energy dispersive spectrometer coupled to an SEM or to a TEM. The energy dispersive X-ray (EDX) spectroscopy allows for elemental mapping

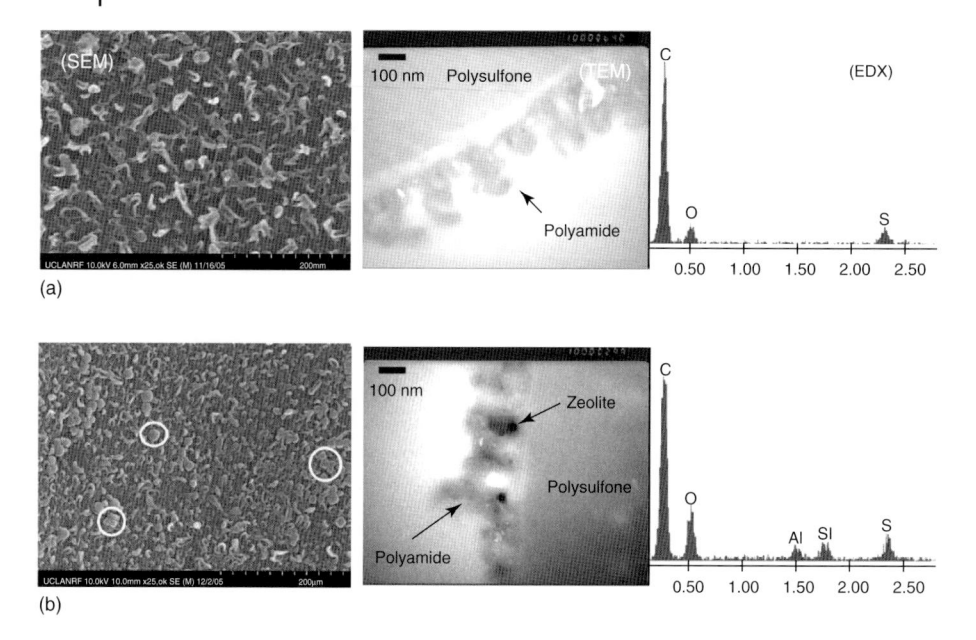

Figure 7.8 Characterization of TFC (a) and TFN (b) polyamide membranes by SEM, TEM, and EDX. (Three white circles in SEM image are drawn around features believed to be zeolite A) [23].

and line scanning, which provides information on elemental composition of the membranes. This technique is effective in confirming the presence of inorganic fillers inside polymer membranes (Figure 7.8), such as zeolite A [23].

The characterization of the membrane surface morphology is also complemented by atomic force microscopy (AFM), which allows the study of the surface of nonconducting materials down to the scale of nanometers [102, 103]. Compared to electron microscopy, other advantages of AFM measurements are that special specimen preparation is not needed and that the imaging can be performed directly in air or liquid [104, 105]. Therefore, AFM is commonly used to determine surface roughness, surface porosity, and pore size distribution without affecting the polymer membrane materials [106].

Surface hydrophilicity and surface charge are crucial parameters that strongly affect membrane filtration performance. Owing to the simplicity, its measurement is most commonly used for the determination of the membrane hydrophilicity, which is closely related to water permeability. Only a small piece of membrane is needed for the measurement [107]. Membrane surface charge is often characterized by the zeta (Z) potential and estimated using the streaming potential measurement. The Z potential values depend on the chemical structure of the membrane material, which is responsible, at least in part, for its ability to reject salts such as sodium chloride [107].

The chemical composition of membrane surfaces can be analyzed by infrared (IR) spectroscopy. In the characterization of PA TFN or TFC membranes, IR spectroscopy provides valuable information on the functional groups (such as aromatic, carboxylic acid, or amide groups) and verifies the successful formation of PA after the interfacial polymerization. Meanwhile, internal reflection spectroscopy (IRS), which combines Fourier transform infrared spectroscopy (FTIR) spectrometry with an attenuated total reflectance (ATR) technique, is one of the most common ways to perform membrane surface analysis. The ATR-FTIR technique allows for the depth profiling of membrane surfaces. ATR-FTIR quantitative analysis of thin films is also possible, which has been demonstrated in a number of studies [108]. Moreover, attempts have been made to develop ATR-FTIR method for estimating the thickness of an organic polymer layer on a polymer surface [109, 110].

X-ray photoelectron spectroscopy (XPS) is also used to study elemental composition of TFN RO membranes near the surface except hydrogen element, confirming the formation of the expected polymer chemical structures after the interfacial polymerization. Both qualitative and quantitative analyses are possible with XPS. Compared with XPS, Rutherford backscattering spectrometry (RBS) can probe to a greater depth into the sample and determine the elemental composition and depth profile. Previous studies suggested that RBS was required to study the elemental composition and depth profile of the entire PA active layers for better understanding of physicochemical properties of RO membrane [111, 112].

7.4
Membrane Properties Tailored by the Addition of Fillers

The wide availability of fillers with different structures and sizes enables us to tailor the morphology, microstructure, permeability, salt rejection, chlorine stability, and fouling resistance of TFN membranes for RO desalination. Other properties such as thermal stability and mechanical strength may also be varied by the addition of inorganic fillers in a polymer matrix.

7.4.1
Water Permeability and Salt Rejection

The separation performance (water permeability and salt rejection) of TFN membranes is related to the intrinsic properties of membranes, including porosity, roughness, thickness, hydrophilicity, and surface charge, which may be tailored by using different types of inorganic particles and varying the membrane preparation conditions. For instance, hygroscopic silica nanoparticles increased membrane hydrophilicity and led to a higher water flux [36]. The presence of Ag nanoparticles in PA matrix led to a slightly increasing salt rejection (~0.2%) and a lower flux (~2%). This was ascribed to the structural compactness of the membranes resulting from the specific interaction between silver and PA matrix [43]. Because of super-hydrophilicity and negatively charged framework of zeolite A, the zeolite

A-PA TFN membranes showed more hydrophilic and negative charged surface and improved water permeability, as compared to conventional TFC membranes [23]. Recent research shows that the TFN membrane preparation methods play a critical role in determining the membrane morphology, thus separation performance. For example, zeolite-PA TFN membranes fabricated by the pre-seeding polymerization method exhibited a compact and flat surface morphology, when compared with PA TFC membranes. Such morphology was also observed in the TFN membranes fabricated by direct interfacial polymerization with adding zeolites in TMC-hexane solution. The pre-seeding polymerization method might lead to a better interfacial contact of polymer and zeolites, thus improving the desalination performance. In contrast, the zeolite-PA TFN membranes prepared after the interfacial polymerization by adding zeolite nanoparticles in TMC-hexane solution exhibited unsatisfactory salt rejection, which might be due to the defects existing between nanoparticles and PA [60, 113]. On the other hand, the TFN membranes prepared by interfacial polymerization with the dispersion of zeolite nanoparticles in MPD aqueous solution exhibited similar morphology to the pure PA TFC membranes [60].

The effects of inorganic particle size and loading on the properties of PA TFN membranes are shown in Figure 7.9. By increasing the sizes of zeolite nanoparticles, the Z potential of the TFN membranes significantly decreased, whereas the contact angle almost remained constant or slightly higher (Figure 7.9a) [33]. In addition, the surface roughness of TFN membranes increased when larger particles were added (Figure 7.9a) [33]. Smoother surface was observed when smaller particles were added (Figure 7.9b) and even with increasing the amount of zeolites [23, 60]. The TFN membranes with higher zeolite loadings possessed more hydrophilic surfaces and more negative Z potential (Figure 7.9b), which further promoted the membrane water uptake resulting in higher water flux [23]. Previous studies suggested that smaller inorganic particles (e.g., ~100 nm) best match the characteristic thin film thickness and thus make greater enhancement of permeability than the use of larger particles (e.g., ~300 nm) [33]. The greater roughness of membrane was expected to decrease the membrane permeation resistance, because of the increased surface area of membrane in contact with the liquid during permeation [43]. This may explain an improvement of water permeability accompanied by worsened membrane fouling resistance. The PA TFN membranes sometimes had a greater thickness than pristine PA TFC membranes (Figure 7.9c) [33, 43, 44], which may partially offset the decrease in the membrane resistance caused by the increasing surface area.

Most studies attributed the great improvement in membrane separation performance, especially water flux, to that the presence of foreign inorganic fillers in PA membranes greatly tailored the membrane structure. The addition of foreign inorganic fillers affected the membrane polymerization kinetics and altered the film structure, thus more or less changing the surface morphology and the molecular packing of polymer chains. In general, more and larger pores were observed in the TFN membranes, as compared to TFC membranes (Figure 7.9d) [33, 44]; larger porous structures allowed the water to permeate easily but at the same time the salt permeation also increased. For example, in the fabrication of zeolite A-PA

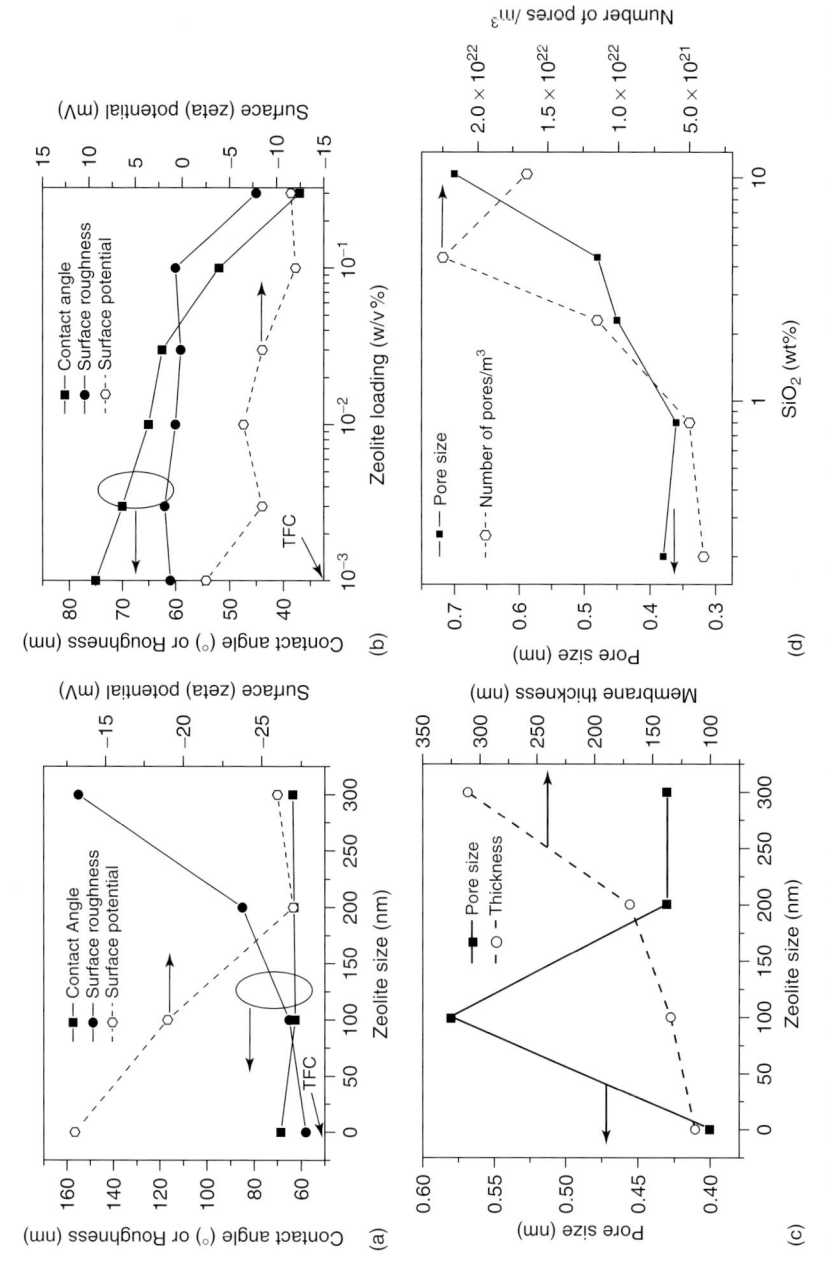

Figure 7.9 (a) Effect of zeolite A particle sizes on the contact angle, roughness, and surface potential of TFN membranes [33]. (b) Effect of zeolite A loading on the contact angle, roughness, and surface (zeta) potential of TFN membranes [23]. (c) Effect of zeolite A particle size on the pore sizes and thickness of TFN membranes; pore sizes were estimated from PEG200 rejection data through a simple steric-exclusion pore transport model [33]. (d) Effect of silica loading on the membrane pore sizes and pore number density of TFN membranes [44].

TFN films, Lind et al. [33] attributed that the presence of super-hydrophilic zeolite A nanocrystals in the organic phase could enhance the miscibility of the MPD aqueous and TMC organic phase during interfacial polymerization; therefore, PA chains appeared less densely packed in the zeolite A-PA TFN membranes. Consequently, these relatively loose membranes exhibited high permeate flux but poor separation. This phenomenon is particularly significant for the TFN membranes with higher loading of inorganic particles. The increased loading resulted in the membranes with higher pore density and larger pore sizes, and thus an abruptly increased flux and sharp decreased salt rejection [44].

Similar to other MMMs for gas separation or pervaporation purposes, the adhesion/bonding between the inorganic fillers and the organic polymer matrix is one of the most important factors in determining the separation properties of TFN membranes. Some inorganic filler particles have surface functional groups that favorably interact with the polymer matrix. For example, TiO_2 nanoparticles were shown to be strongly bonded to the pendant free carboxylic acid on membrane surface by a bidentate coordination and an H bond [42]. There was an interaction between the carboxylic group of modified MWCNTs in the polymer matrix and the amide bond generated in the interfacial polymerization, which made the resultant TFN membranes more stable against chlorine when compared with TFC membranes [34, 35]. However, in some cases, the interfacial voids could result from different surface properties arising from polymers and inorganic fillers; this presumably is the major cause for the more or less deteriorated performance as water and hydrated ion molecules bypass these nonselective and less resistant voids. For instance, as compared to parent PA TFC membranes, Lind and coworkers partially ascribed the dramatic increase of solute permeability in zeolite A-PA TFN membranes to the creation of microporous defects [33]. Some methods have been reported to improve the interfacial interaction between polymer and inorganic particles in the fabrication of MMMs for gas separation [4, 114, 115]. In particular, organic functionalization of particle surfaces has proved to be an effective way in the preparation of MMMs with improved separation performance. These methods could be adopted in the preparation of TFN membranes to achieve desirable desalination properties. So far, the nature of inorganic filler–polymer interaction has not been fully elucidated in the TFN desalination membranes. More experimental and theoretical studies are needed for the fundamental understanding of how inorganic particles interact with polymer matrix during the membrane formation.

As discussed earlier, the modification of membrane separation properties with the addition of inorganic fillers results mainly from the change of pristine membrane structure and sometimes from the permeation properties of porous inorganic fillers. The nonporous inorganic fillers, for example, silver and titanium dioxide, are not permeable to water or hydrated ion molecules (Figure 7.10a). The role of nonporous materials is to affect the PA polymerization process, and modify the PA structure (pore size and pore density); thus, the salt rejection and water flux of nanocomposite membranes would be tuned. In contrast, as microporous inorganic fillers, zeolites have excellent molecular sieving properties

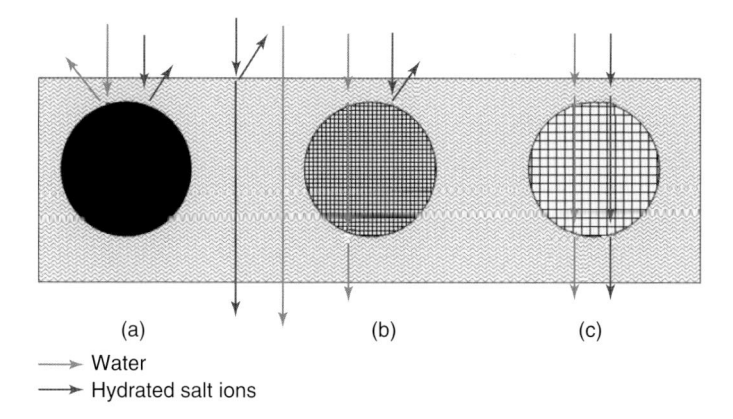

Figure 7.10 Schematic of water and hydrated ions transport through TFN membranes with different fillers: (a) nonporous materials (e.g., titanium dioxide, silica, silver), (b) microporous materials (e.g., zeolite), and (c) other large-pore materials (e.g., mesoporous silica and carbon nanotubes).

(size and shape selectivities) because of their well-defined pore sizes and extremely narrow pore distributions. It is hypothesized that some zeolites having high charge density and super-hydrophilicity provide preferential flow paths for water molecules, while displaying excellent salt rejection (Figure 7.10b). So far, only zeolite A (LTA) and zeolite Y (FAU) have been studied in the development of TFN membranes. Because zeolite A has a pore window size of 4.0–4.2 Å and zeolite Y possesses a pore window size of around 7.3 Å, these hydrophilic zeolites would allow small water molecules (2.8 Å) to permeate through, while blocking hydrated ions [23, 32, 33, 60]. The zeolite (A, Y)-PA TFN membranes have been shown to have improved permeability while retaining a good rejection [23, 32, 33, 60]. However, it is unclear how significant the water preferential flow through zeolite nanocrystals would be. Other large-pore materials, such as mesoporous silica [36] and MWCNTs [34, 35], have also been reported as inorganic fillers of TFN membranes. Because their pore sizes are greater than the sizes of hydrated ions, these filler particles do not function as molecular sieves in the desalination process (Figure 7.10c). In particular, carbon nanotubes have attracted great attention in the separation areas because of the fast water transport arising from their atomic surface smoothness. The proper alignment of carbon nanotubes to the influent is important for the fabrication of high-performance desalination membranes, as shown in Figure 7.11. For instance, the functionalized single-walled carbon nanotubes (SWCNTs) were aligned onto the barrier layer that was formed by conventional interfacial polymerization on a microporous polyethersulfone support [61]. As given in Table 7.2, in the presence of SWCNTs (1.2–1.4 nm in diameter), SWCNTs-PA TFN membranes were observed with a slightly higher salt rejection and almost doubled water flux, as compared with the parent PA TFC membrane [61]. However, more work is required to further understand the transport mechanism of the SWCNTs-PA TFN membranes.

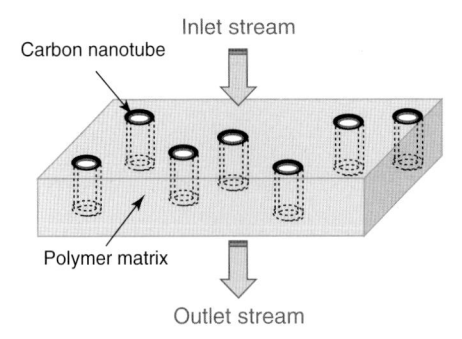

Figure 7.11 Schematic of the target TFN membrane where carbon nanotubes are embedded in a thin film layer with the pores open to the feed direction.

Table 7.2 Desalination results for SWCNTs (1.2–1.4 nm in diameter)-PA TFN membranes (2000 ppm NaCl feed solution at 200 psi) [61].

Membrane samples	Salt rejection (%)	Permeability (m^3 $(m^2 \cdot s \cdot Pa)^{-1}$)
With nanotubes	97.69	5.12×10^{-12}
With nanotubes	99.30	9.9×10^{-12}
Without nanotubes	96.19	2.97×10^{-12}
GE osmonics®[a]	94.91	2.29×10^{-12}

[a] Commerically available salt water reverse osmosis membrane from GE Osmonics® Inc. (Minnetonka, Minn., US), which contains no carbon nanotubes.

7.4.2
Fouling Resistance, Chlorine Stability, and Other Properties

Membrane separation performances, for example, salt rejection and water flux, have been considered as important aspects in the development of TFN membranes. It is noted that other properties, especially chlorine stability and fouling resistance, which are limiting factors in the application of commercial PA TFC desalination membranes, are required for further improvement.

Surface properties, such as roughness, can strongly affect the membrane fouling properties. The Ag-PA TFN membranes were shown to exhibit higher roughness and water flux than plain PA TFC membranes; this was attributed to the increased surface area of membranes [43]. However, rough surfaces may cause a severe fouling problem for TFN membranes because of easy deposition of foulants, when compared with smoother TFC membranes. In other cases, the surfaces of zeolite-PA TFN membranes were observed to be smoother than those of PA TFC membranes [23, 60]. These nanocomposite membranes are potentially more fouling resistant.

Considerable efforts have been made to ease membrane biofouling problem. A significant improvement in antibiofouling has been seen on adding antibacterial nanoparticles, such as silver and titanium dioxide [42, 116–118]. For example, AgA (zeolite LTA in the form of Ag) – PA TFN membranes, exhibiting more hydrophilic and smooth interfaces, inhibited the adhesion of bacteria onto the membrane surface because of bactericidal activity from AgA [32]. In the Ag - PA membranes, the colonies of *Pseudomonas* were significantly suppressed after cultivation, indicating the effectiveness of silver nanoparticles as an antifouling agent [43]. In addition, the sterilization of the TiO_2-PA membranes induced by UV illumination was completed within 4 h, which was more effective than that of the neat PA TFC membranes treated under the same condition [40]. The fouling experiment verified a substantial prevention of the TiO_2-PA TFN membranes against the microbial fouling, demonstrating their potential as antibiofouling membranes [40, 41].

Much attention has been paid to the sensitivity of aromatic PA to chlorine disinfectants that are commonly applied in the desalination processes [6]. The RO membranes with incorporated MWCNTs showed an improved chlorine resistance when compared with the conventional PA RO membranes. For instance, the salt rejection of PA TFC membrane decreased from 98.3 to 73.5% after the membrane was immersed into 3000 ppm NaOCl solution for 4 h. The PA TFN membrane with the incorporation of 1 (w/v)% MWCNTs had a salt rejection 92.5 and 82.4%, before and after the chlorine treatment of membranes, respectively. It was because the interaction between the carboxylic group of the modified MWCNTs and the amide bond made the membranes more stable against chlorine [34, 35].

Other properties such as thermal stability and mechanical strength are also changed after the addition of inorganic particles. Jadav and coworkers reported that the PA TFN membranes incorporated with silica nanoparticles exhibited a higher thermal stability than plain TFC membranes [44]. The composite membranes with embedded MWCNTs also showed higher thermal stability than plain polymer membranes [34]. A lower degree of polymerization and poor mechanical strength were observed in the membranes with higher TiO_2 loadings, ascribed to the significant interference of interfacial polymerization by inorganic TiO_2 nanoparticles [42].

7.5
Commercialization and Future Developments of TFN Membranes

Until now, zeolite-PA TFN membranes capable of single-pass seawater desalination has been demonstrated for the first time by Lind *et al.* with the addition of zeolite A nanoparticles (~250 nm). The TFN membranes showed superior flux and salt rejection (>99.4%) than neat hand-casted TFC membranes and commercially available TFC membranes (SWRO) [59, 119]. NanoH$_2$O, *Inc.* (see *www.Nanoh2o.com*) has developed a QuantumFlux (Qfx) seawater RO element, incorporating nanomaterials (zeolites) into the thin film PA layer of composite membranes for desalination and water purification purposes. Results from their laboratory experiments suggested the potential of fabricating macroscopic samples of TFN RO membranes with significantly increased membrane permeability while matching comparable salt rejection to the commercial TFC RO membranes [120]. The utilization of TFN RO membranes could lead to a 20% decrease in energy consumption [121], which showed substantial reduction in the requirement of feed pressure in the desalination processes [122]. However, there are still more studies required for further development of zeolite-PA TFN membranes. For example, even though in zeolite-PA TFN membranes, improved performance (e.g., water flux) has been reported, more experimental and theoretical studies are needed for fundamental understanding of how zeolite nanocrystals interact with different polymer matrix during membrane formation, and how the types and sizes of zeolite nanocrystals affect TFN membrane structure and performance. Such understanding is crucial for designing and optimizing suitable TFN membranes for water desalination. Despite the incorporation of zeolite A nanoparticles in the PA matrix has led to higher salt rejection and flux than those of a commercial product, long-term operational data are still unavailable.

It is noted that recent experimental and simulation results have revealed considerably fast transport of water molecules through carbon nanotubes thanks to the atomically smooth and hydrophobic walls of carbon nanotubes [123–126]. Thus, carbon nanotubes and other types of nanotubes [127] may have the potential for use in high-flux desalination membranes. In particular, considering the salt and water molecular sizes, nanotubes with very small channel sizes (e.g., <1 nm) [128–131] are highly desirable for high salt rejection. Therefore, the fabrication of small-channel nanotube-polymer TFN membranes deserves more attention. The breakthrough may be seen in the synthesis of such membranes with significantly improved water flux and salt rejection by orienting small-channel carbon nanotubes in the polymer matrix.

In addition to the superior separation performance (e.g., salt rejection and flux), greater attention must be paid on the improvement of certain other properties such as chlorine stability and fouling resistance, which are currently the main limitations of commercial PA TFC films, to develop energy-efficient desalination membranes. The use of nanostructured fillers based on materials, such as silver and titanium oxide, has already been explored in polymeric membranes and imparted enhanced fouling/chlorine resistance to TFN membranes, as compared with the parent PA polymer [40–43]. However, the use of resultant TFN membranes for brackish or seawater desalination is still an emerging area of research, and only limited results have been reported so far. Therefore, more efforts are required to study their separation performance and evaluate their potential for commercialization.

As discussed earlier, the development and commercialization of TFN membranes for water desalination and purification is still in the early stage and requires more systemic studies into the following aspects: better understanding of transport processes, reduction of costs (capital, operating, and materials), productivity of RO membranes, optimization of membrane structures, and integration into a large system level. In particular, the question of how the most promising TFN membrane materials will perform in an actual desalination process must be answered. The engineering economics are needed to be evaluated on scaling-up these membrane materials for industrial application. The ease and costs of large-scale fabrication of TFN membrane materials will be critical in determining their applicability on an industrial scale. All of these are important factors in affecting the development of TFN membranes in desalination industry. The successful implementation of TFN membranes will lead to more energy-efficient desalination process and cost reduction of desalinated water.

7.6 Summary

RO desalination has over half a century of industrial operation, and it has been considered as one of the promising technologies for desalting brackish water and seawater. Over the past decade, in addition to the system-level advancements in RO technology, the development of membrane materials with better performance has undoubtedly made the RO desalination more economic. Currently, PA TFC membranes, which exhibit good separation performance but have limitations in terms of poor chlorine and fouling resistance, have dominated the RO desalination industry.

In recent years, the strategy of forming high-performance separation TFN membranes by incorporating functional nanomaterials into a polymer matrix has been adopted for developing RO membranes. The availability of different types of nanoparticles and advanced knowledge on the integration of the particles in membrane structures has opened a new avenue in the development of novel membrane materials. A wide variety of characterization techniques of TFN membrane materials, such as electron microscopy, AFM, and IR, have been used to correlate membrane properties with their separation performance and multifunctionality. Until now, the use of various types of inorganic fillers, such as silica, silver, titanium dioxide, zeolites, and carbon nanotubes, in the fabrication of inorganic–organic TFN PA desalination membranes with improved properties, such as flux, salt retention, chlorine stability, fouling resistance, and thermal stability has been reported. Nevertheless, the research of TFN membrane materials with superior separation performance and multifunctionality, especially antifouling capacity and chlorine resistance, is still ongoing. More experimental and theoretical studies are required for the fundamental understanding of effects of nanomaterials on the TFN membrane structures and performances. Meanwhile, other aspects, including the ease of synthesis, cost of materials, and their actual performance in industry-scale desalination, need to be looked at before the TFN membranes can be widespread commercialized.

List of Main Abbreviations

ATR	attenuated total reflectance
AFM	atomic force microscopy
CA	cellulose acetate
DLS	dynamic light scattering
EDX	energy dispersive X-ray spectroscopy

FTIR	Fourier transform infrared spectroscopy
MMMs	mixed matrix membranes
MPD	m-phenylenediamine
MWCNTs	multiwalled carbon nanotubes
PS	polysulfone
PA	polyamide
RBS	Rutherford backscattering spectrometry
RO	reverse osmosis
SDAs	structure-directing agents
SEM	scanning electron microscopy
SWCNTs	single-walled carbon nanotubes
TEM	transmission electron microscopy
TMC	trimesoyl chloride
TFC	thin film composite
TFN	thin film nanocomposite
XPS	X-ray photoelectron spectroscopy
XRD	X-ray diffraction

References

1. Basu, S., Khan, A.L., Cano-Odena, A., Liu, C., and Vankelecom, I.F.J. (2010) Membrane-based technologies for biogas separations. *Chem. Soc. Rev.*, **39** (2), 750–768.
2. Greenlee, L.F., Lawler, D.F., Freeman, B.D., Marrot, B., and Moulin, P. (2009) Reverse osmosis desalination: water sources, technology, and today's challenges. *Water Res.*, **43** (9), 2317–2348.
3. Shao, P. and Huang, R.Y.M. (2007) Polymeric membrane pervaporation. *J. Membr. Sci.*, **287** (2), 162–179.
4. Chung, T.-S., Jiang, L.Y., Li, Y., and Kulprathipanja, S. (2007) Mixed matrix membranes (MMMs) comprising organic polymers with dispersed inorganic fillers for gas separation. *Prog. Polym. Sci.*, **32** (4), 483–507.
5. Strathmann, H. (2001) Membrane separation processes: current relevance and future opportunities. *AICHE J.*, **47** (5), 1077–1087.
6. Li, D. and Wang, H. (2010) Recent developments in reverse osmosis desalination membranes. *J. Mater. Chem.*, **20**, 4551–4566.
7. Khawaji, A.D., Kutubkhanah, I.K., and Wie, J.-M. (2008) Advances in seawater desalination technologies. *Desalination*, **221** (1–3), 47–69.
8. Petersen, R.J. (1993) Composite reverse osmosis and nanofiltration membranes. *J. Membr. Sci.*, **83** (1), 81–150.
9. El-Saied, H., Basta, A.H., Barsoum, B.N., and Elberry, M.M. (2003) Cellulose membranes for reverse osmosis part I. RO cellulose acetate membranes including a composite with polypropylene. *Desalination*, **159** (2), 171–181.
10. Younos, T. and Tulou, K.E. (2005) Overview of desalination techniques. *J. Contemp. Water Res. Educ.*, **132**, 3–10.
11. Wethern, M. and Katzaras, W. (1995) Reverse osmosis treatment of municipal sewage effluent for industrial reuse. *Desalination*, **102** (1–3), 293–299.
12. Petersen R. J. and Cadotte J. E. (1990) Thin film composite reverse osmosis membrane, in *Handbook of Industrial Membrane Technology* (ed. M. C. Porter), Noyes Publication.
13. Asano, T. (1998) *Wastewater Reclamation and Reuse*, CRC Press.
14. Liu, M., Yu, S., Tao, J., and Gao, C. (2008) Preparation, structure characteristics and separation properties

of thin-film composite polyamide-urethane seawater reverse osmosis membrane. *J. Membr. Sci.*, **325** (2), 947–956.
15. Kuehne, M.A., Song, R.Q., Li, N.N., and Petersen, R.J. (2001) Flux enhancement in TFC RO membranes. *Environ. Prog.*, **20** (1), 23–26.
16. Glater, J., Hong, S.-K., and Elimelech, M. (1994) The search for a chlorine-resistant reverse osmosis membrane. *Desalination*, **95** (3), 325–345.
17. Konagaya, S. and Watanabe, O. (2000) Influence of chemical structure of isophthaloyl dichloride and aliphatic, cycloaliphatic, and aromatic diamine compound polyamides on their chlorine resistance. *J. Appl. Polym. Sci.*, **76** (2), 201–207.
18. Gabelich, C.J., Frankin, J.C., Gerringer, F.W., Ishida, K.P., and Suffet, I.H. (2005) Enhanced oxidation of polyamide membranes using monochloramine and ferrous iron. *J. Membr. Sci.*, **258** (1–2), 64–70.
19. Belfer, S., Purinson, Y., Fainshtein, R., Radchenko, Y., and Kedem, O. (1998) Surface modification of commercial composite polyamide reverse osmosis membranes. *J. Membr. Sci.*, **139** (2), 175–181.
20. Shannon, M.A., Bohn, P.W., Elimelech, M., Georgiadis, J.G., Marinas, B.J., and Mayes, A.M. (2008) Science and technology for water purification in the coming decades. *Nature*, **452** (7185), 301–310.
21. Hong, S. and Elimelech, M. (1997) Chemical and physical aspects of natural organic matter (NOM) fouling of nanofiltration membranes. *J. Membr. Sci.*, **132** (2), 159–181.
22. Zhu, X. and Elimelech, M. (1997) Colloidal fouling of reverse osmosis membranes: measurements and fouling mechanisms. *Environ. Sci. Technol.*, **31** (12), 3654–3662.
23. Jeong, B.-H., Hoek, E.M.V., Yan, Y., Subramani, A., Huang, X., Hurwitz, G., Ghosh, A.K., and Jawor, A. (2007) Interfacial polymerization of thin film nanocomposites: a new concept for reverse osmosis membranes. *J. Membr. Sci.*, **294** (1–2), 1–7.
24. Wee, S.-L., Tye, C.-T., and Bhatia, S. (2008) Membrane separation process–pervaporation through zeolite membrane. *Sep. Purif. Technol.*, **63** (3), 500–516.
25. Bowen, T.C., Noble, R.D., and Falconer, J.L. (2004) Fundamentals and applications of pervaporation through zeolite membranes. *J. Membr. Sci.*, **245** (1–2), 1–33.
26. Ismail, A.F., Goh, P.S., Sanip, S.M., and Aziz, M. (2009) Transport and separation properties of carbon nanotube-mixed matrix membrane. *Sep. Purif. Technol.*, **70** (1), 12–26.
27. Lin, J. and Murad, S. (2001) A computer simulation study of the separation of aqueous solutions using thin zeolite membranes. *Mol. Phys.*, **99** (14), 1175–1181.
28. Li, L.X., Dong, J.H., Nenoff, T.M., and Lee, R. (2004) Desalination by reverse osmosis using MFI zeolite membranes. *J. Membr. Sci.*, **243** (1–2), 401–404.
29. Caro, J., Noack, M., Kölsch, P., and Schäfer, R. (2000) Zeolite membranes – state of their development and perspective. *Microporous Mesoporous Mater.*, **38** (1), 3–24.
30. Paul, D.R. and Kemp, D.R. (1973) Diffusion time lag in polymer membranes containing adsorptive fillers. *J. Polym. Sci., Part C: Polym. Symp.*, **41**, 79–93.
31. Zornoza, B., Irusta, S., Teĉllez, C., and Coronas, J.N. (2009) Mesoporous silica sphere–polysulfone mixed matrix membranes for gas separation. *Langmuir*, **25** (10), 5903–5909.
32. Lind, M., Jeong, B., Subramani, A., Huang, X., and Hoek, E. (2009) Effect of mobile cation on zeolite-polyamide thin film nanocomposite membranes. *J. Mater. Res.*, **24** (5), 1624–1631.
33. Lind, M.L., Ghosh, A.K., Jawor, A., Huang, X., Hou, W., Yang, Y., and Hoek, E.M.V. (2009) Influence of zeolite crystal size on zeolite-polyamide thin film nanocomposite membranes. *Langmuir*, **25** (17), 10139–10145.
34. Park, J., Choi, W., Cho, J., Chun, B., Kim, S., and Lee, K. (2010) Carbon nanotube-based nanocomposite desalination membranes from layer-by-layer

assembly. *Desalin. Water Treat.*, **15** (1–3), 76–83.
35. Park, J., Choi, W., Kim, S., Chun, B., Bang, J., and Lee, K. (2010) Enhancement of chlorine resistance in carbon nanotube-based nanocomposite reverse osmosis membranes. *Desalin. Water Treat.*, **15** (1–3), 198–204.
36. Park, K., Kim, S., Chun, B., and Bang, J. (2010) Sulfonated poly (arylene ether sulfone) thin-film composite reverse osmosis membrane containing SiO_2 nano-particles. *Desalin. Water Treat.*, **15** (1–3), 69–75.
37. Vu, D.Q., Koros, W.J., and Miller, S.J. (2003) Mixed matrix membranes using carbon molecular sieves: I. Preparation and experimental results. *J. Membr. Sci.*, **211** (2), 311–334.
38. Vu, D.Q., Koros, W.J., and Miller, S.J. (2003) Mixed matrix membranes using carbon molecular sieves: II. Modeling permeation behavior. *J. Membr. Sci.*, **211** (2), 335–348.
39. Lee, K.P., Arnot, T.C., and Mattia, D. (2011) A review of reverse osmosis membrane materials for desalination–development to date and future potential. *J. Membr. Sci.*, **370** (1–2), 1–22.
40. Kwak, S.-Y., Kim, S.H., and Kim, S.S. (2001) Hybrid organic/inorganic reverse osmosis (RO) membrane for bactericidal anti-fouling. 1. Preparation and characterization of TiO_2 nanoparticle self-assembled aromatic polyamide thin-film-composite (TFC) membrane. *Environ. Sci. Technol.*, **35** (11), 2388–2394.
41. Kim, S.H., Kwak, S.-Y., Sohn, B.-H., and Park, T.H. (2003) Design of TiO_2 nanoparticle self-assembled aromatic polyamide thin-film-composite (TFC) membrane as an approach to solve biofouling problem. *J. Membr. Sci.*, **211** (1), 157–165.
42. Lee, H.S., Im, S.J., Kim, J.H., Kim, H.J., Kim, J.P., and Min, B.R. (2008) Polyamide thin-film nanofiltration membranes containing TiO_2 nanoparticles. *Desalination*, **219** (1–3), 48–56.
43. Lee, S.Y., Kim, H.J., Patel, R., Im, S.J., Kim, J.H., and Min, B.R. (2007) Silver nanoparticles immobilized on thin film composite polyamide membrane: characterization, nanofiltration, antifouling properties. *Polym. Adv. Technol.*, **18** (7), 562–568.
44. Jadav, G.L. and Singh, P.S. (2009) Synthesis of novel silica-polyamide nanocomposite membrane with enhanced properties. *J. Membr. Sci.*, **328** (1–2), 257–267.
45. Merkel, T.C., Freeman, B.D., Spontak, R.J., He, Z., Pinnau, I., Meakin, P., and Hill, A.J. (2002) Ultrapermeable, reverse-selective nanocomposite membranes. *Science*, **296** (5567), 519–522.
46. Sridhar, S., Aminabhavi, T.M., Mayor, S.J., and Ramakrishna, M. (2007) Permeation of carbon dioxide and methane gases through novel silver-incorporated thin film composite Pebax membranes. *Ind. Eng. Chem. Res.*, **46** (24), 8144–8151.
47. Matteucci, S., Kusuma, V.A., Kelman, S.D., and Freeman, B.D. (2008) Gas transport properties of MgO filled poly (1-trimethylsilyl-1-propyne) nanocomposites. *Polymer*, **49** (6), 1659–1675.
48. Hosseini, S.S., Li, Y., Chung, T.-S., and Liu, Y. (2007) Enhanced gas separation performance of nanocomposite membranes using MgO nanoparticles. *J. Membr. Sci.*, **302** (1–2), 207–217.
49. Sadeghi, M., Semsarzadeh, M.A., and Moadel, H. (2009) Enhancement of the gas separation properties of polybenzimidazole (PBI) membrane by incorporation of silica nano particles. *J. Membr. Sci.*, **331** (1–2), 21–30.
50. Hibshman, C., Cornelius, C.J., and Marand, E. (2003) The gas separation effects of annealing polyimide-organosilicate hybrid membranes. *J. Membr. Sci.*, **211** (1), 25–40.
51. Merkel, T.C., He, Z., Pinnau, I., Freeman, B.D., Meakin, P., and Hill, A.J. (2003) Sorption and transport in poly (2,2-bis (trifluoromethyl)-4,5-difluoro-1,3-dioxole-co-tetrafluoroethylene) containing nanoscale fumed silica. *Macromolecules*, **36** (22), 8406–8414.
52. Winberg, P., DeSitter, K., Dotremont, C., Mullens, S., Vankelecom, I.F.J., and Maurer, F.H.J. (2005) Free volume and interstitial mesopores in silica

filled poly (1-trimethylsilyl-1-propyne) nanocomposites. *Macromolecules*, **38** (9), 3776–3782.
53. Moaddeb, M. and Koros, W.J. (1997) Gas transport properties of thin polymeric membranes in the presence of silicon dioxide particles. *J. Membr. Sci.*, **125** (1), 143–163.
54. Xu, Z.-l., Yu, L.-Y., and Han, L.-F. (2009) Polymer-nanoinorganic particles composite membranes: a brief overview. *Front. Chem. Eng. Chin.*, **3** (3), 318–329.
55. Li, D., Zhu, H.Y., Ratinac, K.R., Ringer, S.P., and Wang, H.T. (2009) Synthesis and characterization of sodalite-polyimide nanocomposite membranes. *Microporous Mesoporous Mater.*, **126** (1–2), 14–19.
56. Pechar, T.W., Tsapatsis, M., Marand, E., and Davis, R. (2002) Preparation and characterization of a glassy fluorinated polyimide zeolite-mixed matrix membrane. *Desalination*, **146** (1–3), 3–9.
57. Kim, J. and Van der Bruggen, B. (2010) The use of nanoparticles in polymeric and ceramic membrane structures: review of manufacturing procedures and performance improvement for water treatment. *Environ. Pollut.*, **158** (7), 2335–2349.
58. Hoek E. M. V., Jeong B. H. and Yan Y. (2005) US Patent 20110027599.
59. Lind, M.L., Eumine Suk, D., Nguyen, T.-V., and Hoek, E.M.V. (2010) Tailoring the structure of thin film nanocomposite membranes to achieve seawater RO membrane performance. *Environ. Sci. Technol.*, **44** (21), 8230–8235.
60. Kong, C., Shintani, T., and Tsuru, T. (2010) "Pre-seeding"-assisted synthesis of a high performance polyamide-zeolite nanocomposite membrane for water purification. *New J. Chem.*, **34**, 2101–2104.
61. Ratto T. V., Holt J. K. and Szmodis A. W. (2010) US Patent 20100025330.
62. Ebelmen, M. (1846) Sur les combinaisons des acides borique et silicique avec les ethers. *Ann. Chim. Phys.*, **16**, 129.
63. Livage, J. (1997) Sol–gel processes. *Curr. Opin. Solid State Mater. Sci.*, **2** (2), 132–138.
64. Hench, L.L. and West, J.K. (1990) The sol–gel process. *Chem. Rev.*, **90** (1), 33–72.
65. Laurent, S., Forge, D., Port, M., Roch, A., Robic, C., Vander Elst, L., and Muller, R.N. (2008) Magnetic iron oxide nanoparticles: synthesis, stabilization, vectorization, physicochemical characterizations, and biological applications. *Chem. Rev.*, **108** (6), 2064–2110.
66. Pierre, A.C. (1998) *Introduction to Sol–gel Processing*, Kluwer Academic Publishers, Boston, MA.
67. Brinker, C.J. and Scherer, G.W. (1990) *Sol–gel Science: The Physics and Chemistry of Sol–gel Processing*, Academic Press, San Diego, CA.
68. Wright, J.D. and Sommerdijk, N.A.J.M. (2001) *Sol–gel Materials: Chemistry and Applications*, Taylor & Francis, London.
69. Sakka, S. (2002) *Sol–Gel Science and Technology: Topics in Fundamental Research and Applications: Sol Gel Prepared Ferroelectrics and Related Materials*, Kluwer Academic Publishers, Boston, MA.
70. Yan, C., Sun, L., and Cheng, F. (2003) *Handbook of Nanophase and Nanostructured Materials*, Kluwer, New York.
71. Cushing, B.L., Kolesnichenko, V.L., and O'Connor, C.J. (2004) Recent advances in the liquid-phase syntheses of inorganic nanoparticles. *Chem. Rev.*, **104** (9), 3893–3946.
72. Gurav, J.L., Jung, I.-K., Park, H.-H., Kang, E.S., and Nadargi, D.Y. (2010) Silica aerogel: synthesis and applications. *J. Nanomater.*, 1–11.
73. Brinker, C.J. and Sherrer, G.W. (1990) *Sol–gel Science*, Academic Press, New York.
74. Chen, X. and Mao, S.S. (2007) Titanium dioxide nanomaterials: synthesis, properties, modifications, and applications. *Chem. Rev.*, **107** (7), 2891–2959.
75. Cundy, C.S. and Cox, P.A. (2003) The hydrothermal synthesis of zeolites: history and development from the earliest days to the present time. *Chem. Rev.*, **103** (3), 663–702.

76. Tosheva, L. and Valtchev, V.P. (2005) Nanozeolites: synthesis, crystallization mechanism, and applications. *Chem. Mater.*, **17** (10), 2494–2513.
77. Cundy, C.S. and Cox, P.A. (2005) The hydrothermal synthesis of zeolites: precursors, intermediates and reaction mechanism. *Microporous Mesoporous Mater.*, **82** (1–2), 1–78.
78. Jansen, J.C. (1991) The preparation of molecular sieves, in *Studies in Surface Science and Catalysis* (eds H.V. Bekkum, E.M. Flanigen, and J.C. Jansen), Elsevier.
79. Yu, J. (2007) Synthesis of zeolites, in *Introduction to Zeolite Science and Practice* (ed. J. Čejka), Elsevier, Oxford.
80. Szostak, R. (2007) *Molecular Sieves – Principles of Synthesis and Identification*, Springer, New York.
81. Zhan, B.-Z., White, M.A., Lumsden, M., Mueller-Neuhaus, J., Robertson, K.N., Cameron, T.S., and Gharghouri, M. (2002) Control of particle size and surface properties of crystals of NaX zeolite. *Chem. Mater.*, **14** (9), 3636–3642.
82. Valtchev, V.P. and Bozhilov, K.N. (2004) Transmission electron microscopy study of the formation of FAU-type zeolite at room temperature. *J. Phys. Chem. B*, **108** (40), 15587–15598.
83. Larsen, S.C. (2007) Nanocrystalline zeolites and zeolite structures: synthesis, characterization, and applications. *J. Phys. Chem. C*, **111** (50), 18464–18474.
84. Wang, H., Holmberg, B.A., and Yan, Y. (2002) Homogeneous polymer-zeolite nanocomposite membranes by incorporating dispersible template-removed zeolite nanocrystals. *J. Mater. Chem.*, **12** (12), 3640–3643.
85. Wang, H., Holmberg, B.A., and Yan, Y. (2003) Synthesis of template-free zeolite nanocrystals by using in situ thermoreversible polymer hydrogels. *J. Am. Chem. Soc.*, **125** (33), 9928–9929.
86. Wang, H., Wang, Z., and Yan, Y. (2000) Colloidal suspensions of template-removed zeolite nanocrystals. *Chem. Commun.*, (23), 2333–2334.
87. Valtchev, V.P., Tosheva, L., and Bozhilov, K.N. (2005) Synthesis of zeolite nanocrystals at room temperature. *Langmuir*, **21** (23), 10724–10729.
88. Kim, E.-S., Kim, Y.J., Yu, Q., and Deng, B. (2009) Preparation and characterization of polyamide thin-film composite (TFC) membranes on plasma-modified polyvinylidene fluoride (PVDF). *J. Membr. Sci.*, **344**, 71–81.
89. Kim, H.I. and Kim, S.S. (2006) Plasma treatment of polypropylene and polysulfone supports for thin film composite reverse osmosis membrane. *J. Membr. Sci.*, **286** (1–2), 193–201.
90. Steen, M.L., Hymas, L., Havey, E.D., Capps, N.E., Castner, D.G., and Fisher, E.R. (2001) Low temperature plasma treatment of asymmetric polysulfone membranes for permanent hydrophilic surface modification. *J. Membr. Sci.*, **188** (1), 97–114.
91. Pendergast, M.T.M., Nygaard, J.M., Ghosh, A.K., and Hoek, E.M.V. (2010) Using nanocomposite materials technology to understand and control reverse osmosis membrane compaction. *Desalination*, **261** (3), 255–263.
92. Wei, J., Jian, X., Wu, C., Zhang, S., and Yan, C. (2005) Influence of polymer structure on thermal stability of composite membranes. *J. Membr. Sci.*, **256** (1–2), 116–121.
93. Wu, C., Zhang, S., Yang, D., Wei, J., Yan, C., and Jian, X. (2006) Preparation, characterization and application in wastewater treatment of a novel thermal stable composite membrane. *J. Membr. Sci.*, **279** (1–2), 238–245.
94. Wu, C., Zhang, S., Yang, D., and Jian, X. (2009) Preparation, characterization and application of a novel thermal stable composite nanofiltration membrane. *J. Membr. Sci.*, **326** (2), 429–434.
95. Cadotte J. E. (1981) US Patent 4277344.
96. Freger, V. (2003) Nanoscale heterogeneity of polyamide membranes formed by interfacial polymerization. *Langmuir*, **19** (11), 4791–4797.
97. Liu, L.-F., Yu, S.-C., Zhou, Y., and Gao, C.-J. (2006) Study on a novel

polyamide-urea reverse osmosis composite membrane (ICIC-MPD): I. Preparation and characterization of ICIC-MPD membrane. *J. Membr. Sci.*, **281** (1–2), 88–94.

98. Yu, S., Liu, M., Liu, X., and Gao, C. (2009) Performance enhancement in interfacially synthesized thin-film composite polyamide-urethane reverse osmosis membrane for seawater desalination. *J. Membr. Sci.*, **342** (1–2), 313–320.

99. Zhou, Y., Yu, S., Liu, M., and Gao, C. (2005) Preparation and characterization of polyamide-urethane thin-film composite membranes. *Desalination*, **180** (1–3), 189–196.

100. Cong, H., Radosz, M., Towler, B.F., and Shen, Y. (2007) Polymer-inorganic nanocomposite membranes for gas separation. *Sep. Purif. Technol.*, **55** (3), 281–291.

101. Xiao, Y., Low, B.T., Hosseini, S.S., Chung, T.S., and Paul, D.R. (2009) The strategies of molecular architecture and modification of polyimide-based membranes for CO_2 removal from natural gas–A review. *Prog. Polym. Sci.*, **34** (6), 561–580.

102. Hernández, A., Calvo, J.I., Prádanos, P., and Palacio, L. (1999) A multidisciplinary approach towards pore size distributions of microporous and mesoporous membranes, in *Surface Chemistry and Electrochemistry of Membranes* (ed. T.S. Sørensen), Marcel Dekker, New York.

103. Bowen W. R., Hilal N., Lovitt R. W. and Wright V. J. (1999) Atmoic force microscope studies of membrane surfaces. in *Surface Chemistry and Electrochemistry of Membranes* (ed. T. S. Sørensen), Marcel Dekker, New York.

104. Bowen, W.R., Hilal, N., Lovitt, R.W., and Williams, P.M. (1996) Atomic force microscope studies of membranes: surface pore structures of diaflo ultrafiltration membranes. *J. Colloid Interface Sci.*, **180** (2), 350–359.

105. Yalamanchili, M.R., Veeramasuneni, S., Azevedo, M.A.D., and Miller, J.D. (1998) Use of atomic force microscopy in particle science and technology research. *Colloids and Surfaces A: Physicochemical and Engineering Aspects*, **133** (1–2), 77–88.

106. Khulbe, K.C., Feng, C.Y., and Matsuura, T. (2008) *Synthetic Polymeric Membranes*, Springer, Berlin, Heidelberg, pp. 101–139.

107. Kallioinen, M. and Nyström, M. (2008) *Advanced Membrane Technology and Applications*, John Wiley & Sons, Inc., pp. 841–877.

108. Freger, V. and Ben-David, A. (2005) Use of attenuated total reflection infrared spectroscopy for analysis of partitioning of solutes between thin films and solution. *Anal. Chem.*, **77** (18), 6019–6025.

109. Yang, P., Meng, X., Zhang, Z., Jing, B., Yuan, J., and Yang, W. (2005) Thickness measurement of nanoscale polymer layer on polymer substrates by attenuated total reflection infrared spectroscopy. *Anal. Chem.*, **77** (4), 1068–1074.

110. Singh, P.S., Joshi, S.V., Trivedi, J.J., Devmurari, C.V., Rao, A.P., and Ghosh, P.K. (2006) Probing the structural variations of thin film composite RO membranes obtained by coating polyamide over polysulfone membranes of different pore dimensions. *J. Membr. Sci.*, **278** (1–2), 19–25.

111. Mi, B., Coronell, O., Mariñas, B.J., Watanabe, F., Cahill, D.G., and Petrov, I. (2006) Physico-chemical characterization of NF/RO membrane active layers by Rutherford backscattering spectrometry. *J. Membr. Sci.*, **282** (1–2), 71–81.

112. Bartels, C.R. (1989) A surface science investigation of composite membranes. *J. Membr. Sci.*, **45** (3), 225–245.

113. Li, D., He, L., Dong, D., Forsyth, M., and Wang, H. (2012) Preparation of silicalite-polyamide composite membrane for desalination. *Asia-Pac. J. Chem. Eng.*, **7** (3), 434–441.

114. Bernardo, P., Drioli, E., and Golemme, G. (2009) Membrane gas separation: a review/state of the art. *Ind. Eng. Chem. Res.*, **48** (10), 4638–4663.

115. Aroon, M.A., Ismail, A.F., Matsuura, T., and Montazer-Rahmati, M.M. (2010) Performance studies of mixed matrix membranes for gas separation:

a review. *Sep. Purif. Technol.*, **75** (3), 229–242.
116. Choi, O., Deng, K.K., Kim, N.-J., Ross, L. Jr.,, Surampalli, R.Y., and Hu, Z. (2008) The inhibitory effects of silver nanoparticles, silver ions, and silver chloride colloids on microbial growth. *Water Res.*, **42** (12), 3066–3074.
117. Savage, N. and Diallo, M.S. (2005) Nanomaterials and water purification: opportunities and challenges. *J. Nanopart. Res.*, **7** (4), 331–342.
118. Lv, Y., Liu, H., Wang, Z., Liu, S., Hao, L., Sang, Y., Liu, D., Wang, J., and Boughton, R.I. (2009) Silver nanoparticle-decorated porous ceramic composite for water treatment. *J. Membr. Sci.*, **331** (1–2), 50–56.
119. Pendergast, M.M. and Hoek, E.M.V. (2011) A review of water treatment membrane nanotechnologies. *Energy Environ. Sci.*, **4** (6), 1946–1971.
120. Hoek, E.M.V. and Ghosh, A. (2009) Nanotechnology-based membranes for water purification, in *Nanotechnology Applications for Clean Water* (eds N. Savage, M. Diallo, J. Duncan, A. Street, R. Sustich), William Andrew.
121. NanoH$_2$O (2010) Nanotechnology Advances Reverse Osmosis Membrane Performance. (NanoH$_2$O).
122. Subramani, A., Badruzzaman, M., Oppenheimer, J., and Jacangelo, J.G. (2011) Energy minimization strategies and renewable energy utilization for desalination: a review. *Water Res.*, **45** (5), 1907–1920.
123. Hummer, G., Rasaiah, J.C., and Noworyta, J.P. (2001) Water conduction through the hydrophobic channel of a carbon nanotube. *Nature*, **414** (6860), 188–190.
124. Sun, L. and Crooks, R.M. (2000) Single carbon nanotube membranes: a well-defined model for studying mass transport through nanoporous materials. *J. Am. Chem. Soc.*, **122** (49), 12340–12345.
125. Majumder, M., Chopra, N., Andrews, R., and Hinds, B.J. (2005) Nanoscale hydrodynamics: Enhanced flow in carbon nanotubes. *Nature*, **438** (7064), 44–44.
126. Alexiadis, A. and Kassinos, S. (2008) Molecular simulation of water in carbon nanotubes. *Chem. Rev.*, **108** (12), 5014–5034.
127. Won, C.Y. and Aluru, N.R. (2007) Water permeation through a subnanometer boron nitride nanotube. *J. Am. Chem. Soc.*, **129** (10), 2748–2749.
128. Tang, Z.K., Sun, H.D., Wang, J., Chen, J., and Li, G. (1998) Mono-sized single-wall carbon nanotubes formed in channels of AlPO$_{4-5}$ single crystal. *Appl. Phys. Lett.*, **73** (16), 2287–2289.
129. Hulman, M., Kuzmany, H., Dubay, O., Kresse, G., Li, L., Tang, Z.K., Knoll, P., and Kaindl, R. (2004) Raman spectroscopy of single wall carbon nanotubes grown in zeolite crystals. *Carbon*, **42** (5–6), 1071–1075.
130. Karanikolos, G.N., Wydra, J.W., Stoeger, J.A., García, H., Corma, A., and Tsapatsis, M. (2007) Continuous c-oriented AlPO$_{4-5}$ films by tertiary growth. *Chem. Mater.*, **19** (4), 792–797.
131. Deng, S., Dalton, A., Terasaki, O., and Balkus, K.J. Jr. (2004) Carbon nanotubes synthesized in zeolites UTD-1, UTD-18 and UTD-12. *Stud. Surf. Sci. Catal.*, **154**, 903–910.

8
Application of Ceramic Membranes in the Treatment of Water

Weihong Xing, Yiqun Fan, and Wanqin Jin

8.1
Introduction

A membrane can be generally defined as a semipermeable barrier that provides permselective properties in separation processes. Recently, many researchers have focused on inorganic ceramic membranes because of their various properties, including sufficiently high mechanical strength, thermal stability, and resistance to corrosive environments. The development of ceramic membranes, which were first commercialized in the early 1980s, can be divided into three periods: (i) the enrichment of the element U^{235} using ceramic membranes for the Manhattan project during the Second World War, when ceramic membranes were mainly applied in military applications; (ii) the development of microfiltration (MF) (pore diameter >50 nm) and ultrafiltration (UF) (pore diameter between 2 and 50 nm) ceramic membranes in industrial liquid filtration from the 1980s onward; and (iii) the development of nanosized ceramic membranes, which mainly focused on the application of nanofiltration (NF), pervaporation (PV), gas separation (GS), and ceramic membrane reactors.

Ceramic membranes can be classified into dense membranes and porous membranes. Dense membranes are made from solid layers of metals and mixed conducting oxides with considerable thickness, and they possess only a single layer. This type can be used in hydrogen and oxygen separation processes. Porous ceramic membranes usually consist of several layers, with macroporous support, intermediate layers and a top (separation) layer. The macroporous support provides mechanical strength, and the intermediate layers make the surface smooth enough for top layer formation. The materials used for ceramic membranes are numerous, and they include oxide materials such as Al_2O_3, TiO_2, ZrO_2, SiO_2, and combinations thereof. IUPAC classifies ceramic membranes according to pore size as follows: (i) macroporous membranes with a pore size larger than 50 nm; (ii) mesoporous membranes with a pore size in the range of 2–50 nm; and (iii) microporous membranes with a pore size smaller than 2 nm. Membrane processes are classified into MF, UF, NF, PV, GS, reverse osmosis (RO), dialysis, and electrodialysis (ED) depending on the driving force and the physical size of the separated species.

Functional Nanostructured Materials and Membranes for Water Treatment, First Edition.
Edited by Mikel Duke, Dongyuan Zhao, and Raphael Semiat.
© 2013 Wiley-VCH Verlag GmbH & Co. KGaA. Published 2013 by Wiley-VCH Verlag GmbH & Co. KGaA.

8.2
Membrane Preparation

There are various methods for the preparation of ceramic membranes, including plastic extrusion and the sol–gel process [1]. One of the most important objectives of preparation is to obtain defect-free (no cracks or pinholes) ceramic membranes. The process for the fabrication of ceramic membranes usually consists the following steps [2]:

1) Initial layer formation in a precursor solution
2) Growth of film thickness (lyogel state)
3) Drying of the film to a xerogel state
4) Calcination/sintering to obtain the final membrane.

The properties and quality of the final membranes critically depend on the support quality, concentration, and structure of the precursor solutions, as well as the drying and sintering conditions [2]. In addition, ceramic membrane performance (e.g., porosity, pore diameter, mechanical strength, and corrosion resistance) varies depending on the preparation route used.

8.2.1
Extrusion

As a simple and economical technique, extrusion is a promising shaping method for ceramics. This procedure is particularly suitable for the preparation of flat, tubular, or monolithic structures of macroporous ceramic supports. The detailed steps involved are as follows:

1) Mixing the primary ceramic powder, the additives, the plasticizing agent, and the water in a certain mass ratio;
2) Pugging the aforementioned mixture and refining it into plastic clay materials;
3) Extruding the well-pugged ceramic mixture with an extruder and drying the green body afterward; and
4) Calcining the dried samples in a furnace to obtain a sufficiently strong macroporous ceramic support.

8.2.2
Sol–Gel Process

The sol–gel method focuses on hydrolysis and polycondensation with metal alkoxide as precursors. The main advantage of the sol–gel method is that the particle size can be readily controlled by applying either the colloidal or the polymeric route. When the colloidal sol–gel route is adapted, the spaces between colloidal particles are assumed to be pores in ceramic membranes. Spaces within amorphous polymeric networks (mostly of silica-based materials) can be obtained through the polymeric sol–gel route, which can be used to fabricate PV or GS membranes.

Commercial ceramic membrane materials are most commonly employed for MF, UF, NF, and PV. Microporous ceramic membranes for GS remain at the laboratory research and pilot-scale stage. At present, the majority of ceramic membrane companies produce MF and UF membranes. The Inopor Corporation in Germany is the only company in the world that can produce NF membranes. However, over 7000 companies use ceramic membranes in the separation process, and the total area they use for running ceramic membrane equipment is 350 000 m^2.

Since their introduction into commercial applications, MF and UF ceramic membranes have made rapid progress in many areas, including food and beverage processing, biotechnology applications, and water treatment. Many factories can provide MF/UF ceramic membranes and modules, most of which are in the United States (CeraMem, Hillard, and CorpMempro), France (Novasep, Pall Exekia, and TAMI), Germany (Atech Innovations, Inopor, ITN Nanovation, Pall Schumacher, Westfalia Separation, and Membraflow), Japan (Kubota and NGK), and China (Jiuwu).

NF is a new, promising membrane separation process based on the pressure-driven technique. It is considered to be an intermediate between UF and RO. This procedure possesses the advantages of a simple course and energy efficiency, and it does not need a chemical change. Therefore, it can be used in the desalination of seawater, food engineering, and biotechnology. The pore size of an NF membrane is in the nanometer or subnanometer scale with a rejection for solutes in the range of 200–1000 $g \cdot mol^{-1}$, and the charge effect of the membrane surface enables it to exhibit high rejection toward multivalent ions (very low rejection properties can also be observed).

Ceramic NF membranes are important for harsh environments, such as extreme pH (>14 or <1), hydrocarbon solvent (especially aromatic hydrocarbons and oxidants), and corrosive conditions, where polymeric NF membranes are unstable. Gamma-alumina, titanium oxide, and zirconia are the most studied inorganic materials. However, at least two problems still exist for the commercialization of the ceramic NF membranes: (i) increased stability in extreme pH systems and (ii) a broader range of application of ceramic NF membranes. To solve these problems, a more efficient sol–gel fabrication route is needed.

For the porous ceramic membrane used in the PV process, silica membranes are the most popular. Titania, zirconia, silica/titania, and silica/zirconia composite membranes are also common. The pore diameter distribution of microporous ceramic membranes is wider than the pore diameter distribution of zeolite membranes. As a result, microporous ceramic membranes can principally be applied to a boarder range of alcohols during solvent dehydration and separation processes (e.g., methanol separated from methylbenzene (MTBE)). However, the temperature of the operation must be less than 80 °C because of the poor stability of silica-based membranes in the presence of water vapor. Recently, the University of Twente and the Energy Research Center of the Netherlands (ECN) developed a new type of organic–inorganic hybrid silica microporous membrane that provided both an excellent separation factor and a high flux toward the dehydration of butanol. More importantly, the properties of these membranes were stabile at 150 °C over 1000 days, which is very promising for industrial applications [3].

Inorganic-membrane-based GS offers an environment friendly alternative to conventional separation technologies because of its high energy efficiency and high performance (i.e., flux and selectivity at elevated temperatures) [4]. Silica membranes are one of the most promising molecular sieving materials because of their excellent separation performance with small gases such as H_2, N_2, O_2, and CH_4. This promising material has drawn a great deal of attention for its application in the separation of industrially relevant gas streams under harsh environments because of its enhanced stability. However, the modest hydrothermal stability of silica membranes severely restricts its broader industrial application, especially when humid atmospheres are encountered, such as in the steam reforming of hydrocarbons and water–gas shift reactions. Humidity can induce the densification of the microporous silica network, as explained by the breakage of siloxane bonds in the presence of steam, followed by the generation of silanol groups and subsequent recombination and rearrangement into a siloxane network [3].

Several approaches have been proposed for improving the hydrothermal stability of silica membranes [5–7]. First, transition-metal ions can be used as dopants to modify the amorphous silica matrix. For example, Igi et al. [6] found that doping with cobalt hampered the thermally induced molecular motion of the silica matrix, thereby increasing the hydrothermal stability of the silica membrane. Second, organic functional groups can be introduced into the silica network [8–11]. Duke et al. [8, 9] reported improved stability of carbonized silica membranes where the remaining carbon reduced the free motion of the silanol groups. Furthermore, de Vos et al. [10] and Wei et al. [11] also dispersed organic functional groups into a silica matrix to make it more hydrophobic. Another strategy to overcome the instability of silica membranes in the presence of steam is to substitute tetraethyl-ortho-silicate (TEOS) for other precursors, which are more hydrothermally stable (e.g., transition-metal alkoxides or bridged bis-silyl precursors) [12].

8.3
Clarification of Surface Water and Seawater Using Ceramic Membranes

Ceramic membranes are often used in chemical, fermentation, and food industries because of their excellent material properties. Although their high price limits their widespread application for water clarification, ceramic membranes could be effective as a separation technology for water clarification. In fact, Japan has used ceramic membranes to clarify drinking water. For example, NGK's honeycomb ceramic membranes were first used in a drinking water plant [13]. In this section, ceramic membrane applications in surface water and seawater are introduced.

8.3.1
Ceramic Membrane Microfiltration of Surface Water

The demand for the clarification of surface water is continually increasing, but it is difficult to produce high-quality drinking water with traditional methods of sedimentation and filtration. As a result, membrane technology is often used for producing drinking water. Ceramic membranes can be a particularly useful method that usually includes pretreatment and ceramic membrane MF.

Membrane fouling is one of the most significant issues affecting the development of membranes for the treatment of drinking water. Membrane fouling leads to lower membrane performance by decreasing the specific permeate flux. At present, new technologies such as ultrasound and ozonation are being used to reduce membrane fouling [14–16].

8.3.1.1 Pretreatment with Flocculation/Coagulation

Surface water contains small particles and natural organic matter (NOM) that would quickly block a porous membrane. Therefore, these substances should be removed before MF. Flocculation/coagulation is effective for removing suspended particles and NOM in surface water. Flocculation and coagulation can not only have a positive influence on MF permeate flux but also improve the water quality of the permeate. The results for flocculation or coagulation with different membrane pore sizes were investigated using ceramic membranes with pore sizes of 50, 200, 500, and 800 nm [17]. The steady flux of MF was observed to increase when coupled with online coagulation. For example, the steady flux increased from 225 to 640 $l \cdot m^{-2} \cdot h^{-1}$ when a ceramic membrane with a pore size of 500 nm was used. However, the flux increased slightly for the membrane pore size of 800 nm. The coagulation–MF experimental results revealed that the permeate flux and turbidity removal were higher with online coagulation than without.

The dose of the flocculant or coagulant had a significant influence on membrane filtration, and it varies for different water feeds [18]. Membrane fouling is much lower at an optimum dose (i.e., the optimum dose of flocculant for the NOM concentration of feed water). The aggregates formed at a low dose of flocculant are possibly smaller than those formed at a higher dose. However, a coagulant dose that is too low for neutralizing the charge of the organic matter might decrease aggregation and precipitation. This would result in higher pore blockage and lower filterability of the membrane. At a dose of flocculant higher than the optimum, aggregation is improved and material deposition on the membrane surface is increased.

8.3.1.2 Effect of Transmembrane Pressure (TMP) and Cross-Flow Velocity (CFV)

When MF was performed under a very low cross-flow velocity (CFV), the permeate flux decreased dramatically during the whole assay, and a steady-state was not reached [19]. The flux decrease was lower for higher velocities than for lower velocities, and fouling decreased when CFV increased. The effects of flow velocity on DOC removal and UV_{254} reduction were not significant.

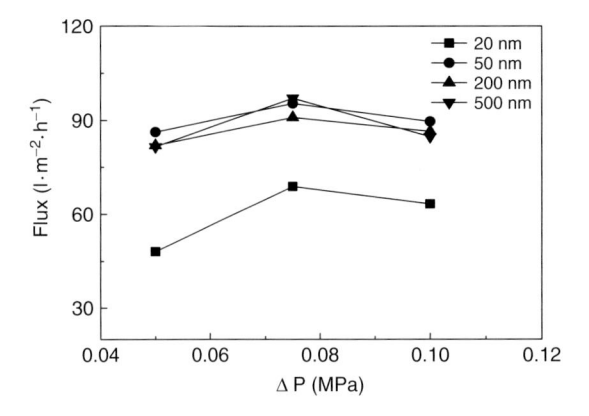

Figure 8.1 The relationship between TMP and pseudo-steady permeate flux at 25 °C.

Figure 8.1 displays the relationship between pseudo-steady permeate flux and transmembrane pressure (TMP) in our study. When the pressure changed from 50 to 75 kPa, the flux increased for all membranes. In contrast, when the pressure changed from 75 to 100 kPa, however, the pseudo-steady permeate flux declined slightly. Therefore, the critical pressure in this experiment was 75 kPa. Because the formed floc cake was compressible, its resistance significantly increased with increasing TMP. Furthermore, Ca^{2+} could make the cake more dense [20]. The concentration of Ca^{2+} in this feed water was 68.65 mg·l^{-1}, which might have resulted in the deposition of calcium.

8.3.1.3 Ultrasound Cleaning

To elucidate the effect of pH on the ultrasonic cleaning process, the zeta potential values of the particles and membranes were determined [18]. These potential values have a significant effect on membrane fouling. Zeta potential data provide a mechanistic insight into the role of solution conditions in ultrasonic cleaning. At the lowest pH investigated (3.5), the particles had the smallest zeta potential (least negative), while the membrane charge was highly positive. Therefore, particle–particle repulsive forces were lower than at higher pH values. It was also likely that the lower particle–particle repulsive forces resulted in a more compact cake structure. Consequently, greater force was required to remove individual particles from the cake layer than for higher pH values. Visual observations during the ultrasonic cleaning at pH 3.5 showed large, millimeter-scale particle aggregates detaching from the membrane surface. Once in solution, the flocs slowly disaggregated owing to the ultrasonically generated turbulence. The stronger particle–particle interactions and denser cake layer structure at this low pH apparently facilitated the patch removal of aggregates from the surface. This removal mechanism resulted in a very effective membrane cleaning with a cleaned flux ratio of 0.94, more than five times than that of the fouled flux.

8.3.1.4 Hybrid Ozonation–Ceramic Ultrafiltration

NOM has been implicated as a major foulant of UF membranes, which are used to treat natural waters. Ozone has a high oxidation potential and high reactivity with NOM and has been shown to reduce membrane fouling [19]. They found that under appropriate operating conditions, the hybrid ozonation–ceramic UF membrane system could significantly reduce membrane fouling in the treatment of natural waters. However, the fouling behavior was strongly dependent on ozone concentration and hydrodynamic conditions. Higher permeate flux was observed at higher levels of cross-flow velocity and ozone concentration and at a lower TMP. For example, the permeate flux decreased to ∼60% of the initial value after 12 h. When ozone was applied to the system, the permeate flux recovered to ∼95% of the initial permeate flux within a period of 30–40 min [20]. In addition, their experimental results showed that continuous rather than intermittent ozonation could be used in full-scale operations and that it resulted in considerable savings in capital and operating costs.

8.3.1.5 Ceramic Membrane Applications for Industrial-Scale Waterworks

In our study, when ceramic membranes were used for the purification of surface water in a full-scale operation, the flux of this process was higher than 100 $l \cdot m^{-2} \cdot h^{-1}$ for 37 days when the temperature decreased from 16 to 5 °C (Figure 8.2). Water viscosity was higher at low temperature, which resulted in a smaller flux. Furthermore, the TMP of this process ranged from 0.05 to 0.25 MPa. Therefore, this membrane process could operate stably over a long term.

8.3.2 Pretreatment of Seawater RO Using Ceramic Membranes

8.3.2.1 Effect of Operational Parameters

CFV Cross-flow velocity has a large effect on membrane fouling and concentration polarization. Some investigations have been conducted on pretreatment for

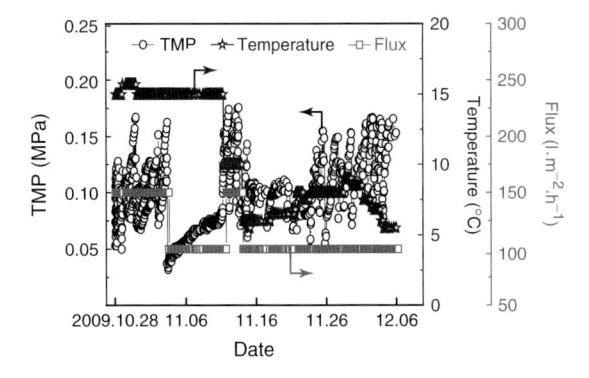

Figure 8.2 Ceramic membrane used for the purification of surface water in a full-scale operation.

the seawater reverse osmosis (SWRO) at the Huangdao power plant (Shandong province of China) [21]. The results indicated that the membrane was more susceptible to fouling at very low or very high CFV. At the optimum CFV of 3.7 m·s^{-1}, the normalized flux decreased to 96.7% of the initial flux after 20 min and then remained stable in the range of 96.5–96.8%. Therefore, proper hydraulic conditions could control the extent of membrane fouling.

TMPs The driving force of the membrane process increased with TMP until the filter cake was compacted and the filtration resistance increased dramatically [21]. The permeate flux increased by 16.9% when TMP increased from 0.1 to 0.18 MPa. However, the increase in permeate flux was negligible when TMP increased from 0.18 to 0.20 MPa. It was also related to the concept of "critical flux," which is used to understand and improve the operation of membrane filtration systems.

The turbidity of the permeate decreased with TMP, whereas the COD_{Mn} of the permeate decreased from 1.2 to 0.85 mg·l^{-1} when TMP increased from 0.10 to 0.18 MPa and then increased to 0.11 mg·l^{-1} above 18 MPa. This likely resulted from the different transport mechanisms for suspended solids and dissolved organic matter in the filtration of seawater. Because the size of suspended solids is larger than dissolved organic matter, the formed cake layer on the membrane surface might become a "prefilter." The density and thickness of the cake layer increased with TMP, which resulted in the lowest turbidity of 0.01 NTU for the permeate at 0.20 MPa.

8.3.2.2 Ceramic Membrane Application for the Industrial-Scale SWRO Plant

Ceramic membrane pretreatment was stably used for SWRO in Shengsi, China. Beyond 2800 h, the TMP of the ceramic membrane process was below 0.1 MPa. In our study, the economic evaluation of the industrial ceramic membrane process used for SWRO pretreatment is shown in Table 8.1. The capital cost of water treatment using a ceramic membrane was 730 RMB·m^{-3}·day^{-1}. The electricity consumption was in the range of 0.078–0.18 kWh with an average of 0.15 kWh. The quality of the water produced by the ceramic membrane was also stable. The reagent and human costs were well controlled below 0.16 RMB·m^{-3}.

8.4
Ceramic Membrane Application in the Microfiltration and Ultrafiltration of Wastewater

Oily wastewater is the main pollutant emitted in industrial and domestic sewage. It includes primarily salts and hydrocarbons that can be toxic to the environment [22–24]. In particular, the metal-rolling and metal-finishing industries produce a large volume of waste-soluble oil emulsions that need to be treated before being discharged into municipal sewage systems. Emulsions are defined as colloidal suspensions of a liquid within another liquid in droplet sizes of typically less than 20 μm, and waste oil emulsions generally contain oil (mineral, vegetable, or

Table 8.1 Economic evaluation for a 1000 $m^3 \cdot day^{-1}$ SWRO pretreatment plant.

Items		Value
Equipment parameters	Average water yield ($m^3 \cdot d^{-1}$)	1000
	Membrane area (m^2)	6.3
	Operating pressure (MPa)	0.04–0.15
Capital cost	Cost of equipment (thousands RMB)	690
	Cost of control system (thousands RMB)	20
	Pump (thousands RMB)	20
	Total cost (thousands RMB)	730
Running cost	Power consumption ($kWh \cdot m^{-3}$)	0.078–0.18
	Cost of labor ($RMB \cdot m^{-3}$)	0.08
	Cost of reagent ($RMB \cdot m^{-3}$)	0.08

synthetic), fatty acids, emulsifiers (anionic and nonionic surfactants), corrosion inhibitors (amines), bactericides, and many other chemicals.

The treatment of these waste emulsions has been a serious problem. So far, there are several techniques for oil separation. They typically include pH adjustment, gravity settling, centrifugal settling, filter coalescence, heating treatment, electrostatic coalesce, and membrane filtration. There are advantages and disadvantages associated with each of these techniques [22–25].

Membrane separation has been developed extensively and is becoming a promising technology. There are many reports on the use of membranes to treat waste emulsions because membranes have a high rate of oil removal and low energy cost compared to traditional treatment methods. This technology has several advantages, including a stable effluent quality and a small area requirement. Many studies on membrane separation for oily wastewater treatment have been reported, particularly in MF, UF, and RO with organic membranes [22–27]. However, inorganic membranes are used in industrial-scale applications because of their advantages of higher flux, higher porosity, and larger hydrophilic surface than organic membranes.

The resistance of ceramic membranes to mechanical, thermal, and chemical stress allows for a better recovery of membrane performance. For example, Yang et al. used a zirconia membrane and three different alumina membranes to separate an oil–water emulsion obtained from steelworks in the evaluation of their permeability and separation. The results showed the same oil rejection efficiency but the flux of zirconia was higher than others [28]. Wang et al. used alumina and zirconia MF membranes to treat waste rolling emulsion. The results showed that the zirconia membranes gave a higher flux, but the rejection efficiency for both was the same [29]. Kwan and Yeung also used alumina and ZrO_2 membranes to treat an oil emulsion. The results showed that there was a similar rejection of oil but a higher flux for ZrO_2 [30]. Furthermore, Muller et al. [31] studied two ceramic membranes (α-Al_2O_3) in the treatment of an oily water Hueneme field in California. The oil removal efficiencies were 98–99%. Moreover, Zhong et al. [32] studied the performance of MF using a 0.2 μm ZrO_2 ceramic membrane combined with traditional chemical flocculation as pretreatment. The resulting oil removal efficiencies were 99.4–99.9%. Finally, Abadi et al. [22] chose an α-Al_2O_3-based ceramic membrane because of its excellent performance in removing oil, acceptable permeate flux, and cheaper price. A review of this body of work is presented in the following sections.

8.4.1
Microstructure of the Membranes

The study of the membrane structure includes two parts: porosity and pore size. As the preparation of ceramic membranes requires a fixed high-temperature sintering procedure, commercial membranes have almost the same porosity. Although Benito et al. [33] investigated the treatment of oily emulsion wastewater using a γ-Al_2O_3 with very high porosity (~70%) and obtained comparable results, the resistance of the chemical in the material hindered the application because chemical cleaning was required. Therefore, investigations mainly focus on the impact of membrane pores during oil–water separation. The mechanism for filtration of oily wastewater by the membrane is the rejection of suspended particles and oil droplets. This functions to guarantee the retention rate, but a higher flux is needed. It is generally believed that the larger the pore size, the smaller the membrane resistance, and the higher the flux. Meanwhile, the smaller the rejection rate, the smaller the pore size and the greater the resistance of the membrane, the lower the flux, and the higher the rejection [34]. However, it has been found that the relationship between particle distribution in the liquid and the membrane pore size is very complex and that there is an optimal membrane pore size for every different treatment system. The choice of the pore size is therefore especially important for the process.

The results for the rejection of alumina membranes with different pore sizes are shown in Table 8.2 [28]. The table shows that pore sizes ranging between 0.2 and 0.01 μm attained 99% oil retention. As 1.0 μm pore size alumina membrane is larger, the small droplets could penetrate the membrane quite easily. Figure 8.3 shows the relationship between the pore size and the permeate flux when the emulsions were treated under a flow rate of 7 m·s^{-1}, a filtration pressure of 0.1 MPa, and a temperature of 12 °C. The 1 μm membrane flux decreased rapidly

Table 8.2 Separation results obtained with different membranes [28].

Material	Pore size (μm)	Oil concentration of permeate (ppm)	Rejection coefficient (%)
α-Al_2O_3	0.2	<10	99.9
γ-Al_2O_3	0.01	<10	99.8
α-Al_2O_3	1.0	313	94.3

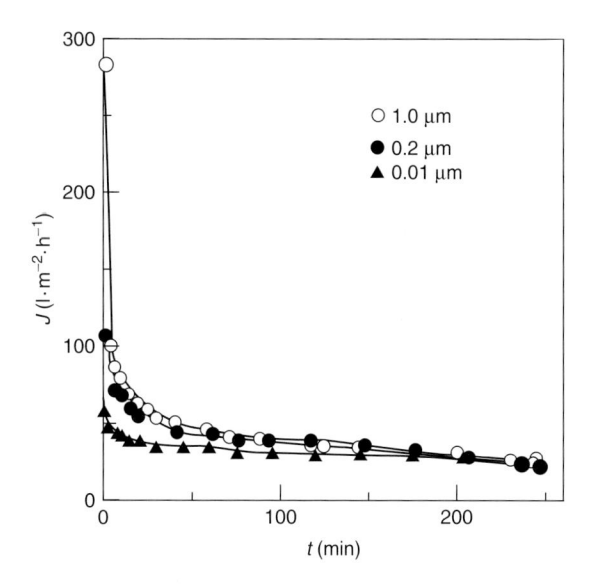

Figure 8.3 The relationship between flux and pore size [28].

owing to the larger pore size and displayed more significant pore clogging. The flux of the membrane with a pore size of 0.01 μm was more stable and less prone to clogging. Furthermore, the three ceramic membranes were stable at ∼25 $l \cdot m^{-2} \cdot h^{-1}$.

Mueller et al. [31] found that a film was formed on the membrane surface during the filtration process and determined that this film affected membrane resistance and flux. In general, the oil layer was tens of micrometers thick, while the thickness of the oil layer was determined by the characteristics of the membrane material. As a result, the layer resistances of the filtration membranes with different pore sizes were close to the overall resistance and reached a similar stable flux. Therefore, the performance of the membrane material is critical in rolling emulsion for ceramic membrane development and design.

8.4.2
Surface Properties of Ceramic Membranes

Ceramic membranes fabricated with different materials have different membrane surface properties, which lead to obvious differences in the force between the droplets and the membrane surface. Currently, the commercially available varieties of ceramic membranes are Al_2O_3, TiO_2, ZrO_2, and composite membranes. These varieties have been investigated for membrane surface properties, focusing on wettability and electrical properties [35, 36].

8.4.2.1 Wettability
The wettability of materials is measured by the contact angle. When the solid–liquid contact angle is larger and the ratio of wetting liquid to solid is smaller, it is more

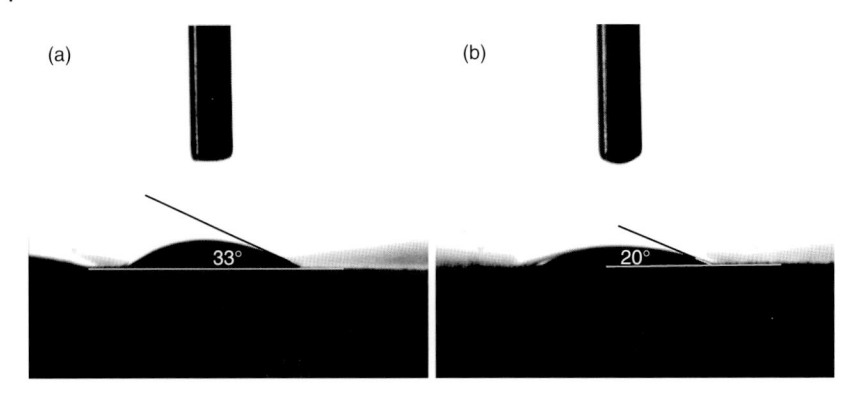

Figure 8.4 Contact angle of the water droplets on the dense Al_2O_3 disk (a) and the ZrO_2 coating (b).

difficult to spread the wetting liquid on the solid surface [37]. When the material is of a contact angle less than 90°, it is defined as a hydrophilic material and is generally used in the separation of O/W emulsions. Zhou et al. [38] reported that commercial Al_2O_3 MF membranes could be modified with a nanosized ZrO_2 coating to reduce membrane fouling by oil droplets and keep the unique character of the inorganic membrane. Figure 8.4 shows the shape of a water droplet on this ZrO_2 coating and the dense alumina disk, respectively. As seen, the apparent contact angle on the ZrO_2 coating is 20°, which is lower than the dense alumina disk (33°). This implies that the modified membrane becomes more hydrophilic when large amounts of surface hydroxyl (–OH) groups exist in the nanosized ZrO_2 coating.

Modified membranes have a stable performance in the oil–water separation of stable emulsions. For example, we determined the wettability of porous ceramic membranes by the dynamic contact angle method (Figure 8.4). The contact angle decreased with the spreading and addition of water droplets on the membrane surface. The Al_2O_3–TiO_2 composite membrane had a lower initial contact angle and a more rapid rate of decrease in the contact angle than the Al_2O_3 membrane (Figure 8.5). The results indicated that TiO_2 doping improved membrane hydrophilicity and provided more hydrophilic groups on the membrane surface. Variations in permeate flux for the different membranes are illustrated in Figure 8.6. The initial permeate flux decreased significantly for both membranes. A higher and more stable flux was observed for the Al_2O_3–TiO_2 composite membrane (605 $l·m^{-2}·h^{-1}$) than for the Al_2O_3 membrane (465 $l·m^{-2}·h^{-1}$). The main reason for this was that the Al_2O_3–TiO_2 composite membrane had more hydrophilic groups, reducing the adsorption of oil droplets.

8.4.2.2 Surface Charge Properties

Porous ceramic membranes made of mineral oxides in an aqueous solution are electrically charged on account of the amphoteric behavior of surface hydroxyl groups [40, 41]. It is now recognized that the surface electrochemical properties of ceramic membranes play a significant role in their separation performance

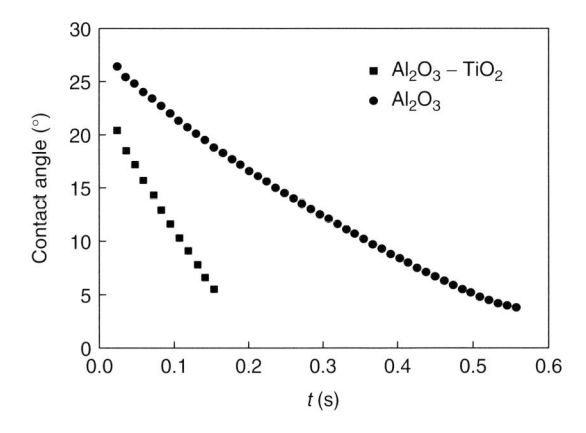

8.4 Ceramic Membrane Application in the Microfiltration and Ultrafiltration of Wastewater

Figure 8.5 Time dependence of the contact angle for Al_2O_3 and Al_2O_3–TiO_2 membranes [39].

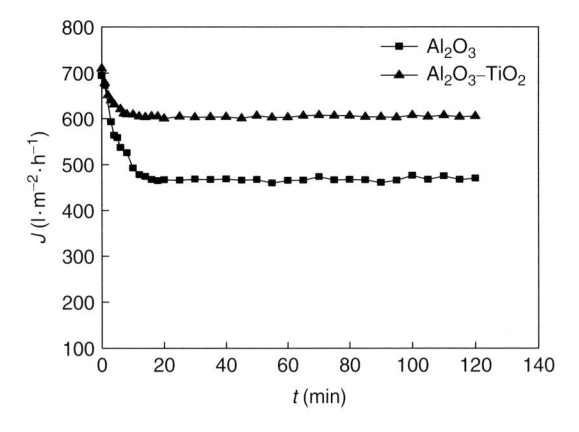

Figure 8.6 Variation of the permeate flux for Al_2O_3 and Al_2O_3–TiO_2 membranes versus filtration time [39].

[42, 43]. The selection of a membrane with suitable charge properties can lead to the optimization of existing processes [44]. The nature of the material determines the surface charge properties of the membrane materials. The electrochemical properties of the membrane surface are frequently characterized in terms of zeta potential, isoelectric point (IEP), and point of zero charge (PZC). Among them, the IEP and PZC are important parameters that reflect the acidic or basic nature of the membrane surface [45]. In the absence of any specific adsorption, the values of IEP and PZC are equal [46, 47]. Figure 8.7 shows the IEP of the Al_2O_3 and ZrO_2 ceramic membranes, 6.6 and 5.8, respectively [48].

Figure 8.8 shows a comparison of filtration using the alumina and zirconia membranes with a pore size of 0.2 μm at 12 °C with a pressure of 0.1 MPa and a membrane surface flux speed of 7 m·s^{-1}. The zirconia membrane maintained a high flux, primarily because the oil in the water emulsion droplet surface is usually

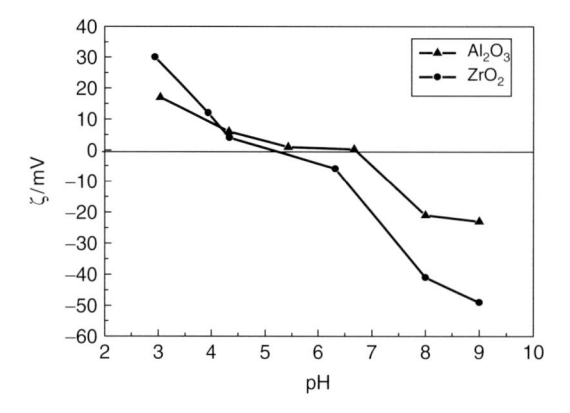

Figure 8.7 pH dependence of ζ of Al_2O_3 and ZrO_2 membranes [48].

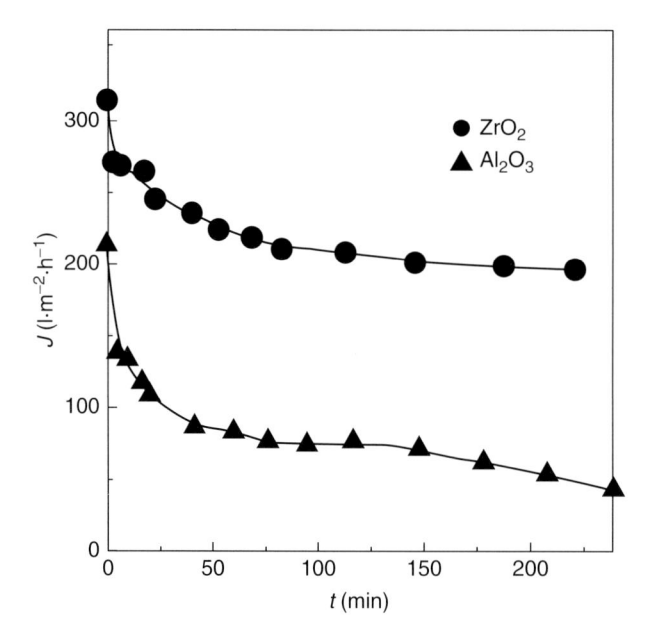

Figure 8.8 Variation of the permeate flux for Al_2O_3 and ZrO_2 membranes versus filtration time [28].

negatively charged [28]. The use of a membrane with a lower isoelectric point helps reduce the oil on the membrane, reducing surface fouling and leading to a sustained higher flux.

8.4.2.3 Technical Process

The pretreatment of raw emulsion has a significant effect on the MF process. As such, Figure 8.9 shows the effect of NaOH pretreatment on the flux and rejection of

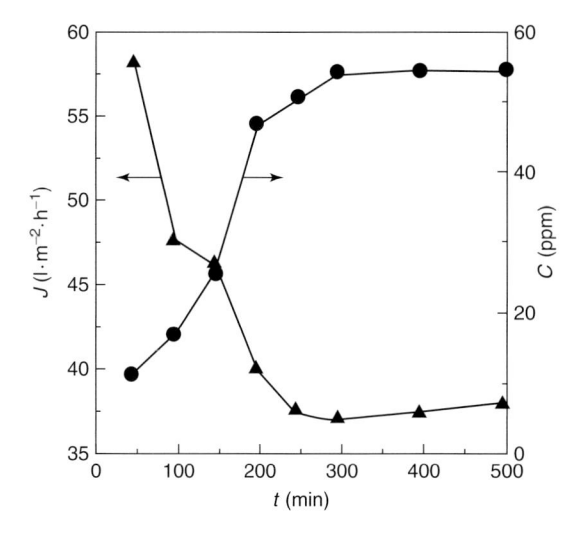

8.4 Ceramic Membrane Application in the Microfiltration and Ultrafiltration of Wastewater

Figure 8.9 The influence of pretreatment on the flux and retention [29].

the membrane [29]. When the amount of the pretreatment was increased up to 300 ppm, the flux clearly increased and the retention improved. The analysis showed that when the emulsion droplet size became larger after pretreatment (from 1 to 10–20 μm), membrane performance improved.

The influence of pressure on the membrane flux under different flow rates is shown in Figure 8.10 [29]. As seen, there is a critical pressure for the filtration of

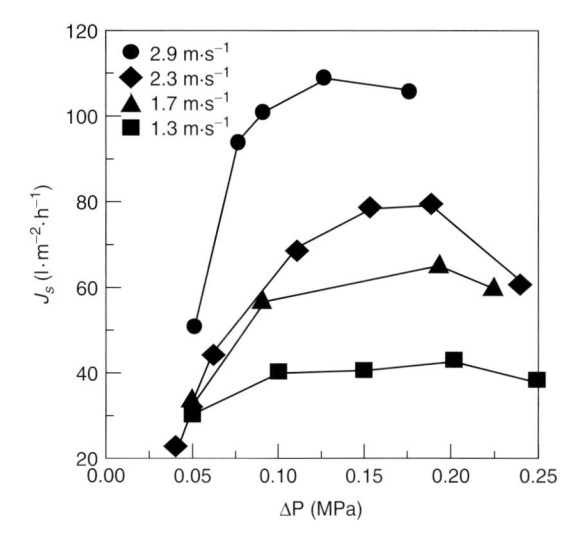

Figure 8.10 The influence of pressure on the membrane flux under different flow rates [29].

Table 8.3 The relationship between pressure and rejection [48].

Pressure (MPa)	Oil content in permeate (ppm)	Retention (%)
0.07	4.0	99.9
0.10	4.7	99.9
0.15	6.7	99.9
0.20	7.4	99.8

the oily wastewater. The flux increased with the pressure when operation pressures were lower than the optimum and the process was pressure controlled. However, aside from the critical drop in pressure, the flux is not significantly improved by pressure increases, as higher pressure makes it easier to push the oil droplets into the membrane pores. This causes pore congestion and increases the mass transfer resistance and decreases the flux.

The oil droplets may be pressed into the membrane pores and then penetrate through the pores into the other side of the membrane, when the pressure is increased above a threshold value; because the oil droplets are compressible, it leads to lower retention rates. However, the pore congestion could reduce the pore size and improve the rejection of the membrane. Therefore, it is necessary to investigate rejection under different pressures. Table 8.3 shows the relationship between the pressure and the rejection after 2 h of filtration. According to the results, gradually increasing pressure to 0.15 MPa is a recommended mode of operation for industrial applications in such systems.

As for the operational mode of the cross-flow filtration, the fluid shear stress could effectively reduce membrane surface deposition and concentration polarization effects. It is generally believed that flux increases as the membrane surface velocity increases, but the influence of the increased velocity and energy consumption on membrane performance must be considered. Figure 8.11 shows the relationship between membrane flux and velocity [29]. When the velocity was below 5 m·s^{-1}, the flux increased almost proportionally to the velocity and then gradually slowed. When the flow rate was more than 6 m·s^{-1}, the flux declined.

The influence of temperature on the filtration process is more complex. With increasing temperature, the viscosity of the liquid drops and the diffusion coefficient increases. The impact is that concentration polarization is reduced. However, increased temperature also changes some of the liquid's properties, such as the solubility of some components in the feed solution, which changes the formation of pollution. Increasing the temperature can reduce the viscosity of the liquid, which increases the flux. The temperature also affects the size distribution of the oil drops, with smaller oil droplets leading to lower flux and creating an optimized operating temperature. Figure 8.12 shows that the flux was proportional to the temperature [29]. The relationship between the temperature and the rejection is

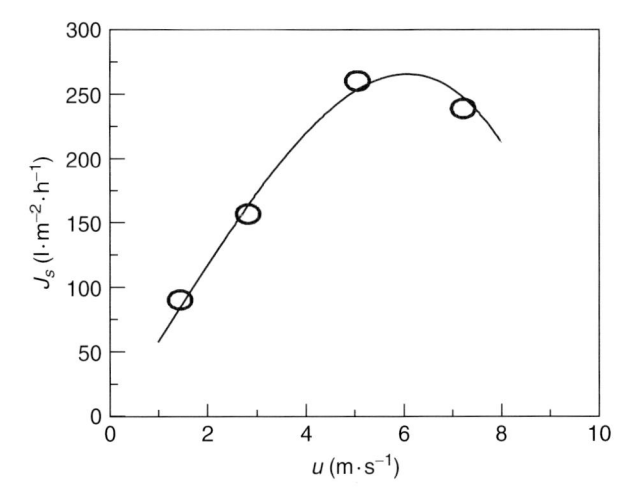

Figure 8.11 The relationship between membrane flux and velocity [48].

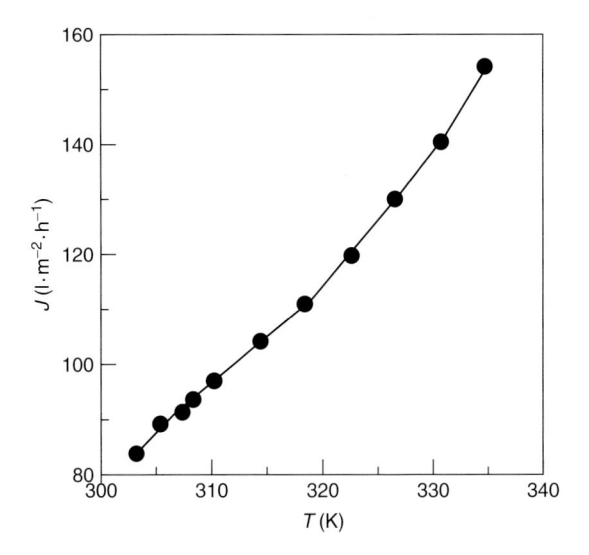

Figure 8.12 The effect of temperature on flux [29].

shown in Table 8.4 [29]. The influence of temperature on retention was low, with rejection decreasing only slightly as temperature increased.

In industrial applications, filtration is usually an enrichment process, so it is necessary to consider the different concentration ratios of the process. Figure 8.13 shows that the concentration had little effect on the flux after 150 min of operation [48]. Furthermore, the relationship between the concentration ratio and the retention rate is shown in Table 8.5 [48]. The rejection rate changed little when the concentration ratio increased, which is more favorable for wastewater treatment processes.

8 Application of Ceramic Membranes in the Treatment of Water

Table 8.4 The influence of temperature on rejection [29].

Temperature (°C)	Oil content in permeate (ppm)	Rejection (%)
16.5	6.7	99.9
25	8.8	99.8
35	7.8	99.8
45	9.2	99.8

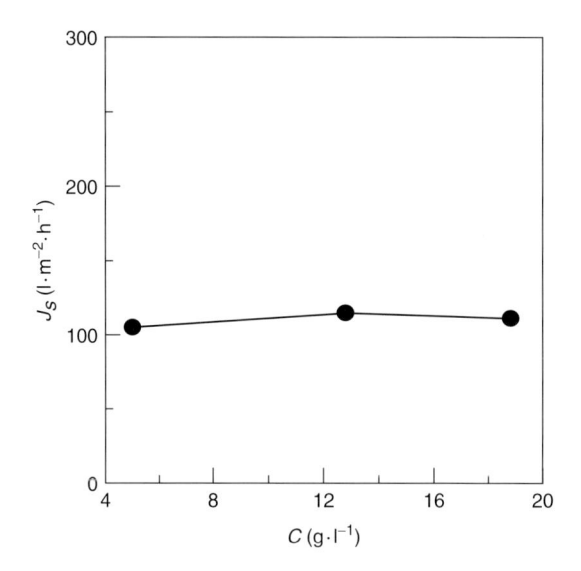

Figure 8.13 The effect of concentration on flux [48].

Table 8.5 The relationship between concentration ratio and rejection [48].

Concentrate ratio	Oil content in permeate (ppm)	Rejection (%)
1	6.7	99.9
2.8	7.4	99.9
3.8	7.8	99.8

8.4.2.4 Cost

Using ceramic membrane equipment to treat emulsion wastewater has an effect comparable to the use of organic membrane equipment. However, organic membrane devices cost four times more to import than ceramic membrane devices. Furthermore, organic membrane devices require 10 times more energy than that of the ceramic membrane devices, and the normal equipment and pharmaceutical costs of organic membrane devices are more than the cost of importing the equipment. The overall cost of treating wastewater with imported devices is thus approximately seven times more for organic membrane devices than ceramic membrane devices if equipment depreciation is taken into account [48].

8.5
Conclusions and Prospects

Increasing attention has been given to the use of ceramic membrane processes for the treatment of industrial wastewater and surface water and the desalination of seawater containing organic matter. These processes have many remarkable advantages, including superior water quality, removal of a wide range of contaminants, easier control of operation, and the capacity to save space. A number of new applications for ceramic membranes are, therefore, likely to arise in the near future. In the short term, it is expected that ceramic NF membranes and ceramic membrane contactors that can be applied to liquid treatment (including biological membrane reactors) will be developed. Ceramic NF membranes with high performance parameters, such as low cutoff values or high fluxes, can only be obtained in an asymmetric multilayer configuration. The requirements for developing such a multilayer configuration includes the shaping of appropriate support material, the formation of mesoporous interlayers, and the synthesis of a microporous top layer with a cutoff value below 1000 Da. Alumina, titania, zirconia, and silica are considered the main ceramic materials for the formation of multilayer structures.

References

1. Li, K. (2007) *Ceramic Membranes for Separation and Reaction*, John Wiley & Sons, Ltd, Chichester.
2. Burggraaf, A.J. and Cot, L. (1996) *Fundamental of Inorganic Membrane Science and Technology*, Elsevier, Amsterdam.
3. Castricum, H.L., Sah, A., Kreiter, R., Blank, D.H.A., Vente, J.F., and ten Elshof, J.E. (2008) Hybrid ceramic nanosieves: stabilizing nanopores with organic links. *Chem. Commun.*, 1103–1105.
4. Ockwig, N.W. and Nenoff, T.M. (2007) Membranes for hydrogen separation. *Chem. Rev.*, **107**, 4078–4110.
5. Fotou, G.P., Lin, Y.S., and Pratsinis, S.E. (1995) Hydrothermal stability of pure and modified microporous silica membranes. *J. Membr. Sci.*, **30**, 2803–2808.
6. Igi, R., Yoshioka, T., Ikuhara, Y.H., Iwamoto, Y., and Tsuru, T. (2008) Characterization of Co-doped silica for improved hydrothermal stability and application to hydrogen separation membranes at high temperatures. *J. Am. Ceram. Soc.*, **91**, 2975–2981.
7. Kanezashi, M. and Asaeda, M. (2006) Hydrogen permeation characteristics and stability of Ni-doped silica membranes in steam at high temperature. *J. Membr. Sci.*, **271**, 86–93.
8. Duke, M.C., da Costa, J.C.D., Do, D.D., Gray, P.G., and Lu, G.Q. (2006) Hydrothermally robust molecular sieve silica for Wet Gas separation. *Adv. Funct. Mater.*, **16**, 1215–1220.
9. Duke, M.C., da Costa, J.C.D., Lu, G.Q.M., Petch, M., and Gray, P. (2004) Carbonised template molecular sieve silica membranes in fuel processing

systems: permeation, hydrostability and regeneration. *J. Membr. Sci.*, **241**, 325–333.

10. de Vos, R.M., Maier, W.F., and Verweij, H. (1999) Hydrophobic silica membranes for gas separation. *J. Membr. Sci.*, **158**, 277–288.

11. Wei, Q., Wang, F., Nie, Z.-R., Song, C.-L., Wang, Y.-L., and Li, Q.-Y. (2008) Highly hydrothermally stable microporous silica membranes for hydrogen separation. *J. Phys. Chem. B*, **112**, 9354–9359.

12. Qi, H., Han, J., Xu, N.P., and Bouwmeester, H.J.M. (2010) Hybrid organic–inorganic microporous membranes with high hydrothermal stability for the separation of carbon dioxide. *ChemSusChem*, **12**, 1375–1378.

13. A Lercha, S Panglisch, P Buchta, Y Tomita, H Yonekawa, K Hattori, R Gimbel. Direct river water treatment using coagulation/ceramic membrane microfiltration, *Desalination*, **179**(1–3): 41–50.

14. Lamminen, M.O., Walker, H.W., and Weavers, L.K. (2006) Effect of fouling conditions and cake layer structure on the ultrasonic cleaning of ceramic membranes. *Sep. Sci. Technol.*, **41**, 3569–3584.

15. Kim, J., Davies, S.H.R., Baumann, M.J., Tarabara, V.V., and Masten, S.J. (2008) Effect of ozone dosage and hydrodynamic conditions on the permeate flux in a hybrid ozonation-ceramic ultrafiltration system treating natural waters. *J. Membr. Sci.*, **311**, 165–172.

16. Karnik, B.S., Davies, S.H.R., Chen, K.C., Jaglowski, D.R., Baumann, M.J., and Masten, S.J. (2005) Effects of ozonation on the permeate flux of nanocrystalline ceramic membranes. *Water Res.*, **39**, 728–734.

17. Li, W.X., Zhou, L.Y., Xing, W.H., and Xu, N.P. (2010) Coagulation-microfiltration for lake water purification using ceramic membranes. *Desalin. Water Treat.*, **18**, 239–244.

18. Meyn, T. and Leiknes, T. (2010) Comparison of optional process configurations and operating conditions for ceramic membrane MF coupled with coagulation/flocculation pre-treatment for the removal of NOM in drinking water production. *J. Water Supply Res. Technol.-Aqua*, **59**, 81–91.

19. de la Rubia, A., Rodriguez, M., and Prats, D. (2006) pH, ionic strength and flow velocity effects on the NOM filtration with TiO_2/ZrO_2 membranes. *Sep. Purif. Technol.*, **52**, 325–331.

20. da Costa, A.R. and Pinho, M.N. (2006) Menachem elimelech mechanisms of colloidal natural organic matter fouling in ultrafiltration. *J. Membr. Sci.*, **281**, 716–725.

21. Xu, J., Chang, C.Y., and Gao, C.J. (2010) Performance of a ceramic ultrafiltration membrane system in pretreatment to seawater desalination. *Sep. Purif. Technol.*, **75**, 165–173.

22. Mohmmadi, T., Kazemimoghadam, M., and Saadabadi, M. (2003) Modeling of membrane fouling and decline in reverse osmosis during separation of oil in water emulsions. *Desalination*, **157**, 369–375.

23. Cui, J., Zhang, X., Liu, H., Liu, S., and Yeung, K.L. (2008) Preparation and application of zeolite/ceramic microfiltration membranes for treatment of oil contaminated water J o M S. Preparation and application of zeolite/ceramic microfiltration membranes for treatment of oil contaminated water. *J. Membr. Sci.*, **325**, 420–426.

24. Hua, F.L., Tsang, Y.F., Wang, Y.J., Chan, S.Y., Chua, H., and Sin, S.N. (2007) Performance study of ceramic microfiltration membrane for oily wastewater treatment. *Chem. Eng. J.*, **128**, 169–175.

25. Abadi, S.R.H. (2011) Ceramic membrane performance in microfiltration of oily wastewater. *Desalination*, **265**, 222–228.

26. Ebrahimi, M., Ashaghi, K.S., Engel, L., Willershausen, D., Mund, P., Bolduan, P., and Czermak, P. (2009) Characterization and application of different ceramic membranes for the oil-field produced water treatment. *Desalination*, **245**, 533–540.

27. Barredo-Damas, S., Alcaina-Miranda, M.I., Bes-Piá, A., Iborra-Clar, M.I., Iborra-Clar, A., and Mendoza-Roca, J.A.

(2010) Ceramic membrane behavior in textile wastewater ultrafiltration. *Desalination*, **250**, 623–628.
28. Yang, C., Zhang, G., Xu, N., and Shi, J. (1998) Preparation and application in oil–water separation of $ZrO_2/\alpha\text{-}Al_2O_3$ MF membrane. *J. Membr. Sci.*, **142**, 235–243.
29. Wang, P., Xu, N., and Shi, J. (2000) A pilot study of the treatment of waste rolling emulsion using zirconia microfiltration membranes. *J. Membr. Sci.*, **173**, 159–166.
30. Kwan, S.M. and Yeung, K.L. (2008) Zeolite micro-fuel cell. *Chem. Commun.*, **31**, 3631–3633.
31. Mueller, J., Cen, J., and Davis, R.H. (1997) Crossflow microfiltration of oily water. *J. Membr. Sci.*, **129**, 221–235.
32. Zhong, J., Sun, X., and Wang, C. (2003) Treatment of oily wastewater produced from refinery processes using flocculation and ceramic membrane filtration. *Sep. Purif. Technol.*, **32**, 93–98.
33. Benito, J.M., Sánchez, M.J., Pena, P., and Rodríguez, M.A. (2007) Development of a new high porosity ceramic membrane for the treatment of bilge water. *Desalination*, **214**, 91–101.
34. Bhave, R.R. and Fleming, H.L. (1988) Removal of oily contaminants in wastewater with microporous alumina membrane. *AIChE Symp. Ser.*, **84**, 19–27.
35. Kawakatsu, T., Trägårdh, G., and Trägårdh, C. (2001) The effect of the hydrophobicity of microchannels and components in water and oil phases on droplet formation in microchannel water-in-oil emulsification. *Colloids Surf., A: Physicochem. Eng. Aspects*, **179**, 29–37.
36. Huisman, I., Trägårdh, G., and Trägårdh, C. (1999) Determining the zeta potential of microfiltration membranes using the electroviscous effect. *J. Membr. Sci.*, **156**, 153–158.
37. Duan, S. and Tan, Y. (1990) *Interface Chemistry*, Higher Education Press, Beijing.
38. Zhou, J.-E., Changa, Q., Wang, Y., Wang, J., and Meng, G. (2010) Separation of stable oil–water emulsion by the hydrophilic nano-sized ZrO_2 modified Al_2O_3 microfiltration membrane. *Sep. Purif. Technol.*, **75**, 243–248.
39. Zhang, Q., Fan, Y., and Xu, N. (2009) Effect of the surface properties on filtration performance of $Al_2O_3\text{-}TiO_2$ composite membrane. *Sep. Purif. Technol.*, **66**, 306–312.
40. Herbig, R., Arki, P., Tomandl, G., and Bräunig, R.E. (2003) Comparison of electrokinetic properties of ceramic powders and membranes. *Sep. Purif. Technol.*, **32**, 363–369.
41. Theodoly, O., Cascão-Pereira, L., Bergeron, V., and Radke, C.J. (2005) A combined streaming-potential optical reflectometer for studying adsorption at the water/solid surface. *Langmuir*, **21**, 10127–10139.
42. Hunter, R.J. (1981) *Zeta Potential in Colloid Science: Principles and Applications*, Academic Press, London.
43. Fievet, P., Szymczyk, A., Aoubiza, B., and Pagetti, J. (2000) Evaluation of three methods for the characterization of the membrane-solution interface: streaming potential, membrane potential and electrolyte conductivity inside pores. *J. Membr. Sci.*, **168**, 87–100.
44. Bowen, W.R. and Cao, X. (1998) Electrokinetic effects in membrane pores and the determination of zeta-potential. *J. Membr. Sci.*, **140**, 267–273.
45. Mullet, M., Fievet, P., Szymczyk, A., Foissy, A., Reggiani, J.-C., and Pagetti, J. (1999) A simple and accurate determination of the point of zero charge of ceramic membranes. *Desalination*, **121**, 41–48.
46. Barthés-Labrousse, M.-G. (2002) Acid–base characterization of flat oxidecovered metal surfaces. *Vacuum*, **67**, 385–392.
47. Franks, G.V. and Meagher, L. (2003) The isoelectric points of sapphire crystals and alpha-alumina powder. *Colloids Surf., A: Physicochem. Eng. Aspects*, **214**, 99–110.
48. Xu, N. (2005) *Process-Oriented Design, Preparation and Application of Ceramic Membranes*, Science Press, Beijing.

9
Functional Zeolitic Framework Membranes for Water Treatment and Desalination

Bo Zhu, Bin Li, Linda Zou, Anita J. Hill, Dongyuan Zhao, Jerry Y. S. Lin, and Mikel Duke

9.1
Introduction

Inorganic membranes are suitable for many applications in the chemical and pharmaceutical industry or in water and wastewater treatments, especially for high temperature, extreme acidity or alkalinity, and high-pressure operations, which preclude the use of existing polymeric membranes. Figure 9.1 shows a simplified schematic of the membrane separation process [1]. When the medium to be separated flows through the channels of the membrane, particles/molecules with a size exceeding the radius of the pores of membranes are retained, while the others permeate through. Pressure or concentration gradient across the membrane is usually the driving force for the separation. Recent studies on inorganic membranes have been mainly focusing on three types of membranes: zeolite membranes, sol-gel-based microporous membranes, and Pd-based and perovskite-like dense membranes [2]. The focus in this chapter is specifically on zeolite membranes.

Zeolites have been applied in many scientific and technological fields because of their unique properties such as ion exchange, absorption, and catalytic properties. Their widespread applications include air components separation, paraffin hydrocarbons separation and recovery, radioactive waste treatment, hydrocarbon catalytic reactions, and air pollutants removal [3].

Zeolites are a group of compounds, naturally formed or synthesized, with open structures, which may incorporate a range of small inorganic and organic species. The frameworks of the zeolitic materials, which form the channels and cavities, are constructed from tetrahedral groups (e.g., AlO_4, SiO_4, PO_4, BeO_4, GaO_4, GeO_4, and ZnO_4) linked to each other by sharing all of the oxygen atoms [4]. The most common zeolites are crystalline, hydrated aluminosilicates. They are based on AlO_4 and SiO_4 tetrahedra linked together to form a three-dimensional network having pores of molecular dimensions comparable to many chemical substances [3–5]. Figure 9.2 presents the structure of MFI-type zeolite, which has orthorhombic crystal symmetry (Figure 9.2a), and nearly cylindrical, 10-member ring channels

Functional Nanostructured Materials and Membranes for Water Treatment, First Edition.
Edited by Mikel Duke, Dongyuan Zhao, and Raphael Semiat.
© 2013 Wiley-VCH Verlag GmbH & Co. KGaA. Published 2013 by Wiley-VCH Verlag GmbH & Co. KGaA.

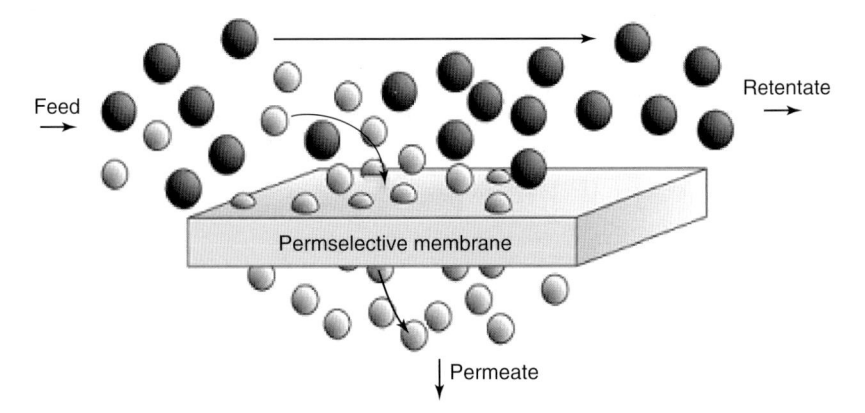

Figure 9.1 Simplified concept schematic of membrane separation [1]. (Source: Reprinted from Journal of Colloid and Interface Science, 314, Lu et al., Inorganic membranes for hydrogen production and purification: a critical review and perspective, 589–603. Copyright (2007), with permission from Elsevier.)

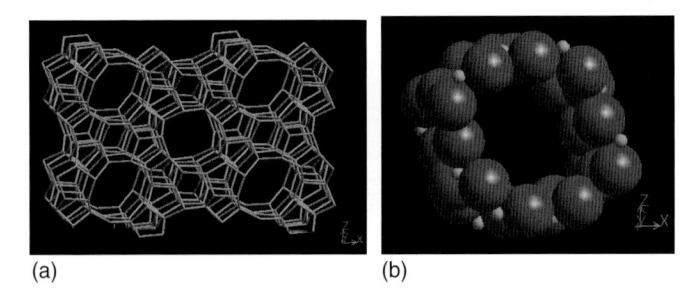

Figure 9.2 Structure of MFI-type zeolite. (a) Framework of MFI-type zeolite viewed along [010] and (b) complex of 10 rings viewed along [010] [6].

(Figure 9.2b) [6]. A pure silica (SiO_2) framework without any defects is neutral, but it becomes negatively charged when aluminum is present in the framework. The negative framework charge is balanced by cations that are relatively mobile and can usually be exchanged by other cations [3, 5]. The remarkably porous crystalline aluminosilicate structure of zeolites has led to their wide application as molecular sieves for the separation of gases, liquids, and so on [7]. For zeolites to be used as molecular sieves, their structure must remain intact after the completion of dehydration [3].

Over the past decade, supported zeolite layers have been studied extensively for many potential applications such as separation membranes, catalytic membrane reactors, and chemical sensors [8]. The crystalline zeolite membranes have a porous structure with well-defined pores at the molecular level and are extremely thermally, chemically, and mechanically stable [9]. The studies of zeolite membranes have been mainly focused on gas separation and liquid pervaporation processes [10].

Molecular sieving and competitive adsorption and diffusion are the general separation mechanisms involved in zeolite membranes for these processes [11–13]. Recently, several research groups have explored the possibility of using zeolite membranes for water treatment. This chapter mainly focuses on the developments and progress in zeolite membranes for water treatment. We hope that this review is beneficial to researchers in the field.

9.2
Preparation of Zeolite Membranes

Figure 9.3 shows the different approaches for the preparation of zeolite membranes [2, 14]. Among these strategies, the most commonly used and the most promising technique is the so-called composite membrane preparation [2]. The main techniques used for the synthesis of composite zeolite membranes are hydrothermal crystallization, which can be classified into two categories: direct *in situ* crystallization and secondary (seeded) growth crystallization [2]. In addition to these two main methods, a "dry" synthesis technique can also be used for the preparation of supported zeolite layers [15–18]. Recently, a microwave-assisted method has been increasingly popular in the synthesis of zeolite membranes. Some novel synthesis

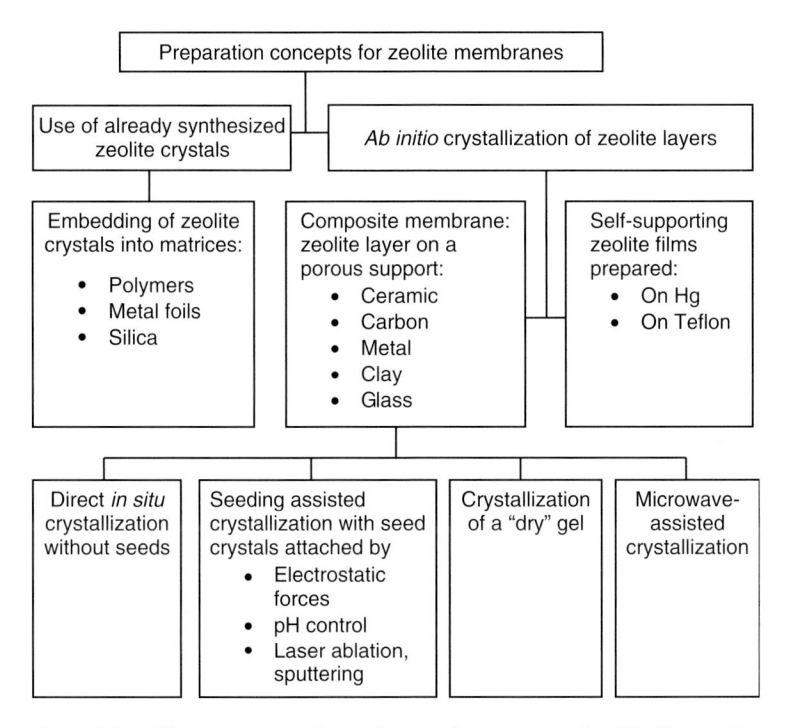

Figure 9.3 Different strategies for zeolite membrane preparation [2]. (Source: Reprinted from Microporous Mesoporous Mater., 38, Caro *et al.*, Zeolite membranes–state of their development and perspective, 3–24. Copyright (2000), with permission from Elsevier.)

routes could be created by combining the microwave technique with the strategies mentioned earlier [14].

The supports used for the preparation of zeolite membrane coatings include ceramic materials, such as alumina, zirconia, and titania, metals such as stainless steel, and other porous materials such as carbon. Among them, alumina is the most commonly used support for zeolite membranes. It is known that alumina has a certain effect on zeolite membrane composition, which has an impact on the separation properties of the membrane [15, 19, 20].

9.2.1
Direct *In situ* Crystallization

The direct *in situ* technique is a one-step method that uses the standard hydrothermal synthesis to crystallize a thin zeolite top layer on top of the support or inside the pores of the support (pore plugging) [2]. The *in situ* crystallization technique is commonly used for the preparation of supported MFI-type zeolite membranes [20–22]. In this approach, an MFI-type zeolite top layer on the surface of a porous support is formed by bringing the support in contact with a precursor solution or gel and followed by hydrothermal treatment under certain concentration and temperature ranges. Figure 9.4a shows a schematic orientation of microcrystals in an MFI-type zeolite top layer synthesized by the direct *in situ* crystallization technique [2]. During direct *in situ* crystallization, the precursor solution or gel for the zeolite crystallization can also be soaked into the pore system of the support to form zeolite plugs [23].

For the direct *in situ* synthesis of MFI-type zeolite membranes, an organic agent, normally tetrapropyl ammonium (TPA), is used as a structure-directing template for the formation of the zeolite crystal structure. Following the membrane synthesis process, sintering treatment at a high temperature is necessary in order to open the pore structure by removing the template from the zeolite membrane. However, the

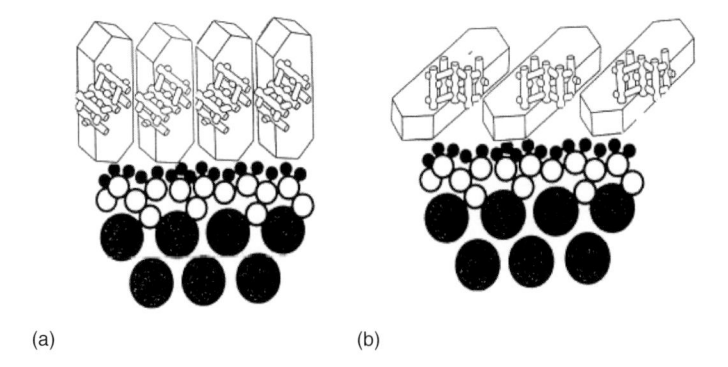

(a)　　　　　　　　　　　　(b)

Figure 9.4 Schematic orientation of microcrystals in an MFI-type zeolite top layer by the direct *in situ* crystallization technique (a) and by seeding supported crystallization (b) [2]. (Source: Reprinted from Microporous Mesoporous Mater., 38, Caro *et al.*, Zeolite membranes–state of their development and perspective, 3–24. Copyright (2000), with permission from Elsevier.)

sintering treatment will also result in microcracks in the zeolite membranes because of thermal stresses mainly caused by (i) the thermal expansion differences between the zeolite layer and the porous support and (ii) the changes in lattice parameters of the zeolite crystals during the removal of the template at high temperature [24–26].

9.2.2
Seeded Secondary Growth

An alternative strategy commonly used for the synthesis of zeolite membranes is seeding-supported crystallization, also referred to as the *seeded secondary growth* [27–29]. In this method, the zeolite layer is formed in several steps. It generally involves (i) the preparation of a colloidal suspension of zeolite seed crystals, (ii) the coating of nanosized zeolite seeds on a suitable support such as a porous ceramic substrate, and (iii) the growth of a continuous zeolite membrane in a subsequent hydrothermal process. The use of zeolite seed crystals can offer a way to control the growth of a zeolite top layer on the surface of the support [8]. Crack propagation could be limited by using nanoparticle seeds (<100 nm) and a small filler layer between seeds during secondary growth [30]. Figure 9.4b shows a schematic orientation of the individual microcrystals in an MFI-type zeolite layer synthesized by seeding supported crystallization [2].

Many researchers [31–34] have reported the recipes for the preparation of the zeolite seed crystals. For the coating of these seeds on the support, there are several methods [2, 8]:

- matching the zeta potentials of the support and the seeds for an electrostatic attachment by changing the pH of the solution [35–37];
- using positively charged cationic polymers to adjust different zeta potentials between the support and the zeolite seeds to be attached by electrostatic forces [38–41];
- electrophoretic deposition of nanosized zeolite seeds on solid supports [42, 43];
- immersion of the dried support into a seed solution followed by thermal treatment to fix the seeds to the support;
- using pulsed laser ablation to prepare a thin zeolite film on a support [44–46].

To avoid the formation of microcracks in zeolite membranes, some researchers have attempted to prepare MFI zeolite membranes using the secondary growth method without organic templates. Mintova and coworkers [39] prepared zeolite ZSM-5 films with thicknesses in the range of 230–3500 nm on quartz substrates from template-free precursor. Colloidal silicalite-1 seed crystals were first coated on the surface-modified dense quartz substrates. These crystals were then grown into continuous films by hydrothermal treatment in a synthesis gel without template. Hedlund et al. [41] used the secondary growth method for synthesis of good quality template-free zeolite ZSM-5 membranes (1.5 μm thick) on a porous alumina support. In their work, the elimination of the calcination step at around 400 °C for template removal has minimized the chances of the formation of cracks. Pan and

Lin [47] also reported the preparation of good quality MFI-type zeolite membranes on porous α-alumina supports by the secondary growth method using pure silica sol without an organic template. In this work, the polished α-alumina supports were first coated with the silicalite suspension with the pH modified to 3–4 by a dip-coating process and were then calcined at different temperatures (450 and 650 °C) to enhance the bonding between the zeolite layer and the support. After sintering treatments, a zeolite seed layer (~2.5 μm) consisting of relatively fine polycrystalline particles was obtained (Figure 9.5a) [47]. The dip-coated samples with the silicalite seed layers were hydrothermally treated with silica sol prepared from template-free precursor, followed by a drying step at low temperature (40 °C). Zeolite membranes were synthesized in this secondary growth step without the calcination and with the thickness increased to about 5.5 μm, as shown in Figure 9.5b [47].

Most recently, Choi et al. [48] developed a calcination technique, named as *rapid thermal processing* (RTP), which was postulated to strengthen grain bonding through condensation of terminal silanol groups (Si–OH) present in zeolite grains. They used RTP to prepare MFI membranes grown on porous α-alumina disks and stainless steel tubes and demonstrated that the use of this calcination technique could eliminate grain boundary defects and substantially improve in *p-/o*-xylene isomer separation. Yoo et al. [30] also used RTP in a single calcination step for the fabrication of MFI zeolite membranes after secondary hydrothermal growth. In their work, confined synthesis [49, 50], as shown in Figure 9.6 [30], was used to prepare zeolite nanoparticles with well-controlled morphology. The obtained zeolite nanoparticles were then deposited as a randomly oriented seed layer on homemade

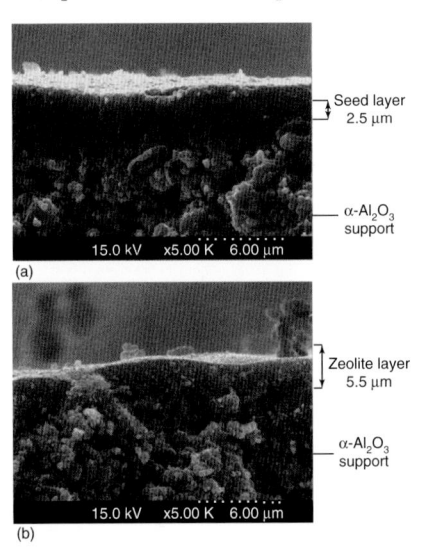

Figure 9.5 SEM images of the silicalite seed layer (a) and the secondary grown zeolite membrane (b) [47]. (Source: Reprinted from Microporous Mesoporous Mater., 43, Pan et al., Template free secondary growth synthesis of MFI type zeolite membranes, 319–327. Copyright (2001), with permission from Elsevier.)

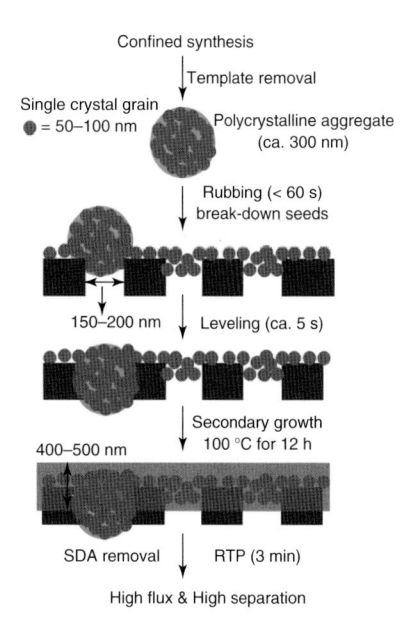

Figure 9.6 Schematic of confined synthesis used for the fabrication of MFI-type zeolite membranes [30]. (Source: Reprinted from Angew. Chem. Int. Ed., 49, Yoo et al., High-performance randomly oriented zeolite membranes using brittle seeds and rapid thermal processing, 8699–8703. 319–327. Copyright (2010), with permission from John Wiley and Sons.)

porous α-alumina supports using a rubbing [51] and leveling method. The rubbing process breaks down the polycrystalline spherical seed particles (about 300 nm in diameter) into smaller grains (50–100 nm in diameter) and deposits them in the pores of the support. The leveling process pushes the unbroken polycrystalline aggregates remaining on the support, after the rubbing process, down deeper into the pores of the support or breaks them down further (Figure 9.6). This simple leveling process can reduce the surface roughness and thus improve the uniformity of the membrane thickness [30]. The combined rubbing and leveling methods can complete the seed-deposition process in 1 min and result in a high-quality seed layer, which can then be used to produce thin zeolite membranes by secondary hydrothermal growth followed by a single RTP step (Figure 9.7). The membranes calcined by the RTP process after rubbing and leveling exhibited increased separation factors (ranged from 123 to 139) for the separation of p-/o-xylene isomers compared to those obtained by the conventional calcinations (separation factor = 28) [30].

9.2.3
Microwave Synthesis

Microwave energy has been widely used for chemical reactions. In the past years, remarkable progress has been made in microwave synthesis of zeolite membranes [14]. Many types of zeolite membranes, such as LTA, MFI, AFI,

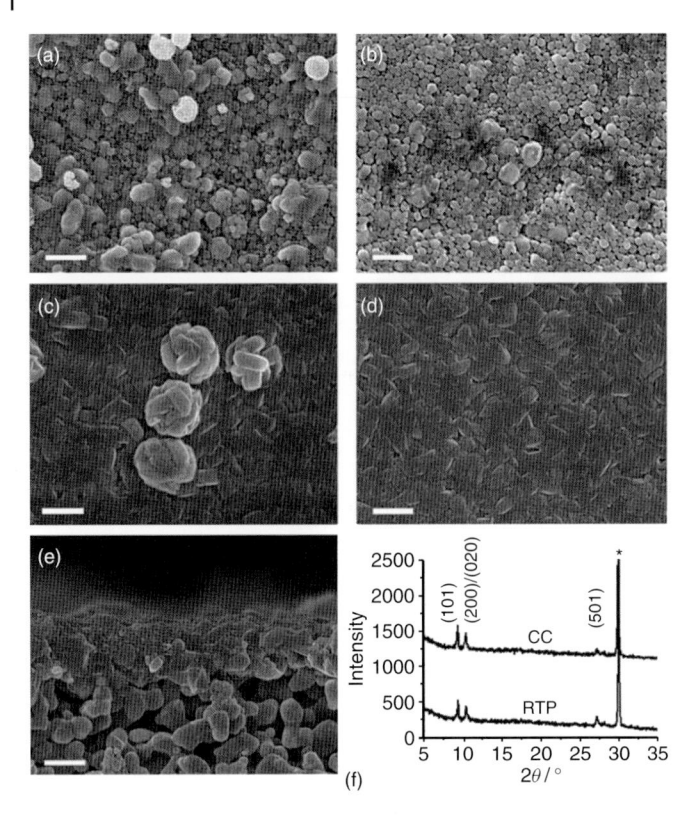

Figure 9.7 Development of the membrane [30]. (a–e) SEM images of seed layers and silicalite-1 seeded membranes on a porous α-alumina support at various stages of the fabrication process. (a) Top view of a seed layer after the rubbing step and (b) after rubbing and leveling. After secondary hydrothermal growth, intergrown MFI–zeolite layers are formed. (c) In membranes prepared without leveling (denoted as R-membranes), several protrusions remain after secondary growth, but (d) they are not prominent in membranes grown from the leveled seed layers as shown in (b) (denoted as RL-membranes). (e) The cross-sectional view of the RL-membrane after secondary growth shows an ultrathin MFI–zeolite layer on top of the α alumina support. All scale bars correspond to 500 nm. (f) X-ray diffraction patterns of zeolite films treated by conventional calcination (CC) and rapid thermal processing (RTP). The XRD patterns were acquired using a PANalytical X-Pert PRO MPD X-ray diffractometer equipped with a Co source ($Co_{K\alpha}$, $\lambda = 1.790$ Å) and an X-Celerator detector. The reflection peak marked with an asterisk (*) originates from the α-alumina support. (Source: Reprinted from Angew. Chem. Int. Ed., 49, Yoo et al., High-performance randomly oriented zeolite membranes using brittle seeds and rapid thermal processing, 8699–8703. 319–327. Copyright (2010), with permission from John Wiley and Sons.)

FAU, SOD, and ETS-4, have been prepared by both *in situ* microwave synthesis [52–56] and microwave-assisted secondary growth [57–61]. The main advantage of the microwave-assisted synthesis is that it can offer a very efficient way for the preparation of a variety of zeolite materials within short times [14, 62, 63]. For example, Xu et al. [64] reported that microwave heating significantly reduced the

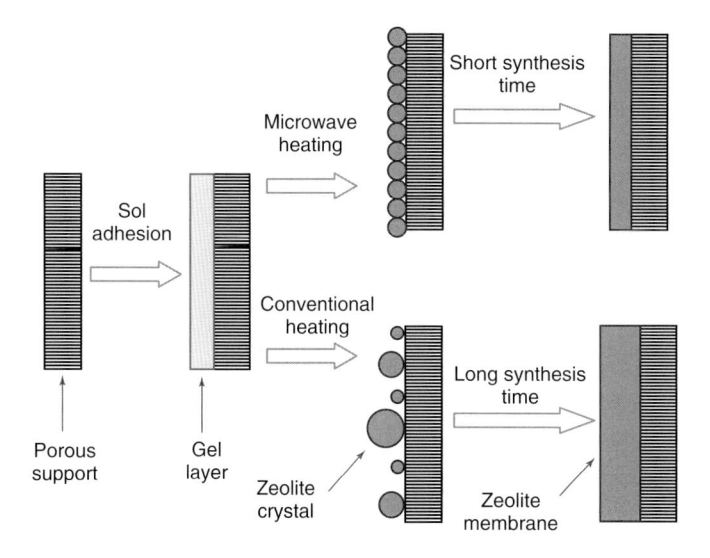

Figure 9.8 Comparative synthesis model of zeolite membrane by microwave heating and conventional heating [14]. (Source: Reprinted from J. Membr. Sci., 316, Li et al., Microwave synthesis of zeolite membranes: a review, 3–17. Copyright (2008), with permission from Elsevier.)

synthesis time (by 8–12 times) for the preparation of NaA zeolite membranes. Figure 9.8 compares the synthesis models of microwave heating and conventional heating for the preparation of zeolite membranes [14].

For the preparation of porous materials, microwave synthesis is also an efficient tool to control the distribution of particle size [65], phase selectivity [66], and macroscopic morphology [67]. In the case of zeolite membranes, the microwave synthesis method could result in different membrane morphology, orientation, composition, and thus different permeation characteristics when compared with those prepared by conventional heating [14]. Figure 9.9 shows recent studies on synthesis of LTA zeolite membranes carried out by Li and Yang [14]. It can be seen that conventional heating produced well-shaped cubic LTA crystals in the

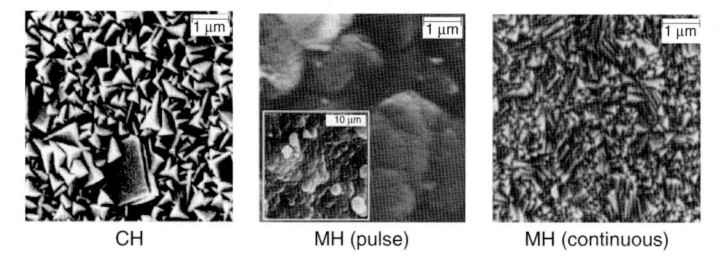

Figure 9.9 SEM images of NaA zeolite membranes synthesized by conventional heating and microwave (both pulsed and continuous) heating [14]. (Source: Reprinted from J. Membr. Sci., 316, Li et al., Microwave synthesis of zeolite membranes: a review, 3–17. Copyright (2008), with permission from Elsevier.)

(a)

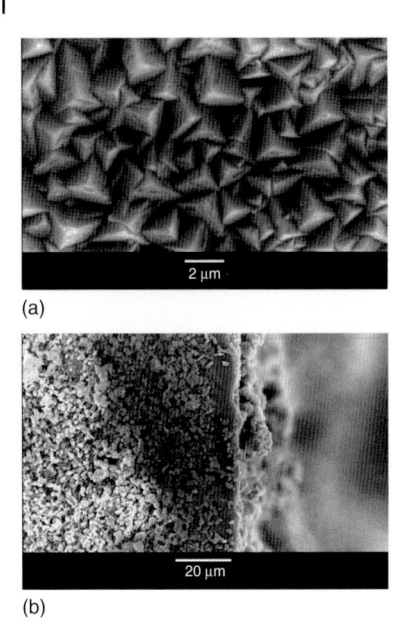

(b)

Figure 9.10 SEM micrographs of NaA membrane synthesized by the microwave technique; (a) surface and (b) thickness [68]. (Source: Reprinted from Desalination, 269, Kunnakorn et al., Performance of sodium A zeolite membranes synthesized via microwave and autoclave techniques for water-ethanol separation: recycle-continuous pervaporation process, 78–83. Copyright (2011), with permission from Elsevier.)

LTA zeolite membrane, whereas pulsed microwave irradiation synthesized LTA zeolite membranes were composed of spherical grains without well-developed crystal faces. However, a tightly packed membrane was formed by bundles of layered crystals under continuous microwave irradiation [14]. Recent work carried out by Kunnakorn et al. [68] showed that the autoclaving (conventional heating) process produced bigger zeolite crystals and gaps in NaA membranes (Figure 9.11a) compared to the microwave technique, which gave well intergrown smaller crystals (Figure 9.10a). The recycle-continuous pervaporation results showed that the NaA membrane synthesized by the microwave technique had a better performance for separation of a water–ethanol mixture [68], which is due to the well intergrown smaller crystals formed by the microwave technique [60]. Changes in composition (Al/Si ratio) of the zeolite membranes were also observed by several research groups [55, 69] when using microwave heating instead of conventional heating.

As mentioned in Sections 9.2.1 and 9.2.2, MFI membranes are classically synthesized by the *in situ* or secondary growth method. Recently researchers have successfully prepared high-quality MFI membranes by combining the seeded secondary growth method with the microwave heating process [57, 61, 70, 71]. Most recently, Xiao et al. [72] developed an effective composite seeding technique in combination with the microwave heating technique to produce thin- and defect-free MFI zeolite membranes. This seeding process includes three steps: (i) premodify the support surface by hot dip coating in a large seed (1.2 µm) suspension, (ii) rub

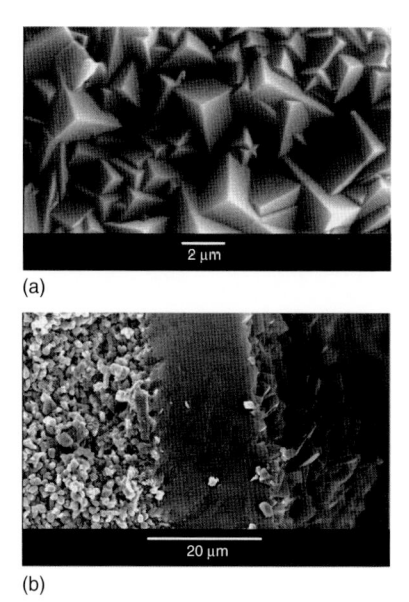

Figure 9.11 SEM micrographs of NaA membrane synthesized by the autoclaving technique; (a) surface and (b) thickness [68]. (Source: Reprinted from Desalination, 269, Kunnakorn *et al.*, Performance of sodium A zeolite membranes synthesized via microwave and autoclave techniques for water–ethanol separation: recycle-continuous pervaporation process, 78–83. Copyright (2011), with permission from Elsevier.)

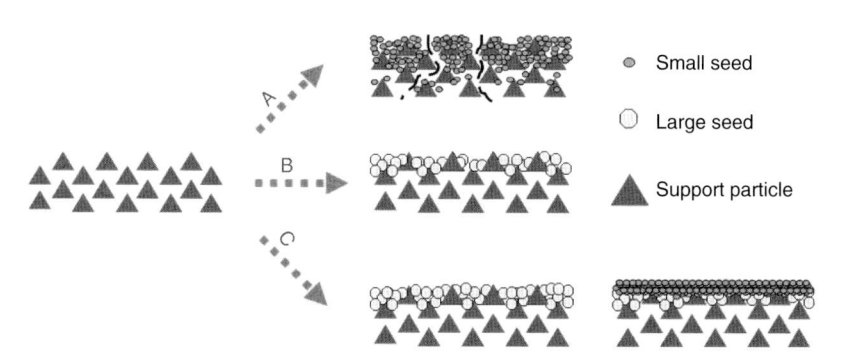

Figure 9.12

off the excess large seeds from the support surface, and (iii) further modify the support by hot dip coating in a small seed (0.4 μm) suspension [72]. Figure 9.12 [72] compares the formation of seed layers by using the conventional dip-coating method and the composite seeding technique developed. This composite seeding process can produce a dense and uniform seed layer, which is preferable for the synthesis of good quality zeolite membranes [72].

9.2.4
Postsynthetic Treatment

Theoretically, a defect-free zeolite membrane can separate components in gaseous and liquid mixtures, primarily by molecular sieving effect, relying on its configurational pores. However, the molecular sieving effect will be spoiled because of the existence of defects in zeolite membranes. To eliminate possible defects in zeolite membranes, significant progress has been achieved through microstructure optimization such as modified secondary growth [73–75], control of grain orientation during film growth [76, 77], and RTP [48].

Apart from these approaches, much effort has also been made toward eliminating the defects in zeolite membranes by postsynthetic treatments [78–84]. Yan et al. [78] used the postsynthetic coking treatment to plug nonzeolitic defects by impregnation of a large aromatic hydrocarbon, 1, 3, 5-triisopropylbenzene (TIPB) followed by heat treatment in air at high temperature. After coking treatment by TIPB, the single gas ratio of n-butane over i-butene at 185 °C increased from 9.7 to 107, but the flux decreased significantly. Surface coating was another method used by Tsapatsis' group [79, 80] to seal the cracks formed during the calcinations of MFI membranes. They applied a mesostructured surfactant–silica layer on the top of the MFI membrane by dip coating the membrane in a mixture of hexadecyltrimethylammonium bromide (CTAB) and polymeric silica sol. After drying, this surfactant–silica layer could be easily peeled off by blowing the membrane surface with compressed air. This method was successful in improving the membrane selectivity but also led to a serious flux decrease owing to zeolitic-pore blocking. Nomura et al. [81] modified silicalite membranes using a counter-diffusion chemical vapor deposition (CVD) technique. Tetraethyl orthosilicate (TEOS) was used as the silica source and O_3 as the oxidizing agent. By using this technique, an amorphous silica layer was formed on the membrane surface, and the separation selectivity for an n/i-butane mixture was increased from 9.1 to 87.8 at 15 °C.

Recently, Zhang et al. [82] developed a novel counterdiffusion chemical liquid deposition (CLD) technique for selective defect patching of zeolite membranes. By using dodecyltrimethoxysilane (DMS) as a silane coupling agent, a protective layer was coated on the membrane surface to keep the zeolitic pores intact during the subsequent reparation. The separation factor of the membrane for a 50/50 n/i-butane gas mixture increased from 4.4 to 35.8 after reparation. Most recently, a simple modification method with hydrolysis of silanes was proposed by Hong et al. [83] to heal nonzeolitic pores of MFI silicalite membranes. They compared three silane precursors, TEOS, tetramethoxysilane (TMOS), and dimethoxydimethylsilane (DMDS), and found that DMDS was more suitable as a precursor for healing nonzeolitic pores. The hydrolysis of silanes in nonzeolitic pores could obviously improve separation performance of the MFI zeolite membranes. The p-/o-xylene separation factor of the membrane increased from 2.34 to 10.84 after DMDS modification. It was also reported by Chen and coworkers [84] that functional defect patching (FDP) can eliminate grain boundary defects of a mordenite zeolite membrane (MZM) by grafting functional polymer groups (poly(4-vinylpyridine)) onto grain boundaries and thus improving the pervaporation performance for the dehydration of acetic acid.

9.3 Zeolite Membranes for Water Treatment

9.3.1 Zeolite Membranes for Desalination

Reverse osmosis (RO) has been widely used in a range of applications in water treatment such as desalination of seawater or brackish water and drinking water purification. To date, polymer membranes have been deployed in the state-of-the-art RO membrane desalination plants. In addition, the commercially available polymeric membranes have high fouling tendency caused by natural organic matter (NOM) and are prone to degradation because of their low resistance to chlorine and other oxidants. Therefore, RO membranes made from ceramic materials have attracted increasing attention in recent years. They potentially offer many advantages over commercial polymeric membranes for RO applications, including excellent chemical, thermal, and mechanical stability, long reliable working lifetime, and high resistance to chlorine, oxidants, radiation, and solvents. Thus, a ceramic desalination membrane offers great potential in expanding the versatility of membrane-based desalination, for example, for "aggressive" industry wastewaters.

Zeolite membranes are of great interest and have been shown to be potential candidates for desalination. Zeolites provide a rigid ceramic structure suitable for separating the small water molecules from large salt ions. Lin and Murad [7] recently carried out a molecular dynamic simulation of ion diffusion in zeolite. Two thin membranes used for their simulation studies were cut from a cubic cell of pure silica ZK-4 zeolite. The uniform pores of a perfect pure silica zeolite are small enough to restrict the entry and transport of hydrate ions, but sufficiently large for water molecules to diffuse through [7]. As shown in Figure 9.13 [7], the region between two membranes was filled with water molecules, and aqueous NaCl solution was filled into the regions to the right of the right membrane and left of the left membrane. In most cases, the simulation system consisted of 2240 molecules in the basic cyclically replicated parallelepiped of which 704 molecules constitute the two membranes. Both osmosis and RO phenomena were observed from the study. The computer simulation results have demonstrated the feasibility of using ZK-4 zeolite membranes to separate water from aqueous NaCl solutions. The simulation study showed that 100% rejection of Na^+ could be achieved using a perfect (single crystal), pure silica ZK-4 zeolite membrane by RO. The size exclusion of hydrated ions is the separation mechanism of the perfect ZK-4 zeolite membrane. The aperture of the ZK-4 zeolite (diameter 0.42 nm) is significantly smaller than the kinetic sizes of hydrated ions (0.8–1.0 nm for $[Na(H_2O)_x]^+$) [85].

Following the computer simulation study by Lin and Murad [7], several research groups have explored the possibility of using MFI-type zeolite membranes for

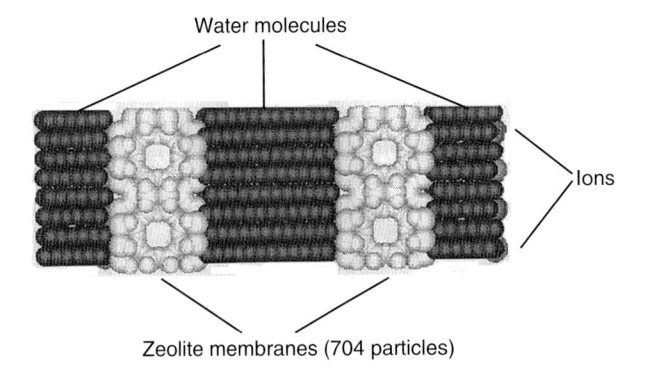

Figure 9.13 The structure of the membrane and a schematic of the solvent/solution compartments [7]. (Source: Reprinted from Mol. Phys., 99, Lin et al., A computer simulation study of the separation of aqueous solutions using thin zeolite membranes, 1175–1181. Copyright (2001), with permission from Taylor & Francis.)

desalination [10, 29, 86–88]. The MFI-type zeolite has an orthorhombic crystal symmetry with nearly cylindrical, 10-member ring channels. The diameter of the nanopores of the MFI-type zeolite is around 0.56 nm, which is smaller than the sizes of hydrated ions. Li and coworkers [10] carried out the first experimental study on α-alumina-supported MFI-type zeolite membranes for RO desalination (Figure 9.14) of aqueous solutions containing a single cation and multiple cations. The α-alumina-supported MFI-type zeolite membranes used for the study were synthesized from Al-free precursors by double hydrothermal treatments using the *in situ* technique [20]. Their results indicated that the prepared MFI-type membranes had high ion rejection rates. A Na^+ rejection of 76.7% with a water flux of about 0.112 kg m^{-2} h^{-1} was achieved for a single cation feed solution

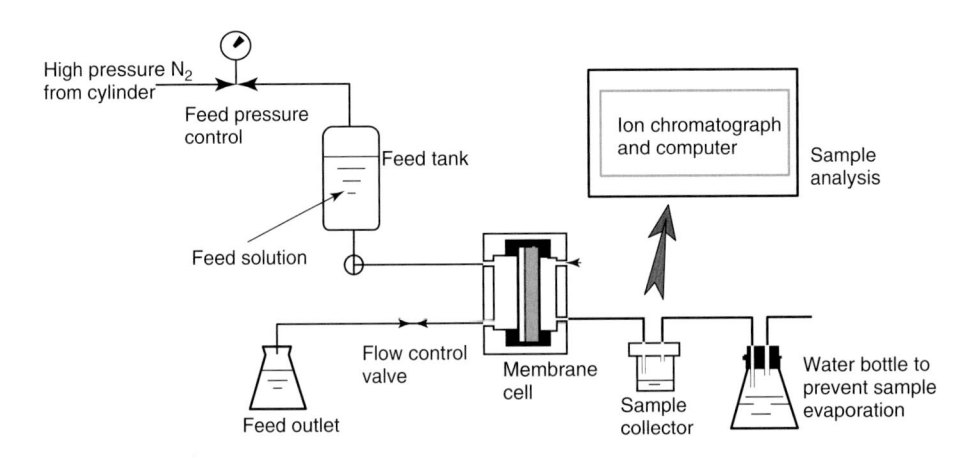

Figure 9.14 Schematic diagram of the RO system [10]. (Source: Reprinted from J. Membr. Sci., 243, Li et al., Desalination by reverse osmosis using MFI zeolite membranes, 401–404. Copyright (2004), with permission from Elsevier.)

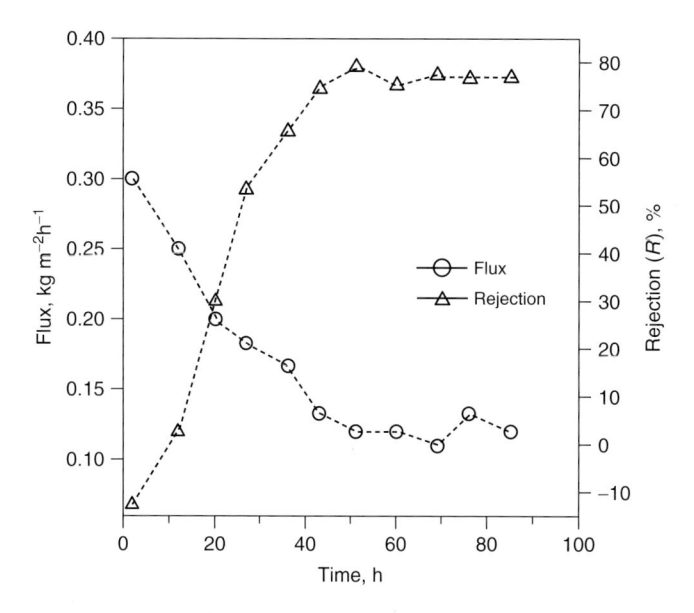

Figure 9.15 Water flux and Na$^+$ rejection as functions of RO operation time for 0.1 M NaCl solution [10]. (Source: Reprinted from J. Membr. Sci., 243, Li et al., Desalination by reverse osmosis using MFI zeolite membranes, 401–404. Copyright (2004), with permission from Elsevier.)

(0.1 M NaCl) under an applied pressure of 2.07 MPa (Figure 9.15) [10]. For a complex feed solution containing multiple cations (0.1 M NaCl + 0.1 M KCl + 0.1 M NH$_4$Cl + 0.1 M CaCl$_2$ + 0.1 M MgCl$_2$), different rejections were obtained for different cations (58.1, 62.6, 79.9, 80.7, and 88.4% for Na$^+$, K$^+$, NH$_4^+$, Ca^{2+}, and Mg^{2+}, respectively) after 145 h of operation at an applied pressure of 2.4 MPa, with a stabilized water flux of 0.058 kg m^{-2} h^{-1} (Figure 9.16) [10]. It can be seen that the Na$^+$ rejection significantly reduced when the ionic strength was increased, from around 77% for the single cation solution to about 58% for the solution containing multiple cations. The divalent cations showed a higher rejection than monovalent ions in a feed containing mixed ion species. As the sizes of hydrated ions are virtually independent of the ion concentration [89], the size exclusion separation mechanism is insufficient to explain the real membrane process.

To understand the mechanisms of ion and water transport in real membranes, Li et al. [86] investigated the effects of charge density, size, and hydration number of the ion on RO process through MFI-type zeolite membranes. The MFI-type zeolite membranes were developed on the disk-shaped alumina substrates by a single hydrothermal treatment through the *in situ* crystallization [20]. The RO separation results (Table 9.1) [86] showed that as ion rejection increased, the fluxes of ions and water decreased in the order of 0.10 M NaCl, 0.10 M MgCl$_2$, and 0.10 M AlCl$_3$. It was found that the ion rejection and water flux depended on the charge density of the ion and the dynamic size and diffusivity of the hydrated ion. Their experimental results also revealed that the transport of ions and water was controlled by different

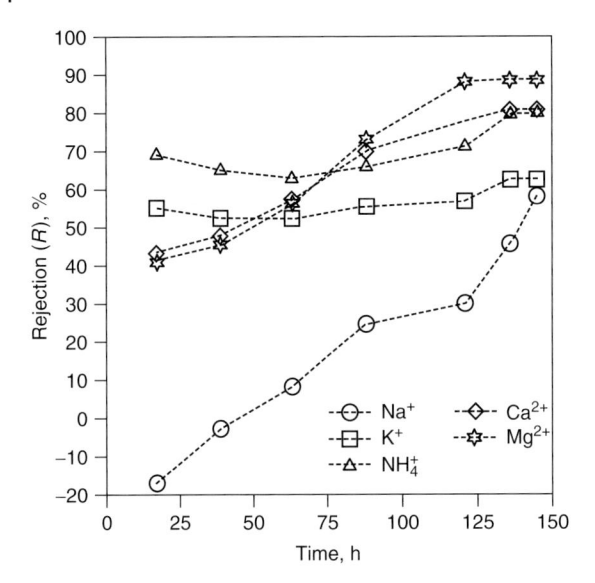

Figure 9.16 Ion rejection as a function of operation time for the multicomponent feed solution containing 0.1 M NaCl, 0.1 M KCl, 0.1 M NH$_4$Cl, 0.1 M CaCl$_2$, and 0.1 M MgCl$_2$ [10]. (Source: Reprinted from J. Membr. Sci., 243, Li et al., Desalination by reverse osmosis using MFI zeolite membranes, 401–404. Copyright (2004), with permission from Elsevier.)

Table 9.1 RO separation results for five single-salt solutions at 25 °C under an applied pressure of 2.1 MPa [86].

Solution	Water flux (kg·m^{-2}·h^{-1})	Permeance (kg·MPa^{-1}·m^{-2}·h^{-1})	Ion rejection (%)
0.10 M NaCl	0.162	0.078	21.6
0.10 M KCl	0.174	0.084	21.9
0.10 M MgCl$_2$	0.081	0.039	68.6
0.10 M CaCl$_2$	0.096	—	57.6
0.10 M AlCl$_3$	0.057	0.028	96.2

mechanisms. Ion rejection on the zeolitic channel can be described by the size exclusion mechanism, whereas ion separation through the intercrystal micropores was attributed to the strong interactions between the ion and the charged double layers, which restricted ion diffusion into the microporous space.

Further studies on the transport of ions and water through MFI-type zeolite membranes during the RO process were carried out by Li et al. [87]. High rejection rates (>95%) were achieved for all alkali metal ions studied. The restricted access of hydrated ions to the zeolitic pores and the competitive diffusion of water and ions in the zeolite channels were found to have a great impact on the RO separation of ions through the MFI zeolite membranes.

Since the first experimental study [10] on MFI-type zeolite membranes for RO desalination showed that the rejection of salt was not high enough and the water flux was too low for practical applications. State-of-the-art polymer membrane RO exhibits fluxes in the order of 20 kg m^{-2} h^{-1} and salt rejections exceeding 99%, so further development of ceramic desalination materials is needed. Work has been carried out on the modification of zeolite structure by introduction of Al content to enhance the membrane performance [88, 90]. The addition of Al content to the membrane can alter the surface hydrophilicity [91, 92] and surface charge [93], which should have a great impact on diffusion of electrolytes. Li et al. [90] recently reported that the addition of Al content can lead to a remarkable improvement in zeolite growth and membrane performance in RO separation. A zeolite membrane with a thickness of 2 μm and a Si/Al ratio of 50 : 50 achieved a Na$^+$ rejection of 92.9% with a water flux of about 1.129 kg m^{-2} h^{-1} at 2.76 MPa [90]. Most recently, Duke et al. [88] explored the desalination of sea salt solutions through the MFI-type zeolite membranes (silicalite and ZSM-5) by the thermally based pervaporation (membrane distillation) and RO modes. The α-alumina-supported zeolite membranes were prepared using the template-free secondary growth method [47] with varying Si/Al ratios. The template-free secondary growth method minimized the defects in the crystal structure of the zeolite membranes. The results obtained for pervaporation of seawater (Figure 9.17) [88] indicated that the addition of alumina into the zeolite membranes appeared to improve equivalent rejection compared to the silicalite (pure silica zeolite) membrane. The membranes with alumina achieved high equivalent salt rejections (>99%), while silicalite showed a salt rejection of around 97%. The proposed mechanism of water and ion diffusion during pervaporation is shown in Figure 9.18 [88]. Salts are deposited on the permeate side after diffusing through the defects or grain boundaries by virtue of the water leaving the permeate as vapor. These salts are removed by periodic permeate flushing with fresh water. According to their knowledge, this is the first reported evidence of seawater desalination by pervaporation (or membrane

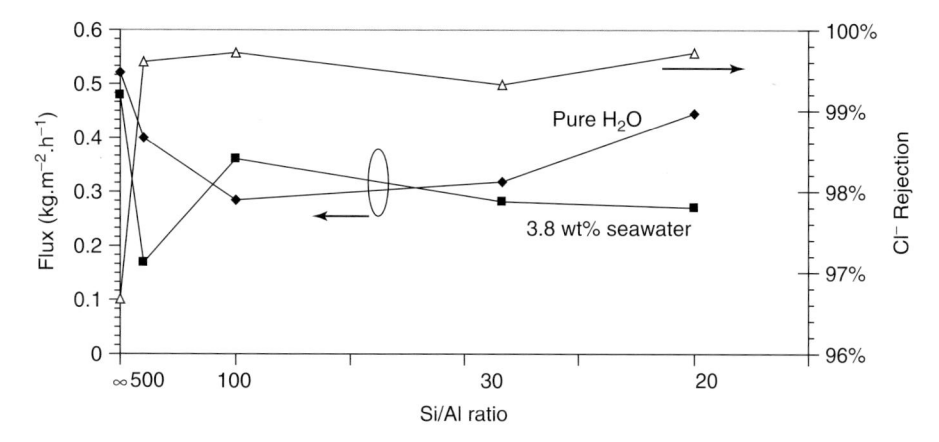

Figure 9.17 Pervaporation results for flux and equivalent Cl$^-$ rejection with pure water and 3.8 wt% seawater feeds [88]. (Source: Reprinted from Sep. Purif. Technol., 68, Duke et al., Seawater desalination performance of MFI type membranes made by secondary growth, 343–350. Copyright (2009), with permission from Elsevier.)

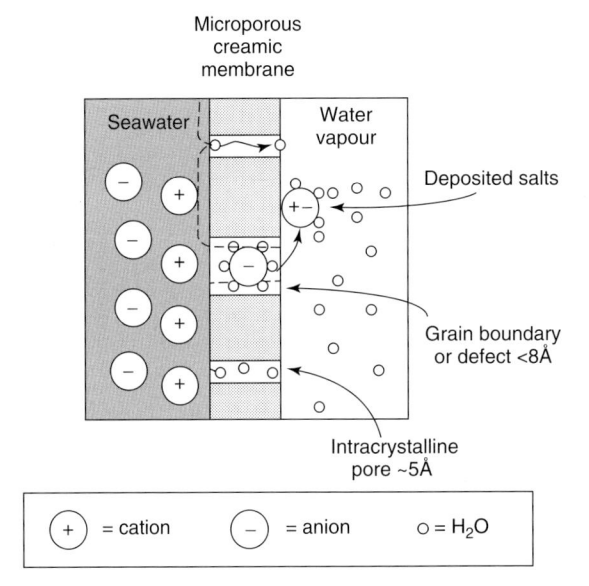

Figure 9.18 Schematic of water and ion diffusion within negatively charged microporous ceramic (zeolite) membranes [88]. (Source: Reprinted from Sep. Purif. Technol., 68, Duke et al., Seawater desalination performance of MFI type membranes made by secondary growth, 343–350. Copyright (2009), with permission from Elsevier.)

distillation) through zeolite membranes. On the basis of the results of the near-perfect rejection for all ions, they concluded that ZSM-5 zeolite membranes are ideal candidates for further research into this application. The studies carried out by the same group [88] on the RO setup showed that the silicalite membrane exhibited the highest ion rejections, and MFI-type zeolite membranes with tailored ratio of Si/Al demonstrated potential applications for ion-selective desalination.

Although there have been some studies to date applying zeolite membranes in the application of desalination [7, 10, 29, 86–88], little work has been carried out to explore the influence of seawater exposure and ion exchange on the intrinsic lattice dimensions and the intracrystalline and intercrystalline porosity of MFI zeolites. Recently, Zhu et al. [94] reported that complex solutions of ions sourced from a synthetic seawater solution in turn have complex interactions with silicalite's inter- and intracrystalline spaces. In their study, MFI framework zeolite (silicalite) was exposed to seawater salts and the relative uptake of cations was explored. Synchrotron X-ray powder diffraction (the Australian Synchrotron) was then used to measure the lattice changes associated with seawater exposure. The N_2 adsorption and positron annihilation lifetime spectroscopy (PALS) were also utilized to develop a better understanding of the effect of seawater exposure on the structure of silicalite powder. The results obtained from ion adsorption testing and synchrotron X-ray powder diffraction showed that the zeolite powder interacted with water, monovalent and divalent cations, and these interactions altered the crystal dimensions of the zeolite. Most exposures to ions and water led to increases

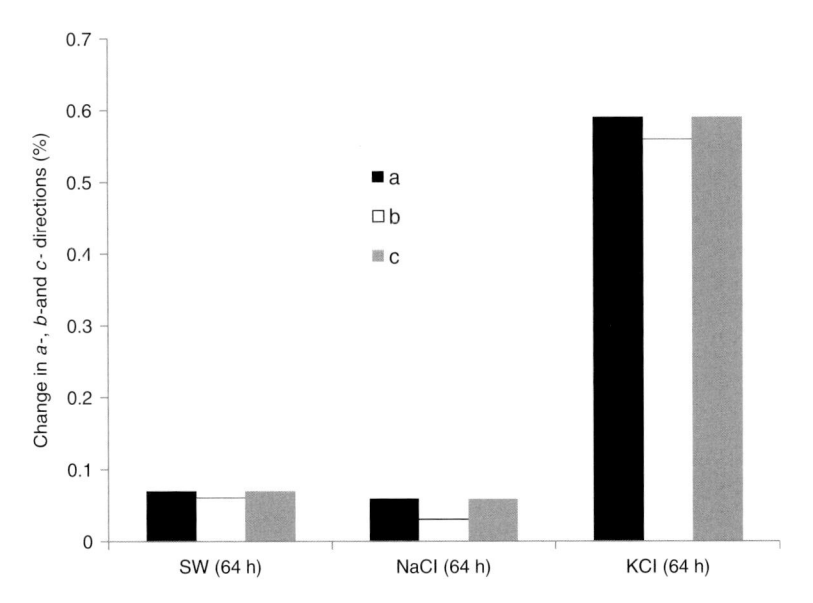

Figure 9.19 Unit cell dimension changes in a-, b-, and c-directions after exposure to seawater ions [94]. (Source: Reprinted from J. Mater. Chem., 20, Zhu et al., Investigation of the effects of ion and water interaction on structure and chemistry of silicalite MFI type zeolite for its potential use as a seawater desalination membrane, 4675–4683. Copyright (2010), with permission from The Royal Society of Chemistry.)

in crystal volumes of <0.1%, but exposure to KCl solution was observed to make the biggest contribution to crystal expansion, at 0.59% (Figure 9.19), particularly in the a- and c-directions as determined by Rietveld refinement (Figure 9.20). On closer inspection of seawater exposed silicalite, both N_2 adsorption and PALS showed 80% micropores in the original material, but the proportion of micropores decreased more than 20% within 1 h exposure to seawater owing to ion adsorption and ion exchange. At 8 h of exposure, a sharp uptake of Ca^{2+} caused an increase in micropore fraction. After 8 h, the monovalent cations were gradually exchanged with divalent ions, resulting in a gradual decrease in micropore fraction (Figure 9.21). This is the first reported evidence of the effect of ions in sea salts solution on the lattice structure, intracrystalline porosity, and intercrystalline porosity of silicalite. The change in structure and porosity will govern transport behavior when silicalite is developed as a membrane for seawater desalination.

9.3.2
Zeolite Membranes for Wastewater Treatment

Zeolite membranes may be used as an alternative to polymeric membranes for treatment of complex wastewater containing organic solvents and radioactive elements [10, 29]. It was also suggested by Li et al. [87] that the separation efficiency could be enhanced in RO by zeolite membranes when operated at

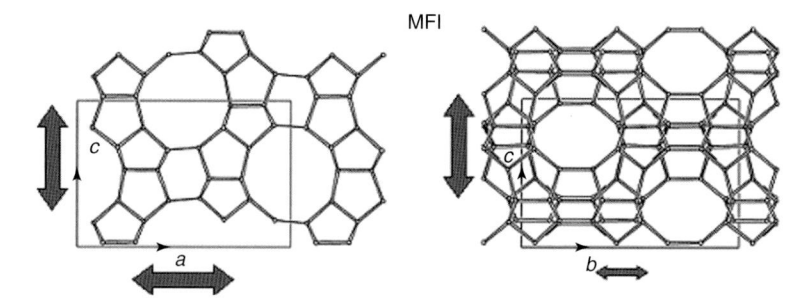

Figure 9.20 Unit cell content in MFI projected along the (b) – left and (a) – right axes. Arrow sizes indicate relative expansion of the crystal measured in Figure 9.19 due to uptake of ions or water [94]. (Source: Reprinted from J. Mater. Chem., 20, Zhu et al., Investigation of the effects of ion and water interaction on structure and chemistry of silicalite MFI type zeolite for its potential use as a seawater desalination membrane, 4675–4683. Copyright (2010), with permission from The Royal Society of Chemistry.)

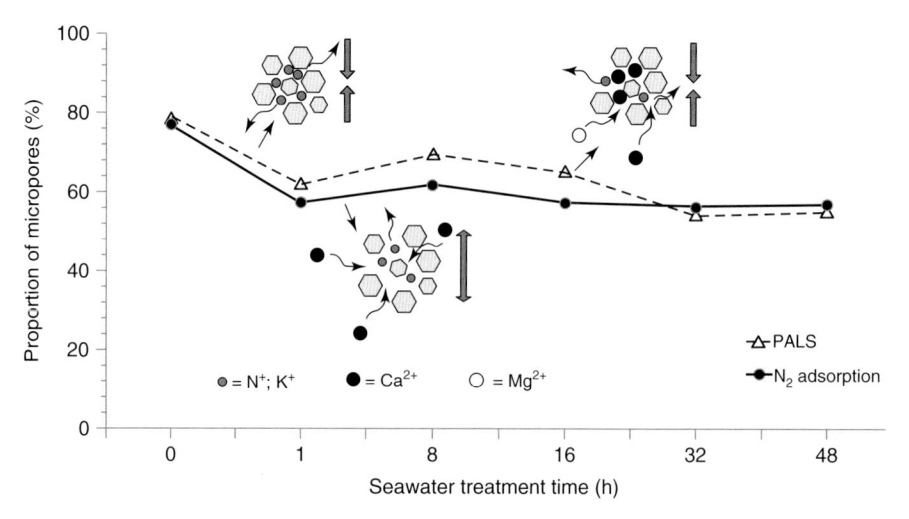

Figure 9.21 N_2 adsorption and PALS measurements of changes in micropore proportion of silicalite powders and schematic illustration of these changes caused by ion interactions over seawater treatment time [94]. (Source: Reprinted from J. Mater. Chem., 20, Zhu et al., Investigation of the effects of ion and water interaction on structure and chemistry of silicalite MFI type zeolite for its potential use as a seawater desalination membrane, 4675–4683. Copyright (2010), with permission from The Royal Society of Chemistry.)

elevated temperatures and high hydraulic pressures. They found in their study [87] that both water and ion fluxes increased significantly when raising the feed temperature from 10 to 50 °C. The Na^+ rejection rate increased continuously with increasing the feed pressure. Both ion and water fluxes increased linearly with the transmembrane pressure, with the water flux being enhanced by more than five times when increasing the transmembrane pressure from 0.71 to 2.65 MPa.

Over the past couple of years, some researchers have attempted to treat oily water, produced water and radioactive solutions, and so on, using different zeolite membranes [95–98]. Cui and coworkers [95] prepared tube-shaped α-Al_2O_3-supported NaA zeolite microfiltration (MF) membranes with average interparticle pore sizes of 1.2, 0.4, and 0.2 μm using *in situ* hydrothermal crystallization. Investigations were carried out on these NaA/α-Al_2O_3 MF membranes for water separation and recovery from oily water. High oil rejection (>99%) was achieved and water containing less than 1 mg l^{-1} oil was produced at 85 l m^{-2} h^{-1} under an applied membrane pressure of 50 kPa. α-Al_2O_3-supported NaA zeolite membranes synthesized by the secondary growth method were also reported by Malekpour *et al.* [96] for desalination of radioactive solutions by a pervaporation process. The results demonstrated a high separation factor (>5000) with a total flux of 1.2 kg m^{-2} h^{-1}. Ion rejection factors exceeding 99% were obtained for the ionic solutions including 0.001 M of Cs^+, Sr^{2+}, and MoO_4^{2-}. It was concluded that zeolitic membranes can be used to treat low-level radioactive wastes through the pervaporation process.

Produced water from the oil and gas industry could be a potential resource for potable water [97]. Indeed, in the United States alone, an estimated 7 billion liters per day of produced water is generated as a by-product of oil and gas exploration and production [99]. Typical oilfield brines contain a high concentration of NaCl and other cations such as K^+, Ca^{2+}, Mg^{2+}, and anions such as SO_4^{2-}. The presence of these cations and anions can reduce the efficiency of RO desalination [97]. Li and coworkers [97] investigated the effects of other co-ions, operating pressure, and feed ion concentration on the performance of RO desalination of NaCl solutions through MFI-type zeolite membranes. They found that the solution chemical compositions, operating pressure, and feed ion concentration had a strong impact on ion and water transport through the zeolite membranes. The counteranions (Br^- and SO_4^{2-}) had minimal influence on the membrane performance. However, both Na^+ rejection and water flux decreased when the multivalent cations (Ca^{2+} and Mg^{2+}) were present in the NaCl solution. This was explained by the changes in surface charge or micropore structure of zeolite membranes in the presence of other cations (e.g., Ca^{2+} and Mg^{2+}) [97]. It was also found that desirable high hydraulic pressures and feed ion concentration could considerably enhance the performance of RO desalination through MFI-type zeolite membranes. The α-Al_2O_3-supported MFI-type zeolite membranes were also used by Liu *et al.* [98] to remove organics from produced water by RO. The results indicated that the MFI-type zeolite membranes had great potential for separation of dissolved organics from aqueous solution. The separation efficiency was determined by the ionic species and dynamic size of dissolved organics. The membrane demonstrated high separation efficiency for electrolytes (e.g., pentanoic acid) and nonelectrolyte organics with larger molecular dynamic size (e.g., toluene) [98]. In a recent report [100] from the same group, a ZSM-5 zeolite membrane (Figure 9.22) with a thickness of around 0.7 μm prepared by seeding and secondary growth achieved excellent rejection for both organic (around 99%) and salt (around 97%). The water flux was also significantly improved through this thin membrane. Apart from the MFI zeolite membranes, Kazemimoghadam *et al.* [101, 102] also developed nanopore hydroxysodalite (HS)

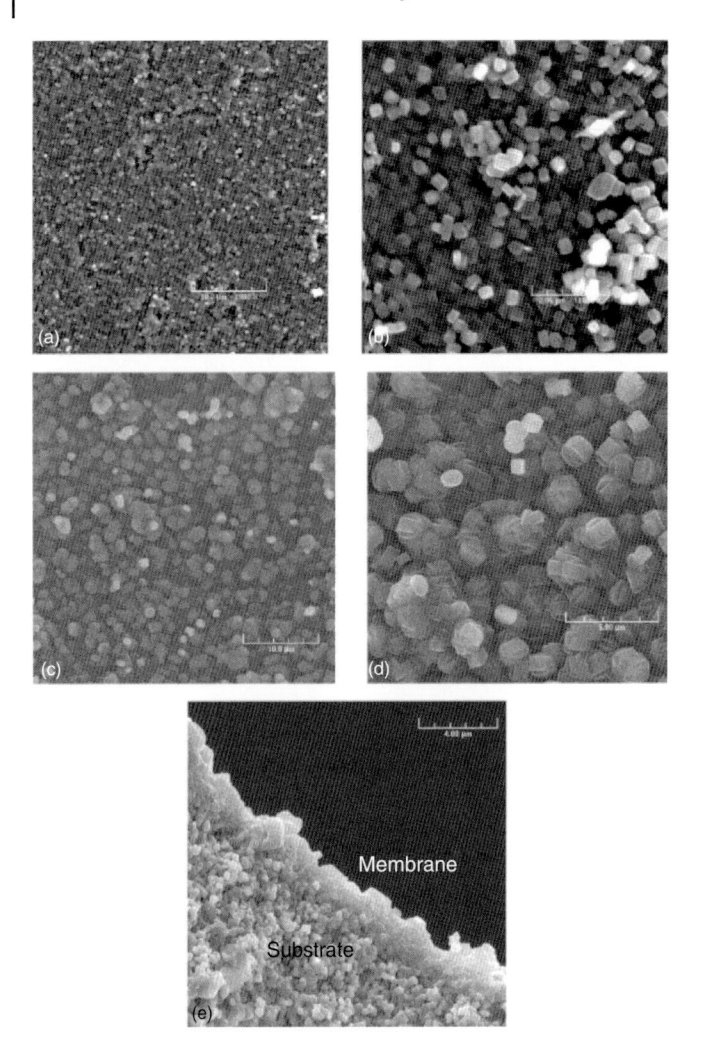

Figure 9.22 SEM images of the synthesized ZSM-5 membrane: (a,b) seed layer; (c,d) secondary growth membrane; (e) cross-section of ZSM-5 membrane [100]. (Source: Reprinted from Sep. Purif. Technol., 72, Lu et al., Organic fouling and regeneration of zeolite membrane in wastewater treatment, 203–207. Copyright (2010), with permission from Elsevier.)

zeolite (Figure 9.23) membranes, which may be used to simultaneously separate ions and dissolved organic compounds from water by RO.

9.3.3
Zeolite Membrane-Based Reactors for Wastewater Treatment

The combination of membrane technology with chemical or biological processes is attractive for wastewater treatment [103]. In these processes, membranes are

Truncated octahedrons (β-cages)

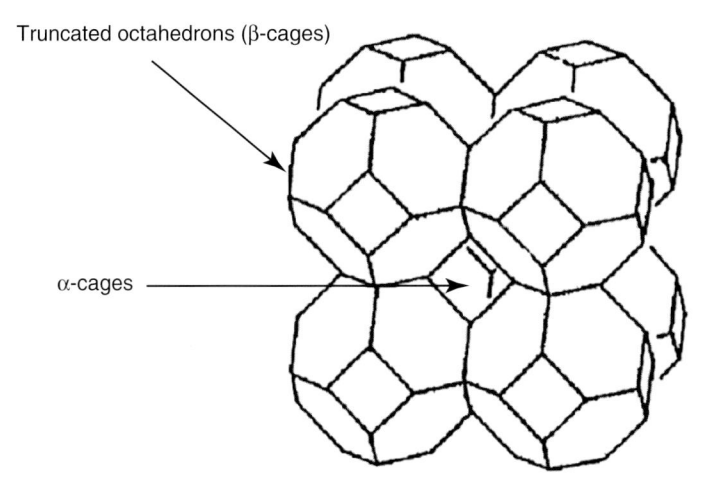

α-cages

Figure 9.23 Repeating unit of zeolite HS [101]. (Source: Reprinted from Desalination, 262, Kazemimoghadam et al., Mechanisms and experimental results of aqueous mixtures pervaporation using nanopore HS zeolite membranes, 273–279. Copyright (2010), with permission from Elsevier.)

often used for sparging gaseous oxidants such as air, oxygen, and ozone for the decontamination of water and degradation of organic pollutants or for the separation of organics [104]. For example, a novel hybrid treatment process described by Baus et al. [105] combined technology with advanced oxidation to treat problematic pollutants. This process employed membrane pervaporation to separate and remove volatile organic compounds from an industrial wastewater, and the pervaporated pollutants were then destroyed by gas phase, UV photolysis. Shanbhag et al. [106] used a capillary silicon membrane contactor to increase the ozone mass transfer rate during the ozonolysis of organic compounds. It was reported [107, 108] that inorganic membranes, being more durable than polymeric membranes, are more suitable for use in ozone contactors.

Most recently, a novel membrane reactor based on ZSM-5 zeolite was developed by Heng and coworkers [104] for ozone water treatment. The reactor consisted of a membrane contactor (alumina capillary membrane) and a separator membrane (ZSM-5 zeolite pervaporation membrane) (Figure 9.24) [104]. By using the ZSM-5 zeolite pervaporation membrane, a 47% higher TOC degradation was achieved in the novel membrane reactor compared to the system using the capillary membrane contactor only, while clean water (TOC < 20 ppm) was produced from the membrane permeate outlet at the same time. The enhancement in TOC degradation was attributed to the increase in the concentration of the organic pollutants in the reaction zone as a result of water withdrawal by membrane pervaporation [104]. Heng et al. [109] also successfully combined multiple functions of the membrane (distributor, contactor, and separator) in a single compact reactor unit to enhance the performance for ozone water treatment. The designed

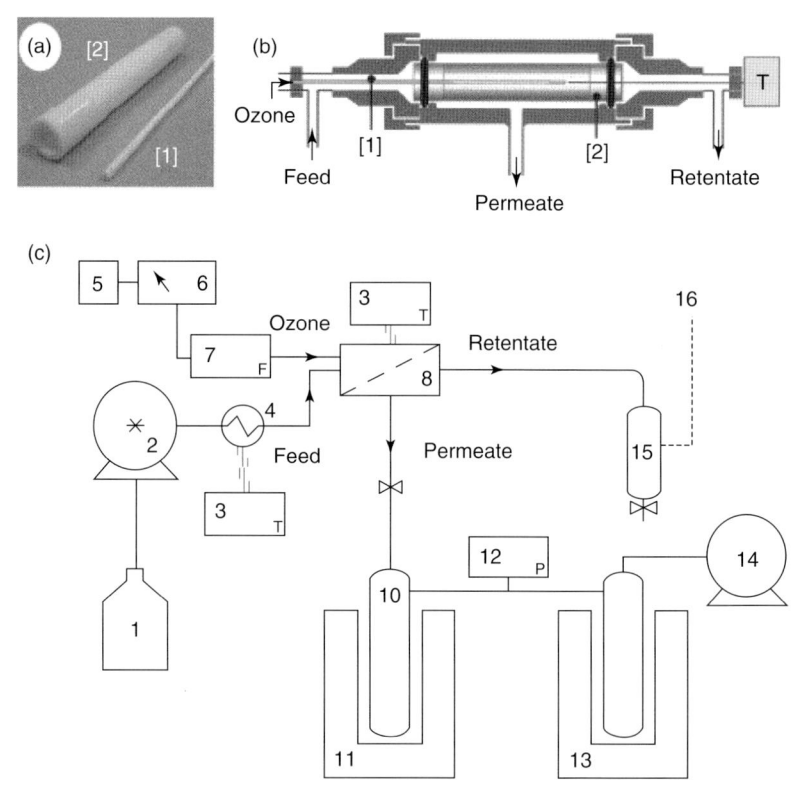

Figure 9.24 (a) A picture of the alumina capillary membrane [1] and zeolite membrane [2]. Schematic drawings of (b) the membrane module and (c) experimental setup. (1) Feed solution; (2) peristaltic pump; (3) thermocouple/heat supplier; (4) heating coil; (5) ozone generator; (6) ozone analyzer; (7) ozone flowmeter; (8) membrane reactor; (9) needle valve; (10) permeate condenser; (11) liquid nitrogen trap #1; (12) digital vacuum meter; (13) liquid nitrogen trap #2; (14) vacuum pump; (15) liquid/gas separator; and (16) ozone destructor [104]. (Source: Reprinted from J. Membr. Sci., 289, Heng et al., A novel membrane reactor for ozone water treatment, 67–75. Copyright (2007), with permission from Elsevier.)

reactor used a membrane distributor to affect a uniform ozone distribution along the reactor length and employed a composite membrane separator/contactor to provide a large contact surface for the gas–liquid reaction and to separate clean water by membrane pervaporation. The ZSM-5 zeolite-based composite membrane used for the study was prepared by coating high surface area γ-Al_2O_3 on the porous ceramic support of the ZSM-5 membrane. Ozone treatment of phthalate contaminated water showed a significantly enhanced TOC degradation (40%) by combining the three different membrane processes compared to those of other treatment processes (1.1% for the traditional sparger in a batch reactor, 20% for a membrane distributor instead of the sparger, and 25% for the combination of a membrane distributor and a membrane contactor).

9.4
Conclusions and Future Perspectives

A remarkably high level of progress in the preparation of zeolite membranes and in the understanding of the separation processes has been achieved in the past few years. Pioneering research published recently showed that zeolite membranes grown on porous ceramic supports can desalinate water; however, unusual ion specific behavior and transient performance over time has not yet been well explained or understood [10]. Although computational predictions of the transport of dissolved ions have been made and related to zeolite structures [7], there is a clear lack of fundamental, experimental research on the various zeolite materials and their structure–property relationships for membrane desalination. Questions such as, "What effect do traditional zeolite modification methods, growth time, and crystal orientation have on desalination?" are yet to be answered. With such an understanding, tailored structures that improve performance are likely to be discovered.

While previous studies have been mainly focused on the MFI-type zeolite membranes for desalination and wastewater treatment, great attention should also be paid to other types of zeolite membranes. Scientific innovations are needed to translate the performance of the zeolite materials into practice as useful membranes.

Over the past two decades, research in molecular scale separation for ceramic membranes has moved from laboratory tests to the development of multiple tube modules with high selectivity for separations of realistic mixtures of gases (e.g., syngas) and liquids (e.g., ethanol). Zeolite membrane ethanol pervaporation units are in active use in demonstration units, so it is likely that their scale-up will increase in the next 15 years. For RO applications, current polyamide membranes for desalination are extremely cost efficient. Desalination could become a major market for zeolite membranes if the membranes can compete on energy costs, reduced membrane cleaning costs, and enhanced service lifetime – as already demonstrated in ceramic membrane nanofiltration. Zeolite membranes have great promise, with exciting opportunities for growth in the next decade. However, improving water flux and reducing production cost will remain to be the key challenges in the coming years for zeolite-based membranes to be practically applied for RO desalination.

Acknowledgments

The financial support provided by the Australian Research Council (ARC) through a Discovery Project (DP0986192) is gratefully acknowledged. This study is also a part of the 2010 Endeavour Executive Award awarded to Mikel Duke by the Australian Federal Government and a travel award awarded to Bo Zhu by the Ian Potter Foundation, Australia.

References

1. Lu, G.Q., Diniz da Costa, J.C., Duke, M., Giessler, S., Socolow, R., Williams, R.H., and Kreutz, T. (2007) Inorganic membranes for hydrogen production and purification: a critical review and perspective. *J. Colloid Interface Sci.*, **314** (2), 589–603.
2. Caro, J., Noack, M., Kölsch, P., and Schäfer, R. (2000) Zeolite membranes – state of their development and perspective. *Microporous Mesoporous Mater.*, **38** (1), 3–24.
3. Breck, D.W. (1974) *Zeolite Molecular Sieves – Structure, Chemistry, and Use*, John Wiley & Sons, Inc., New York.
4. Weller, M.T. (1994) *Inorganic Materials*, Oxford University Press, Inc., New York.
5. Kaduk, J.A. and Faber, J. (1995) Crystal structure of zeolite Y as a function of ion exchange. *Rigaku J.*, **12** (2), 14–34.
6. Baerlocher, C. and McCusker, L.B. (2011) Database of Zeolite Structures, http://www.iza-structure.org/databases/ (accessed October 2011).
7. Lin, J. and Murad, S. (2001) A computer simulation study of the separation of aqueous solutions using thin zeolite membranes. *Mol. Phys.*, **99** (14), 1175–1181.
8. Caro, J. and Noack, M. (2008) Zeolite membranes - recent developments and progress. *Microporous Mesoporous Mater.*, **115** (3), 215–233.
9. McLeary, E.E., Jansen, J.C., and Kapteijn, F. (2006) Zeolite based films, membranes and membrane reactors: progress and prospects. *Microporous Mesoporous Mater.*, **90** (1–3), 198–220.
10. Li, L.X., Dong, J.H., Nenoff, T.M., and Lee, R. (2004) Desalination by reverse osmosis using MFI zeolite membranes. *J. Membr. Sci.*, **243** (1–2), 401–404.
11. Liu, Q., Noble, R.D., Falconer, J.L., and Funke, H.H. (1996) Organics/water separation by pervaporation with a zeolite membrane. *J. Membr. Sci.*, **117** (1–2), 163–174.
12. Kusakabe, K., Kuroda, T., and Morooka, S. (1998) Separation of carbon dioxide from nitrogen using ion-exchanged faujasite-type zeolite membranes formed on porous support tubes. *J. Membr. Sci.*, **148** (1), 13–23.
13. Dong, J.H., Lin, Y.S. and Liu, W. (2000) Multicomponent hydrogen/hydrocarbon separation by MFI-type zeolite membranes. *AIChE J.*, **46** (10), 1957–1966.
14. Li, Y. and Yang, W. (2008) Microwave synthesis of zeolite membranes: a review. *J. Membr. Sci.*, **316** (1–2), 3–17.
15. Kikuchi, E., Yamashita, K., Hiromoto, S., Ueyama, K., and Matsukata, M. (1997) Synthesis of a zeolitic thin layer by a vapor-phase transport method: appearance of a preferential orientation of MFI zeolite. *Micropor. Mater.*, **11** (3–4), 107–116.
16. Matsukata, M., Ogura, M., Osaki, T., Hari Prasad Rao, P.R., Nomura, M., and Kikuchi, E. (1999) Conversion of dry gel to microporous crystals in gas phase. *Top. Catal.*, **9** (1–2), 77–92.
17. Matsufuji, T., Nishiyama, N., Ueyama, K., and Matsukata, M. (2000) Permeation characteristics of butane isomers through MFI-type zeolitic membranes. *Catal. Today*, **56** (1–3), 265–273.
18. Alfaro, S., Arruebo, M., Coronas, J., Menéndez, M., and Santamaría, J. (2001) Preparation of MFI type tubular membranes by steam-assisted crystallization. *Microporous Mesoporous Mater.*, **50** (2–3), 195–200.
19. Sano, T., Yanagishita, H., Kiyozumi, Y., Mizukami, F., and Haraya, K. (1994) Separation of ethanol/water mixture by silicalite membrane on pervaporation. *J. Membr. Sci.*, **95** (3), 221–228.
20. Dong, J.H., Wegner, K., and Lin, Y.S. (1998) Synthesis of submicron polycrystalline MFI zeolite films on porous ceramic supports. *J. Membr. Sci.*, **148** (2), 233–241.
21. Geus, E.R., Exter, M.J., and Bekkum, H. (1992) Synthesis and characterization of zeolite (MFI) membranes on porous ceramic supports. *J. Chem. Soc., Faraday Trans.*, **88**, 3101–3109.
22. Yan, Y., Davis, M.E., and Gavalas, G.R. (1995) Preparation of zeolite ZSM-5 membranes by in-situ crystallization on

porous α-Al$_2$O$_3$. *Ind. Eng. Chem. Res.*, **34** (5), 1652–1661.
23. Piera, E., Giroir-Fendler, A., Dalmon, J.A., Moueddeb, H., Coronas, J., Menéndez, M., and Santamaría, J. (1998) Separation of alcohols and alcohols/O$_2$ mixtures using zeolite MFI membranes. *J. Membr. Sci.*, **142** (1), 97–109.
24. Gues, E.R. and Bekkum, H. (1995) *Zeolites*, **15**, 333.
25. Exter, M.J., Bekkum, H., Rijn, C.J.M., Kapteijn, F., Moulijn, J.A., Schellevis, H., and Beenakker, C.I.N. (1997) *Zeolites*, **19**, 13.
26. Dong, J.H., Lin, Y.S., Hu, M.Z.C., Peascoe, R.A., and Payzant, E.A. (2000) Template-removal-associated microstructural development of porous-ceramic-supported MFI zeolite membranes. *Microporous Mesoporous Mater.*, **34** (3), 241–253.
27. Lovallo, M.C., Tsapatsis, M., and Okubo, T. (1996) Preparation of an asymmetric zeolite L film. *Chem. Mater.*, **8** (8), 1579–1583.
28. Boudreau, L.C. and Tsapatsis, M. (1997) A highly oriented thin film of zeolite A. *Chem. Mater.*, **9** (8), 1705–1709.
29. Kazemimoghadam, M. and Mohammadi, T. (2007) Synthesis of MFI zeolite membranes for water desalination. *Desalination*, **206** (1–3), 547–553.
30. Yoo, W.C., Stoeger, J.A., Lee, P.-S., Tsapatsis, M., and Stein, A. (2010) High-performance randomly oriented zeolite membranes using brittle seeds and rapid thermal processing. *Angew. Chem. Int. Ed.*, **49** (46), 8699–8703.
31. Lovallo, M.C. and Tsapatsis, M. (1996) Preferentially oriented submicron silicalite membranes. *AIChE J.*, **42** (11), 3020–3029.
32. Schoeman, B.J., Sterte, J., and Otterstedt, J.E. (1994) Colloidal zeolite suspensions. *Zeolites*, **14** (2), 110–116.
33. Persson, A.E., Schoeman, B.J., Sterte, J., and Otterstedt, J.E. (1994) The synthesis of discrete colloidal particles of TPA-silicalite-1. *Zeolites*, **14** (7), 557–567.
34. Tsapatsis, M., Lovallo, M.C., Okubo, T., Davis, M.E., and Sadakata, M. (1995) Characterization of zeolite L nanoclusters. *Chem. Mater.*, **7** (9), 1734–1741.
35. Lovallo, M.C., Gouzinis, A., and Tsapatsis, M. (1998) Synthesis and characterization of oriented MFI membranes prepared by secondary growth. *AIChE J.*, **44** (8), 1903–1913.
36. Gouzinis, A. and Tsapatsis, M. (1998) On the preferred orientation and microstructural manipulation of molecular sieve films prepared by secondary growth. *Chem. Mater.*, **10** (9), 2497–2504.
37. Xomeritakis, G., Gouzinis, A., Nair, S., Okubo, T., He, M., Overney, R.M., and Tsapatsis, M. (1999) Growth, microstructure, and permeation properties of supported zeolite (MFI) films and membranes prepared by secondary growth. *Chem. Eng. Sci.*, **54** (15–16), 3521–3531.
38. Mintova, S., Hedlund, J., Schoeman, B., Valtchev, V., and Sterte, J. (1997) Continuous films of zeolite ZSM-5 on modified gold surfaces. *Chem. Commun.*, **1**, 15–16.
39. Mintova, S., Hedlund, J., Valtchev, V., Schoeman, B.J., and Sterte, J. (1998) ZSM-5 films prepared from template free precursors. *J. Mater. Chem.*, **8** (10), 2217–2221.
40. Hedlund, J., Schoeman, B.J., and Sterte, J. (1997) Synthesis of ultra thin films of molecular sieves by the seed film method, *Studies in Surface Science and Catalysis*, **105**, 2203–2210.
41. Hedlund, J., Noack, M., Kölsch, P., Creaser, D., Caro, J., and Sterte, J. (1999) ZSM-5 membranes synthesized without organic templates using a seeding technique. *J. Membr. Sci.*, **159** (1–2), 263–273.
42. Seike, T., Matsuda, M., and Miyake, M. (2002) Preparation of FAU type zeolite membranes by electrophoretic deposition and their separation properties. *J. Mater. Chem.*, **12** (2), 366–368.
43. Shan, W., Zhang, Y., Yang, W., Ke, C., Gao, Z., Ye, Y., and Tang, Y. (2004) Electrophoretic deposition of nanosized zeolites in non-aqueous medium

44. Muñoz, T. Jr., and Balkus, K.J. Jr., (1999) Preparation of oriented zeolite UTD-1 membranes via pulsed laser ablation. *J. Am. Chem. Soc.*, **121** (1), 139–146.
45. Balkus, K.J. Jr., and Scott, A.S. (1999) Zeolite coatings on three-dimensional objects via laser ablation. *Chem. Mater.*, **11** (2), 189–191.
46. Balkus, K.J., Munoz, T., and Gimon-Kinsel, M.E. (1998) Preparation of zeolite UTD-1 films by pulsed laser ablation: evidence for oriented crystal growth. *Chem. Mater.*, **10** (2), 464–466.
47. Pan, M. and Lin, Y.S. (2001) Template-free secondary growth synthesis of MFI type zeolite membranes. *Microporous Mesoporous Mater.*, **43** (3), 319–327.
48. Choi, J., Jeong, H.-K., Snyder, M.A., Stoeger, J.A., Masel, R.I., and Tsapatsis, M. (2009) Grain boundary defect elimination in a zeolite membrane by rapid thermal processing. *Science*, **325** (5940), 590–593.
49. Yoo, W.C., Kumar, S., Penn, R.L., Tsapatsis, M., and Stein, A. (2009) Growth patterns and shape development of zeolite nanocrystals in confined syntheses. *J. Am. Chem. Soc.*, **131** (34), 12377–12383.
50. Yoo, W.C., Kumar, S., Wang, Z., Ergang, N.S., Fan, W., Karanikolos, G.N., McCormick, A.V., Penn, R.L., Tsapatsis, M., and Stein, A. (2008) Nanoscale reactor engineering: hydrothermal synthesis of uniform zeolite particles in massively parallel reaction chambers. *Angew. Chem. Int. Ed.*, **47** (47), 9096–9099.
51. Lee, J.S., Kim, J.H., Lee, Y.J., Jeong, N.C., and Yoon, K.B. (2007) Manual assembly of microcrystal monolayers on substrates. *Angew. Chem. Int. Ed.*, **46** (17), 3087–3090.
52. Madhusoodana, C.D., Das, R.N., Kameshima, Y., and Okada, K. (2006) Preparation of ceramic honeycomb filter supported zeolite membrane modules by microwave-assisted in-situ crystallization. *Key Eng. Mater.*, **317–318**, 697–700.
53. Zhang, L. and Wang, H. (2002) In situ synthesis of AlPO4-14, CoAPO-44 and ZnAPO-34 films on alumina substrates. *J. Mater. Sci.*, **37** (8), 1491–1496.
54. Girnus, I., Pohl, M.M., Richter-Mendau, J., Schneider, M., Noack, M., Venzke, D., and Caro, J. (1995) Synthesis of AlPO4-5 aluminumphosphate molecular sieve crystals for membrane applications by microwave heating. *Adv. Mater.*, **7** (8), 711–714.
55. Li, Y., Liu, J., and Yang, W. (2006) Formation mechanism of microwave synthesized LTA zeolite membranes. *J. Membr. Sci.*, **281** (1–2), 646–657.
56. Julbe, A., Motuzas, J., Cazevielle, F., Volle, G., and Guizard, C. (2003) Synthesis of sodalite/α-Al$_2$O$_3$ composite membranes by microwave heating. *Sep. Purif. Technol.*, **32** (1–3), 139–149.
57. Julbe, A., Motuzas, J., Arruebo, M., Noble, R.D. and Beresnevicius, Z.J. (2005) Synthesis and properties of MFI zeolite membranes prepared by microwave assisted secondary growth, from microwave derived seeds *Studies in Surface Science and Catalysis*, **158** (Part A), 129–136.
58. Coutinho, D., Losilla, J.A., and Balkus, K.J. Jr., (2006) Microwave synthesis of ETS-4 and ETS-4 thin films. *Microporous Mesoporous Mater.*, **90** (1–3), 229–236.
59. Chen, X., Yang, W., Liu, J., and Lin, L. (2005) Synthesis of zeolite NaA membranes with high permeance under microwave radiation on mesoporous-layer-modified macroporous substrates for gas separation. *J. Membr. Sci.*, **255** (1–2), 201–211.
60. Xu, X., Bao, Y., Song, C., Yang, W., Liu, J., and Lin, L. (2004) Microwave-assisted hydrothermal synthesis of hydroxy-sodalite zeolite membrane. *Microporous Mesoporous Mater.*, **75** (3), 173–181.
61. Motuzas, J., Julbe, A., Noble, R.D., van der Lee, A., and Beresnevicius, Z.J. (2006) Rapid synthesis of oriented

silicalite-1 membranes by microwave-assisted hydrothermal treatment. *Microporous Mesoporous Mater.*, **92** (1–3), 259–269.

62. Cundy, C.S. (1998) Microwave techniques in the synthesis and modification of zeolite catalysts. A review. *Collect. Czech. Chem. Commun.*, **63** (11), 1699–1723.

63. Xu, X., Yang, W., Liu, J., and Lin, L. (2000) Fast formation of NaA zeolite membrane in the microwave field. *Chin. Sci. Bull.*, **45** (13), 1179–1181.

64. Xu, X., Yang, W., Liu, J., and Lin, L. (2001) Synthesis of NaA zeolite membrane by microwave heating. *Sep. Purif. Technol.*, **25** (1–3), 241–249.

65. Serrano, D.P., Uguina, M.A., Sanz, R., Castillo, E., Rodríguez, A., and Sánchez, P. (2004) Synthesis and crystallization mechanism of zeolite TS-2 by microwave and conventional heating. *Microporous Mesoporous Mater.*, **69** (3), 197–208.

66. Jhung, S.H., Chang, J.S., Hwang, J.S., and Park, S.E. (2003) Selective formation of SAPO-5 and SAPO-34 molecular sieves with microwave irradiation and hydrothermal heating. *Microporous Mesoporous Mater.*, **64** (1–3), 33–39.

67. Park, S.E., Chang, J.S., Young, K.H., Dae, S.K., Sung, H.J., and Jin, S.H. (2004) Supramolecular interactions and morphology control in microwave synthesis of nanoporous materials. *Catal. Surv. Asia*, **8** (2), 91–110.

68. Kunnakorn, D., Rirksomboon, T., Aungkavattana, P., Kuanchertchoo, N., Atong, D., Kulprathipanja, S., and Wongkasemjit, S. (2011) Performance of sodium A zeolite membranes synthesized via microwave and autoclave techniques for water-ethanol separation: recycle-continuous pervaporation process. *Desalination*, **269** (1–3), 78–83.

69. Weh, K., Noack, M., Sieber, I., and Caro, J. (2002) Permeation of single gases and gas mixtures through faujasite-type molecular sieve membranes. *Microporous Mesoporous Mater.*, **54** (1–2), 27–36.

70. Motuzas, J., Heng, S., Ze Lau, P.P.S., Yeung, K.L., Beresnevicius, Z.J., and Julbe, A. (2007) Ultra-rapid production of MFI membranes by coupling microwave-assisted synthesis with either ozone or calcination treatment. *Microporous Mesoporous Mater.*, **99** (1–2), 197–205.

71. Tang, Z., Kim, S.-J., Gu, X., and Dong, J.H. (2009) Microwave synthesis of MFI-type zeolite membranes by seeded secondary growth without the use of organic structure directing agents. *Microporous Mesoporous Mater.*, **118** (1–3), 224–231.

72. Xiao, W., Zhou, L., Yang, J., Lu, J., and Wang, J. (2011) A simple seeding method for MFI zeolite membrane synthesis on macroporous support by microwave heating. *Microporous Mesoporous Mater.*, **142** (1), 154–160.

73. Miachon, S., Ciavarella, P., van Dyk, L., Kumakiri, I., Fiaty, K., Schuurman, Y., and Dalmon, J.A. (2007) Nanocomposite MFI-alumina membranes via pore-plugging synthesis: specific transport and separation properties. *J. Membr. Sci.*, **298** (1–2), 71–79.

74. Zhao, Q., Wang, J., Chu, N., Yin, X., Yang, J., Kong, C., Wang, A., and Lu, J. (2008) Preparation of high-permeance MFI membrane with the modified secondary growth method on the macroporous α-alumina tubular support. *J. Membr. Sci.*, **320** (1–2), 303–309.

75. Xiao, W., Yang, J., Lu, J., and Wang, J. (2009) Preparation and characterization of silicalite-1 membrane by counter-diffusion secondary growth. *J. Membr. Sci.*, **345** (1–2), 183–190.

76. Lai, Z., Bonilla, G., Diaz, I., Nery, J.G., Sujaoti, K., Amat, M.A., Kokkoli, E., Terasaki, O., Thompson, R.W., Tsapatsis, M., and Vlachos, D.G. (2003) Microstructural optimization of a zeolite membrane for organic vapor separation. *Science*, **300** (5618), 456–460.

77. Kuzniatsova, T.A., Mottern, M.L., Chiu, W.V., Kim, Y., Dutta, P.K., and Verweij, H. (2008) Synthesis of thin, oriented zeolite a membranes on a macroporous support. *Adv. Funct. Mater.*, **18** (6), 952–958.

78. Yan, Y., Davis, M.E., and Gavalas, G.R. (1997) Preparation of highly selective zeolite ZSM-5 membranes by a post-synthetic coking treatment. *J. Membr. Sci.*, **123** (1), 95–103.
79. Xomeritakis, G., Lai, Z., and Tsapatsis, M. (2000) Separation of xylene isomer vapors with oriented MFI membranes made by seeded growth. *Ind. Eng. Chem. Res.*, **40** (2), 544–552.
80. Nair, S., Lai, Z., Nikolakis, V., Xomeritakis, G., Bonilla, G., and Tsapatsis, M. (2001) Separation of close-boiling hydrocarbon mixtures by MFI and FAU membranes made by secondary growth. *Microporous Mesoporous Mater.*, **48** (1–3), 219–228.
81. Nomura, M., Yamaguchi, T., and Nakao, S.-I. (1997) Silicalite membranes modified by counterdiffusion CVD technique. *Ind. Eng. Chem. Res.*, **36** (10), 4217–4223.
82. Zhang, B., Wang, C., Lang, L., Cui, R., and Liu, X. (2008) Selective defect-patching of zeolite membranes using chemical liquid deposition at organic/aqueous interfaces. *Adv. Funct. Mater.*, **18** (21), 3434–3443.
83. Hong, Z., Zhang, C., Gu, X., Jin, W., and Xu, N. (2011) A simple method for healing nonzeolitic pores of MFI membranes by hydrolysis of silanes. *J. Membr. Sci.*, **366** (1–2), 427–435.
84. Chen, Z., Yin, D., Li, Y., Yang, J., Lu, J., Zhang, Y., and Wang, J. (2011) Functional defect-patching of a zeolite membrane for the dehydration of acetic acid by pervaporation. *J. Membr. Sci.*, **369** (1–2), 506–513.
85. Murad, S., Oder, K., and Lin, J. (1998) Molecular simulation of osmosis, reverse osmosis, and electro-osmosis in aqueous and methanolic electrolyte solutions. *Mol. Phys.*, **95** (3), 401–408.
86. Li, L.X., Dong, J.H., Nenoff, T.M., and Lee, R. (2004) Reverse osmosis of ionic aqueous solutions on a MFI zeolite membrane. *Desalination*, **170** (3), 309–316.
87. Li, L.X., Dong, J.H., and Nenoff, T.M. (2007) Transport of water and alkali metal ions through MFI zeolite membranes during reverse osmosis. *Sep. Purif. Technol.*, **53** (1), 42–48.
88. Duke, M., O'Brien-Abraham, J., Milne, N., Zhu, B., Lin, Y.S., and Diniz da Costa, J.C. (2009) Seawater desalination performance of MFI type membranes made by secondary growth. *Sep. Purif. Technol.*, **68**, 343–350.
89. Chowdhuri, S. and Chandra, A. (2001) Molecular dynamics simulations of aqueous NaCl and KCl solutions: effects of ion concentration on the single-particle, pair, and collective dynamical properties of ions and water molecules. *J. Chem. Phys.*, **115** (8), 3732–3741.
90. Li, L., Liu, N., McPherson, B., and Lee, R. (2007) Enhanced water permeation of reverse osmosis through MFI-type zeolite membranes with high aluminum contents. *Ind. Eng. Chem. Res.*, **46** (5), 1584–1589.
91. Jareman, F., Hedlund, J., and Sterte, J. (2003) Effects of aluminum content on the separation properties of MFI membranes. *Sep. Purif. Technol.*, **32** (1–3), 159–163.
92. Noack, M., Kölsch, P., Seefeld, V., Toussaint, P., Georgi, G., and Caro, J. (2005) Influence of the Si/Al-ratio on the permeation properties of MFI-membranes. *Microporous Mesoporous Mater.*, **79** (1–3), 329–337.
93. Noack, M., Kölsch, P., Dittmar, A., Stöhr, M., Georgi, G., Eckelt, R., and Caro, J. (2006) Effect of crystal intergrowth supporting substances (ISS) on the permeation properties of MFI membranes with enhanced Al-content. *Microporous Mesoporous Mater.*, **97** (1–3), 88–96.
94. Zhu, B., Zou, L., Doherty, C.M., Hill, A.J., Lin, Y.S., Hu, X.R., Wang, H.T., and Duke, M. (2010) Investigation of the effects of ion and water interaction on structure and chemistry of silicalite MFI type zeolite for its potential use as a seawater desalination membrane. *J. Mater. Chem.*, **20**, 4675–4683.
95. Cui, J., Zhang, X., Liu, H., Liu, S., and Yeung, K.L. (2008) Preparation and application of zeolite/ceramic microfiltration membranes for treatment of oil contaminated water. *J. Membr. Sci.*, **325** (1), 420–426.

96. Malekpour, A., Millani, M.R., and Kheirkhah, M. (2008) Synthesis and characterization of a NaA zeolite membrane and its applications for desalination of radioactive solutions. *Desalination*, **225** (1–3), 199–208.
97. Li, L., Liu, N., McPherson, B., and Lee, R. (2008) Influence of counter ions on the reverse osmosis through MFI zeolite membranes: implications for produced water desalination. *Desalination*, **228** (1–3), 217–225.
98. Liu, N., Li, L., McPherson, B., and Lee, R. (2008) Removal of organics from produced water by reverse osmosis using MFI-type zeolite membranes. *J. Membr. Sci.*, **325** (1), 357–361.
99. Clark, C.E. and Veil, J.A. (2009) Produced Water Volumes and Management Practices in the United States, ANL/EVS/R-09/1, prepared by the Environmental Science Division, Argonne National Laboratory for the U.S. Department of Energy, Office of Fossil Energy, National Energy Technology Laboratory, September.
100. Lu, J., Liu, N., Li, L., and Lee, R. (2010) Organic fouling and regeneration of zeolite membrane in wastewater treatment. *Sep. Purif. Technol.*, **72** (2), 203–207.
101. Kazemimoghadam, M. and Mohammadi, T. (2010) Mechanisms and experimental results of aqueous mixtures pervaporation using nanopore HS zeolite membranes. *Desalination*, **262** (1–3), 273–279.
102. Kazemimoghadam, M. (2010) New nanopore zeolite membranes for water treatment. *Desalination*, **251** (1–3), 176–180.
103. Sanchez, J.G. and Tsotsis, T.T. (2002) *Catalytic Membranes and Membrane Reactors*, Wiley-VCH Verlag GmbH, Weinheim.
104. Heng, S., Yeung, K.L., Djafer, M., and Schrotter, J.C. (2007) A novel membrane reactor for ozone water treatment. *J. Membr. Sci.*, **289** (1–2), 67–75.
105. Baus, C., Schaber, K., Gassiot-Pintori, I., and Braun, A. (2002) Separation and oxidative degradation of organic pollutants in aqueous systems by pervaporation and vacuum-ultraviolet-photolysis. *Sep. Purif. Technol.*, **28** (2), 125–140.
106. Shanbhag, P.V., Guha, A.K., and Sirkar, K.K. (1995) Single-phase membrane ozonation of hazardous organic compounds in aqueous streams. *J. Hazard. Mater.*, **41** (1), 95–104.
107. Janknecht, P., Wilderer, P.A., Picard, C., Larbot, A., and Sarrazin, J. (2000) Investigations on ozone contacting by ceramic membranes. *Ozone: Sci. Eng.*, **22** (4), 379–392.
108. Mitani, M.M., Keller, A.A., Sandall, O.C., and Rinker, R.G. (2005) Mass transfer of ozone using a microporous diffuser reactor system. *Ozone: Sci. Eng.*, **27** (1), 45–51.
109. Heng, S., Yeung, K.L., Julbe, A., Ayral, A., and Schrotter, J.C. (2008) Preparation of composite zeolite membrane separator/contactor for ozone water treatment. *Microporous Mesoporous Mater.*, **115** (1–2), 137–146.

10
Molecular Scale Modeling of Membrane Water Treatment Processes

Harry F. Ridgway, Julian D. Gale, Zak E. Hughes, Matthew B. Stewart, John D. Orbell, and Stephen R. Gray

10.1
Introduction

Membrane processes have become commonplace in the water industry since the 1990s, with low-pressure microfiltration (MF) and ultrafiltration (UF) membranes used for the removal of particles and bacteria, whereas high-pressure membranes, such as reverse osmosis (RO) and nanofiltration (NF) membranes, are used for desalination and removal of color and trace organics of interest. Water transport and salt/color rejection in RO membranes occurs at the molecular scale, and molecular models have been used to develop an understanding of these processes. For MF/UF membranes, organic fouling is an operational problem that lends itself to modeling at the molecular scale, but there has been little emphasis on this approach to date.

This chapter examines three distinct areas relevant to the molecular modeling of membrane-based water treatment processes. First, Section 10.2 critically discusses the progress and methods used to study, at the molecular level, the structure and properties of high-pressure polymeric membranes, as well as the deficiencies currently present in this approach. Second, Section 10.3 analyzes the current state of research on zeolites as the most promising candidate for inorganic membrane material and describes the application of molecular modeling in studying the transport of water and ions through zeolite materials, as well as providing further insights into ion selectivity. Finally, Section 10.4 investigates the potential of molecular modeling to be applied to the problem of organic fouling of polymeric membranes. Such an approach could provide insights into the new methods to ameliorate organic fouling.

10.2
Molecular Simulations of Polymeric Membrane Materials for Water Treatment Applications

Given the global relevance of water purification membranes to manifest a sustainable water future for mankind, it is perhaps surprising that until very recently molecular simulations have not played a more significant role in membrane

Functional Nanostructured Materials and Membranes for Water Treatment, First Edition.
Edited by Mikel Duke, Dongyuan Zhao, and Raphael Semiat.
© 2013 Wiley-VCH Verlag GmbH & Co. KGaA. Published 2013 by Wiley-VCH Verlag GmbH & Co. KGaA.

materials design, synthesis, and characterization. This is especially true when one considers that state-of-the-art aromatic cross-linked polyamide (PA) membranes were discovered and commercialized in the early to mid-1970s. Moreover, sophisticated molecular simulation algorithms and techniques, as well as the necessary computational resources to carry them out, have since been utilized to probe and delineate the structural and functional aspects of many far more complex biological systems, including aquaporin [1], rhodopsin [2], and cytoplasmic membranes [3].

Molecular simulations refer to computational models of molecules or molecular systems that emulate the structure and/or dynamic behavior of real systems. Two broad levels of theory are used to simulate how atoms and molecules interact with one another: (i) molecular mechanics (MM) and molecular dynamics (MD) are primarily used for large numbers of atoms and are based solely on classical physics and (ii) molecular orbital theory based on quantum mechanical (QM) calculations. QM computations are required to calculate the electronic configurations of atoms and molecules, but they are computationally expensive and can therefore be applied only to small numbers of atoms. Systems comprising thousands of atoms, such as a model of a PA membrane, can only be efficiently simulated using MM. The dynamic aspects of atoms and molecules, that is, how they interact with one another over time, may be simulated by integrating Newton's laws of motion over small time steps on the order of 1 fs (10^{-15} s) or less. This technique is referred to as *molecular dynamics* simulation, and it may be applied using MM, QM, or mixed MM/QM methods to describe the spatial/temporal evolution of a chemical system.

It has only been in the past decade or so that molecular simulations of synthetic membrane materials have begun to play an increasingly significant role, not only in advancing our understanding of the atomic-scale structure and mechanics of polymeric membrane materials but also in assisting in the design of novel membranes with improved performance and durability. Along with this enhanced interest in using molecular simulation methods, there has been noteworthy innovation in designing membrane building algorithms and computational code to better achieve correspondence of model property predictions with experimental analysis of membrane materials and performance properties.

The following sections briefly review some key recent developments in polymer membrane modeling and simulations with the aim of providing broad guidance to those who may be interested in using membrane simulations to assist in their development of new membrane materials, especially for water treatment applications.

10.2.1
RO Membranes: Synthesis, Structure, and Properties

RO is a membrane-based water treatment technique gaining worldwide acceptance in a number of different areas, especially desalination and water reclamation [4]. RO is a pressure-driven process, whereby water is forced through a semipermeable, nonporous membrane, which rejects other constituents that are dissolved in the

10.2 Molecular Simulations of Polymeric Membrane Materials for Water Treatment Applications | 251

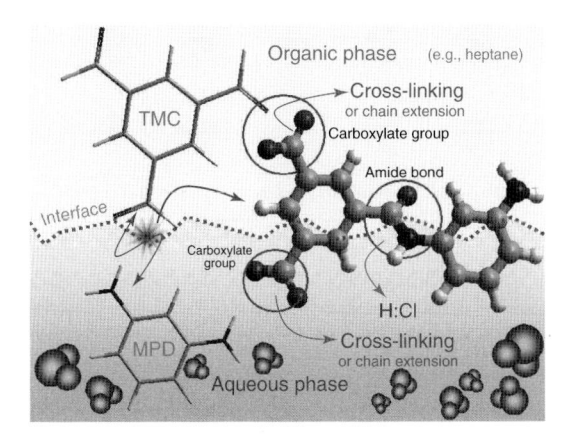

Figure 10.1 Interfacial polycondensation synthetic pathway for an aromatic cross-linked polyamide RO membrane material.

feed water [4] such as ions and dissolved organic molecules. This rejection is achieved by size exclusion, charge exclusion, and physicochemical interactions between the solute, solvent, and membrane surface [5, 6].

The most widely used type of RO membrane material for water purification applications (e.g., seawater desalination, brackish groundwater demineralization, and wastewater reuse) is currently the thin-film composite (TFC) fully aromatic cross-linked PA membrane, discovered in the mid-1970s by John Cadotte, then at North Star Research [7–9]. PA TFC membranes are produced by an interfacial polycondensation process, involving the reaction of an aqueous-phase primary diamine, (e.g., *meta*-phenylenediamine (MPD)), with a hydrophobic trifunctional acid chloride, for example, trimesolylchloride (TMC), dissolved in a suitable organic solvent (e.g., heptane, Figure 10.1). The reaction between TMC and MPD is extremely rapid and occurs at the water–solvent interface where a thin ($\sim$1000–2000 Å) water/solvent-insoluble PA network (i.e., membrane or film) with permselective properties is produced [10]. Because TMC is trifunctional, the membrane is partially cross-linked via MPD residues forming an amorphous polymeric network. Three markedly different diffusion-controlled kinetic regimes have been observed during synthesis, which determine the density, charge, roughness, thickness, and permselective characteristics of the PA membrane [11]. Differential staining techniques suggest that polymer density and charged carboxylic acid and amine groups are distributed across the active PA layer in a nonuniform (asymmetric) manner, with a negatively charged outermost (feed) surface positioned above an innermost (permeate) surface that bears a slight positive charge [12]. However, more recent studies employing near-surface ($\sim$6 nm) and volume-averaged analysis of the PA layer by X-ray photoelectron spectrometry (XPS) and Rutherford backscattering spectrometry (RBS), respectively, suggest that some membrane types (e.g., ESPA3, Hydranautics, Inc.) are spatially more homogeneous than others (e.g., NF90) with respect to the distribution of carboxylate groups, cross-linking density, elemental composition, and ionization behavior (Table 10.1).

Table 10.1 Summary of experimentally derived properties of PA membranes.

Parameter	Value	Method	References
Surface morphology	Rough and bubblelike	TEM of thin sections	Freger [12]
Water content	23 wt%	RBS spectrometry	Mi et al. [13]
Density (of hydrated PA film)	1.38 g cc^{-1}	RBS spectrometry	Mi et al. [13]
Carboxylate charge density (concentration of –R–COO– groups)	0.432 M	RBS spectrometry	Coronell et al. [14]
Amine charge density (–NH$_3^+$ group concentration)	0.036 M	RBS spectrometry	Coronell et al. [15]
pK_{a1} of –NH$_3^+$ groups	3.63 + 0.35–4.62 + 0.28	RBS spectrometry	Coronell et al. [14, 15]
pK_{a1} of –R–COO– groups	3.91 + 1.06–5.72 + 0.06	RBS spectrometry	Coronell et al. [14, 15]
pK_{a2} of –R–COO– groups	5.86 + 0.10–9.87 + 0.23	RBS spectrometry	Coronell et al. [14, 15]
Degree of cross-linking	94%	Calculated from number of free –R–COO– groups	Coronell et al. [15]
Degree of cross-linking	20–50%	Calculation from ATR-FTIR spectrometry	K. Ishida, Orange County Water District, CA (unpublished) and [16]
Pore size	2–4 Å	PALS	Kim et al. [17]
	4.4 Å	PALS	Sharma et al. [18]
Water self-diffusion coefficient	0.01 Å^2 ps^{-1}	Water permeability experiments	Lipp et al. [19]

The fine structure of a typical PA RO membrane is illustrated in Figure 10.2. The feed surface of the PA layer is structurally heterogeneous, comprising many fingerlike extensions or blebs that project into the bulk liquid (feed) phase. Despite several ultrastructural studies, it remains unclear if the fingerlike projections are solid PA or hollow "bubbles," as the transmission electron micrograph (TEM) images might suggest, because of an apparently reduced amount of electron-dense material in the central region of these structures [12, 20]. It is also unclear where exactly water–solute separations take place, a consideration that is especially troubling for materials simulation research. On the basis of the TEM studies of PA membranes differentially stained with uranyl cations or tungstate salts, Freger [12] has argued that separations occur in a dense PA sublayer beneath the bubblelike region. Recently, high-resolution TEM images published by Pacheco and coworkers [20] revealed that the sublayer itself is not of uniform density but rather comprises a distinctly "nodular" morphology that may influence water and solute transport behavior. However, the ultrastructural studies remain somewhat ambiguous and the notion that separations occur at least in part across the enhanced surface area

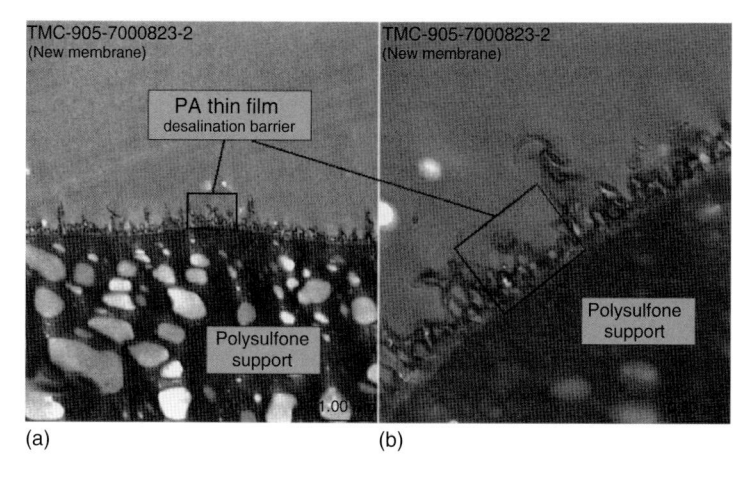

10.2 Molecular Simulations of Polymeric Membrane Materials for Water Treatment Applications | 253

Figure 10.2 Ultrastructure of a typical aromatic cross-linked polyamide RO membrane. Note the rough convoluted surface morphology. (Transmission electron micrographs provided courtesy of Robert L. Riley, Separation Systems Technology, San Diego, California, USA.)

of the "bubble" walls (which would account for the comparatively high flux of PA membranes) has not been entirely ruled out.

A number of analytically derived parameters of PA membranes are summarized in Table 10.1. Of particular relevance for modeling membrane materials is the protonation status of the membrane at neutral or close to neutral pH values. There appear to be at least two populations of ionizable carboxylate groups that can be distinguished by titration with positively charged cationic probes, such as Ba^{2+}, which can be in turn quantified by RBS spectrometry [21, 14, 22].

Until recently, free carboxylic acid groups (R-COOH/R-COO-) residing in the PA separation layer were believed to be uniformly accessible to protons and dissolved cations. Thus, all modeling studies to date have assumed a uniform state of membrane protonation (i.e., either complete protonation or complete deprotonation of all carboxylate groups). However, recent experimental work has proved this assumption to be false. Using RBS analysis, Coronell et al. [21] demonstrated that the proportion of free (i.e., non-cross-linked) carboxylate groups in the PA active layer that were accessible by Ba^{2+} counterions varied substantially in membrane type from about 0.4 to 0.81. Similarly, the ionization behavior of free carboxylate groups in the FT30-like membrane matrix were determined to be biphasic with measured pK_a values in the range of 5–9 [15]. Curiously, in comparison, the ionization behavior of the free amine groups (R-NH_3^+/R-NH_2) in PA active layers could be satisfactorily explained by a single dissociation constant (pK_a = 4.74). The physical basis for the biphasic ionization (and cation accessibility) behavior is not entirely understood, but may be related to the distribution of pore size and shape throughout the membrane matrix and other ultrastructural heterogeneities described by Freger [12] and others [23]. Consonant with this hypothesis are the recent studies of De Baerdemaeker et al. [24] in which positron annihilation lifetime spectroscopy (PALS) with slow positron beam analysis indicated a bimodal pore size distribution in NF membranes, with pore radii for the dual populations ranging from about 1.25 to 1.55 and 3.2 to 3.95 Å. The authors suggest that somewhat smaller pore radii might be expected for RO membranes. These data suggest that in order to more accurately model the PA active layer of an FT30-like membrane material, it may be necessary to protonate only the most ion-accessible carboxylate groups in a manner consistent with experimental findings.

Evidently, no experimental data are available regarding the relative strengths of association of positively charged counterions (e.g., Ba^{2+}, Na^+, K^+, Ca^{2+}, Mg^{2+}, etc.) with ionized carboxylate groups and it remains uncertain if some cations might remain tightly bound to membrane functional groups in actual hydrated membranes. If they do, then such strongly associated cations must also be taken into account to accurately simulate charge–charge interactions (also referred to as *long-range electrostatics*) in membrane models.

Accurate modeling of PA and other RO membrane materials also requires knowledge of the state of water in the membrane polymer matrix. Water in an RO membrane is primarily confined to nanopores that range in size from about 2 to 5 Å. Using quasielectric neutron scattering (QENS), Sharma and coworkers [18] demonstrated that the water in a PA membrane matrix strongly interacts with membrane atoms, thereby reducing water diffusion rates relative to those of bulk phase water. Such intermolecular interactions result in slower self-diffusion of water inside the membrane ($\sim 1.9 \pm 0.4 \times 10^{-5}$ cm^2 s^{-1}) compared to bulk water ($\sim 2.2 \times 10^{-5}$ cm^2 s^{-1}) and an increased residence time ($\sim 2.8 \pm 0.6$ ps). On the basis of QENS analysis, it was further concluded that water translates by a jump diffusion mechanism with a mean jump length of about 1.8 ± 0.5 Å. Finally, using differential scanning calorimetry, water was found to be more highly structured (and, consequently, less mobile) inside the membrane interior resulting in elevation of the freezing point by ~ 2.3 °C. This observation is consistent with strong intermolecular interactions between water and membrane atoms but does not describe the nature of such interactions.

On the basis of the ultrastructural and analytical investigations described earlier, PA membranes clearly possess a previously unanticipated degree of nanoscale physicochemical and morphological heterogeneity. Efforts to build accurate models of thin-film semipermeable PA membrane materials are hindered by such heterogeneity, which cannot be readily captured and simulated atomistically at relevant temporal and spatial scales without exceeding or challenging the computational capacity of modern CPUs. Implicit in the atomistic models of PA membranes that have so far been attempted (see subsequent text) is the notion that the intrinsic structural and permselective properties of these membrane materials can be faithfully captured and computed from models of relatively homogeneous polymer networks. However, it may be necessary to resort to mesoscale or multiscale modeling regimes, for example, dissipative particle dynamics (e.g., see He et al. [25]) or similar approaches to better simulate the influence of larger scale features on membrane structure and function.

10.2.2
Strategies for Modeling Polymer Membranes

Since the mid-1990s, several innovative approaches have been investigated for modeling synthetic cross-linked polymer membrane materials. Kotelyanskii *et al.* [26–28] first introduced the concept of using a random-walk self-avoiding polymer chain built over a three-dimensional (3D) cubic lattice as the initial structure for modeling Cadotte-style cross-linked membrane materials, such as fully aromatic PA membranes (Figure 10.3). In this scenario, the starting configuration was an annealed self-avoiding Markov chain that was sequentially "decorated" with the monomer building blocks of a PA membrane, namely, MPD + TMC dimeric subunits. The software used to create the self-avoiding chain was that of Biosym Technologies, now known as Accelrys, Inc., San Diego, CA. The fully decorated polymer comprised a total of 62 repeat units, each consisting of a covalently bonded MPD + TMC pair. This mode of construction resulted in a $5 \times 5 \times 5$ matrix of alternating MPD/TMC residues packed into a cubic box $\sim$30 Å on a side with imposed periodic boundary conditions.

It is noteworthy that all nonbonded free carboxylic acid groups remained fully protonated in the Kotelyanskii model. However, subsequent experimental work (briefly described earlier; see also Table 10.1) demonstrated that a significant population of carboxylic acid groups become ionized at pH values above about 6.0 or less [14, 15]. Thus, for many water treatment applications where the pH is close to neutrality or sometimes higher, the membrane is expected to be negatively charged. The TMC residues (as well as the MPD + TMC dimeric subunits) were modeled a priori in the Kotelyanskii studies with the assumption of complete protonation, that is, the monomeric building blocks were uncharged. It may be reasonably assumed that partial atomic charges calculated for these monomers

Step 1... Creation of an annealed self-avoiding chain on a three-dimensional lattice matrix

↓

Step 2... Decoration of lattice sites with alternating PA repeat units (i.e., MPD + TMC)

↓

Step 3... Energy minimization (<0.1 kcal mol^{-1}-Å)

↓

Step 4... Cross-linking performed via distance hueristic ($<7 \pm$ Å separation)

↓

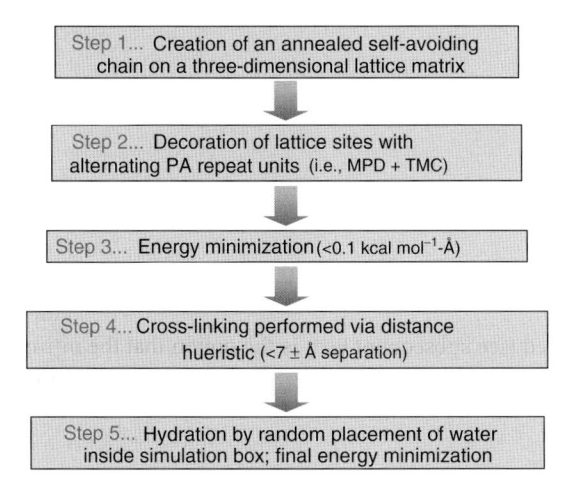

Step 5... Hydration by random placement of water inside simulation box; final energy minimization

Figure 10.3 Lattice decoration strategy used by Kotelyanskii *et al.* [26–28] to build atomistic models of the PA active layer.

would have been significantly different from those of the corresponding ionized monomers. How this factor might have affected the overall system electrostatics and therefore the secondary structural conformation of the polymer matrix was not determined. Nor was it determined how the extent of carboxylate protonation affected the outcome of subsequent MD simulations of these systems.

Using the partially proprietary Biosym CVFF Newtonian force field, Kotelyanskii and coworkers [27] geometry optimized the initial decorated lattice structure to less than 0.1 kcal mol^{-1}·Å using a conjugate-gradient method. Intrachain cross-linking was introduced into their models by "bridging" free carboxylic acid groups separated by no more than 7 ± 2 Å with a difunctional MPD residue. In the Kotelyanskii membrane model, a maximum of six cross-links were introduced to obtain what was believed to be a reasonable estimate of the experimentally derived cross-link density. However, at the time of the 1998 publication, the cross-link density had not yet been independently confirmed outside of Dow-FilmTec Corporation, the sole manufacturer of the FT30 PA membrane.

The Kotelyanskii model was hydrated by insertion of 285 water molecules at random positions throughout the periodic simulation cell. Solvation of the polymer matrix was followed by a final round of energy minimization that served to eliminate bad (i.e., energetically unfavorable) contacts between water molecules and polymer atoms (Figure 10.3). Lattice periodicity inherited from the initial chain conformation (built over the 3D lattice) was found to be rapidly annealed out of the system within the first 100 ps of NVT simulation resulting in a completely amorphous cross-linked polymer network. Experimental X-ray diffraction data indicate that the FT30-like polymer membranes are indeed random structures with no detectable chain periodicity [27]. As a final step in model construction, Na$^+$ and Cl$^-$ ions were added to the polymer matrix at random positions followed by energy minimization.

The Kotelyanskii model was reasonably successful in describing the water permeation characteristics and salt rejection mechanisms of PA membranes (Figure 10.3). As such, this model represents the first successful effort to model the water and solute transport properties of FT30-style RO membrane materials and established a new paradigm for subsequent modeling efforts in this area.

As outlined in Figure 10.4, Harder *et al.* [29] have more recently employed a somewhat different approach for modeling the active layer of cross-linked PA membranes. They begin with a periodic simulation cell containing a spatially randomized mixture of TMC and MPD monomers and then gradually reduce the box dimensions until a target density of about 1.38 g cc^{-1} (for the solvated system) is reached. The entire system is energy minimized and subjected to 1000 steps of MD after each box size readjustment. Subsequently, in what is referred to as *Stage I*, amide bonds are introduced according to a separation distance criterion in which TMC and MPD residues closer than 3.25 Å become covalently linked. Over time, the rate of amide bond formation approaches zero as monomers gradually run out of potential bonding partners that fall within the initial distance heuristic. MD is carried out during this process with bond formation implemented at 1 ps intervals. A second stage of amide bond formation is subsequently initiated using a more

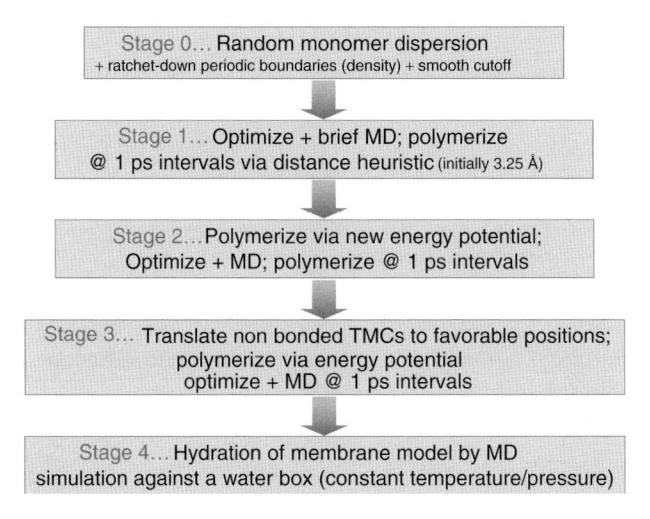

```
┌─────────────────────────────────────────────────────────────┐
│  Stage 0... Random monomer dispersion                       │
│  + ratchet-down periodic boundaries (density) + smooth cutoff│
└─────────────────────────────────────────────────────────────┘
                              ↓
┌─────────────────────────────────────────────────────────────┐
│  Stage 1... Optimize + brief MD; polymerize                 │
│  @ 1 ps intervals via distance heuristic (initially 3.25 Å) │
└─────────────────────────────────────────────────────────────┘
                              ↓
┌─────────────────────────────────────────────────────────────┐
│  Stage 2... Polymerize via new energy potential;            │
│  Optimize + MD; polymerize @ 1 ps intervals                 │
└─────────────────────────────────────────────────────────────┘
                              ↓
┌─────────────────────────────────────────────────────────────┐
│  Stage 3... Translate non bonded TMCs to favorable positions;│
│  polymerize via energy potential                            │
│  optimize + MD @ 1 ps intervals                             │
└─────────────────────────────────────────────────────────────┘
                              ↓
┌─────────────────────────────────────────────────────────────┐
│  Stage 4... Hydration of membrane model by MD               │
│  simulation against a water box (constant temperature/pressure)│
└─────────────────────────────────────────────────────────────┘
```

Figure 10.4 Random monomer dispersion approach used by Harder et al. [29] to build fully atomistic models of the PA active layer.

favorable interaction potential and slightly increased distance criterion. As in Stage I, bond formation in Stage II is cycled at 1 ps intervals during MD simulation. The third and final stage of amide bond formation commences after bond formation in the second stage begins to approach zero. In Stage III, still free TMC residues are translated to more favorable bonding positions that are proximal to "free" MPD monomers that possess at least one nonbonded amine functional group.

Rather than randomly populate their membrane model with individual water molecules, as was done by Kotelyanskii et al. [27, 28], Harder et al. [29] used a preequilibrated orthogonal box of pure TIP3P water (~55 M) to solvate the membrane system. Owing to the cubic nature of the generated membrane model, the orthogonal equilibrated water box can form three interfaces with the cubic membrane model, that is, the faces of the membrane in the x, y, and z Cartesian axes. These three arrangements were prepared and water was allowed to self-diffuse into the adjacent membrane in a series of MD simulations in order to gain better spatial averaging of the polymerized material. These simulations were carried out under periodic boundary conditions and at a constant temperature and pressure (NPT ensemble) of 300 °K and 1 bar, respectively. Using this approach, the authors were able to calculate a water partition coefficient for the membrane of about 0.21, which agreed reasonably well with the reported experimental value of 0.29 [13]. In addition, the calculated density of the fully solvated membrane interior region was 1.38 g cm^{-3}, which was also in close agreement with experimental findings [13].

Ridgway used the Hyperchem modeling environment to develop an automated (i.e., iterative) algorithm for building multiple variants of atomistic models of PA RO membrane materials [30, 31] (Figure 10.5). MPD and TMC monomers are initially constructed using suitable classical force fields, as well as QM and/or semiempirical methods (e.g., PM3 or RM1) to establish partial atomic charges. However, unlike the Kotelyanskii and Harder models described earlier, which used the protonated

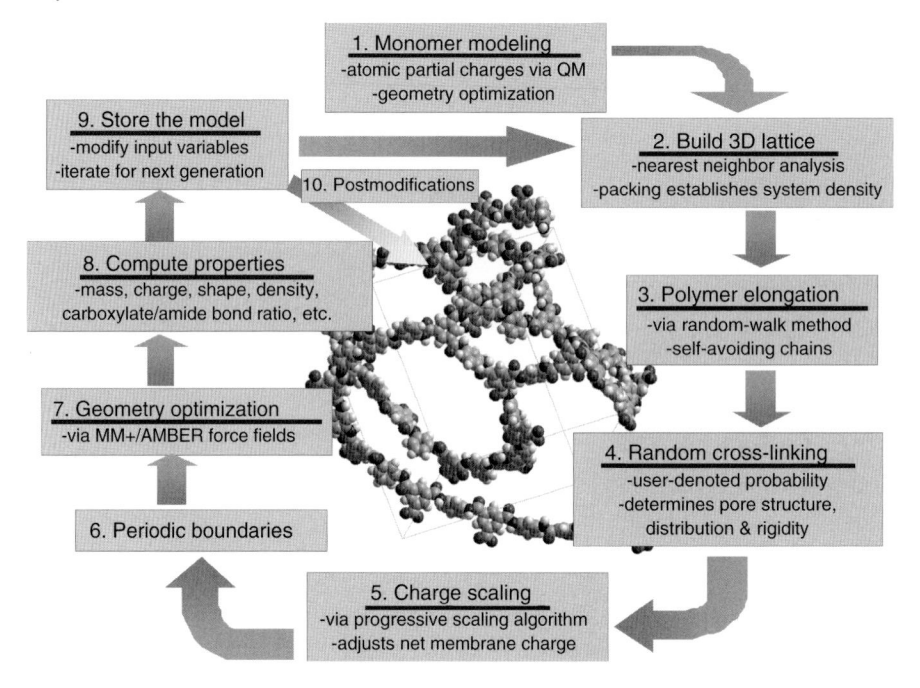

Figure 10.5 Outline of automated membrane model build algorithm of Ridgway. The program generates a series of PA membrane variants, each differing from the one before because of randomization functions for chain elongation, cross-linking probability, density, and so on. Each model is stored and its structural and electronic properties computed before the cycle is iterated.

and halogenated forms of the TMC monomer, respectively, the nonprotonated form of TMC with a net charge of -3.0 e was used by Ridgway to compute monomer atomic partial charges. This decision was based on the experimental zeta potential measurements of PA membrane materials that indicated carboxylate groups are largely deprotonated at pH values above $\sim$6.5 [23, 32]. However, recent ion probe studies by Coronell et al. [15] suggest that a significant proportion of carboxylate functional groups in the PA active layer may possess pK_a values in the range of 6–9 (Table 10.1). To date, the conformation and MD of the PA membrane models have not been determined as a function of monomer partial atomic charges. Short- and long-range electrostatic interactions are extremely important with respect to intermolecular and intramolecular hydrogen-bonding and other Coulomb interactions that undoubtedly influence polymer conformation, dynamics, and function. Atom partial charges can differ substantially between models depending on the method chosen for their computation and the protonation state of the compounds, that is, whether the monomers are charged or neutral. A rigorous sensitivity analysis should be performed to determine how the structural (e.g., density and chain conformations) and the functional (e.g., water and solute transport) properties of various membrane models are affected by systematic adjustment of specific atom partial charges of the monomers.

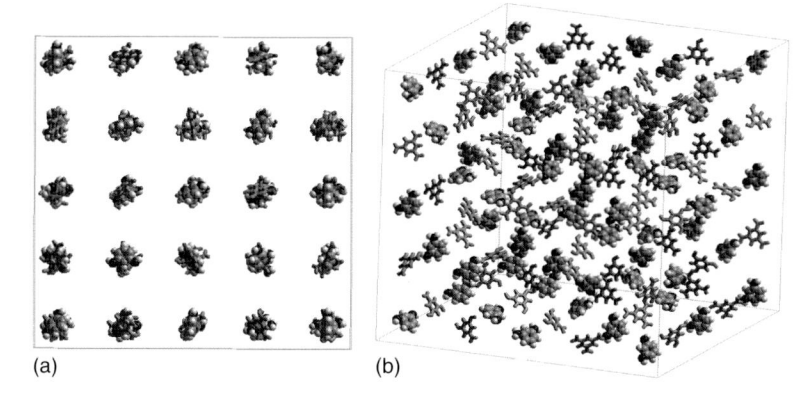

10.2 Molecular Simulations of Polymeric Membrane Materials for Water Treatment Applications | 259

Figure 10.6 End-on (a) and oblique (b) views of initial nonbonded 3D monomer lattice before chain elongation and cross-linking. Monomer spacing in the lattice determines the density of the membrane model. MPD = space filled; TMC = tubes.

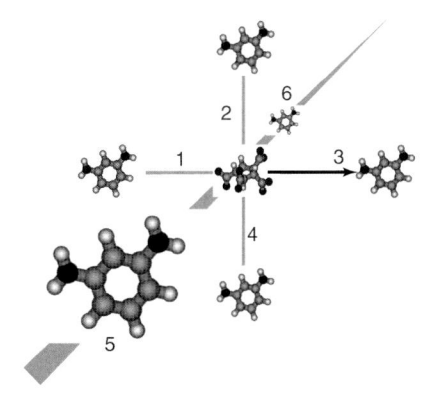

Figure 10.7 Monomer neighborhood analysis in the 3D lattice. Each embedded monomer has six nearest neighbors, which are randomly selected for bonding to form a self-avoiding polymer chain. Central molecule = TMC; others = MPD.

In the Ridgway algorithm, MPD and TMC residues are alternately introduced to build a nonbonded cubic or rectilinear 3D lattice (i.e., matrix) of $N \times R \times L$ dimensions, where N = number of residues per row, R = number of rows, and L = number of row layers (Figure 10.6). In this way, membrane models of differing overall thicknesses (i.e., layer number) may be constructed to better simulate a membrane surface (e.g., for simulations of surface adsorption phenomena). The 3D monomer matrix is somewhat analogous to a uniformly mixed solution of MPD and TMC residues. The distance between the monomers (i.e., the lattice packing density) partially determines the eventual membrane model density. After the 3D matrix is built, a nearest neighbor analysis is performed to identify each of the potential bonding partners for every MPD and TMC residue in the matrix (Figure 10.7). Starting at one corner of the rectilinear lattice, a random walk is then performed through the residue matrix with MPD–TMC amide bonding occurring at each step of the walk, thus creating a randomly folded self-avoiding polymeric chain, not unlike the approach described by Kotelyanskii et al. [26, 27]. Chain termination occurs when all potential bonding partners have been (by random chance) exhausted. The entire length of the chain is subsequently searched to identify any chain-resident TMC monomer pairs that remain separated by a common as-yet nonbonded (free) MPD monomer. A probability function then determines whether a cross-link bond, which bridges the two TMC monomers by the shared proximal MPD residue, is formed. The cross-linking function can be set to any probability ranging from zero to unity to create membrane models with any degree of cross-linking. Following chain elongation and cross-link formation, atom partial charges are progressively adjusted (scaled) to compensate for the loss of partial charges resulting from atoms that were deleted during monomer bond formation. The overall membrane model charge is determined by the number of free (i.e., nonprotonated) carboxylate groups remaining after all monomer bonding and cross-linking operations are completed, with each such group contributing a formal charge of -1.0 e owing to the loss of one proton. Therefore, the net membrane charge, which is averaged over all of the atoms comprising the membrane model, corresponds to the sum of the number of ionized carboxylate groups. Atomic partial charges may also be assigned before scaling, based on values derived from semiempirical QM computations on relatively small solvated polymer systems containing up to several hundred atoms, including Na^+ or other counterions. All nonbonded monomers are deleted from the system. Periodic boundary conditions may be imposed at any time before or following lattice building. Progressive energy minimization is carried out in stages following the monomer bonding and cross-linking, after charge scaling, and subsequent to the addition of water and/or solutes of interest to the membrane model. The program automatically builds as many models as the user chooses, each differing from the one before in terms of cross-link density, net charge, overall matrix dimensions, density, and chain conformation. Each model is stored in memory and its properties (e.g., charge, number of cross-links, density, etc.) are computed and recorded before the next build iteration.

Completed membrane models resemble the image shown in Figure 10.8, in which the partially energy-minimized polymer matrix has been packed into a periodic cell at a specified density. Periodicity may be implemented before matrix formation or at some later step. However, introduction of periodic boundaries later in the building process may result in less stable (i.e., higher energy) configurations that can result when dihedral bond rotations become more restricted because of intrachain cross-linking.

Among the membrane model parameters computed is the ratio of free carboxylate groups to amide bonds in the membrane model (i.e., the $COO-/Am_{II}$ ratio), which can be measured in the actual PA membrane samples by attenuated total reflection Fourier transform infrared (ATR-FTIR) spectrometry [16, 33] (Figure 10.9). As the interfacial polymerization reaction progresses, the intensity of the amide I and II bands increase (i.e., ~1650 cm^{-1} –C=O stretching and ~1550 cm^{-1} –N–H bending, respectively). At the same time, the intensity of

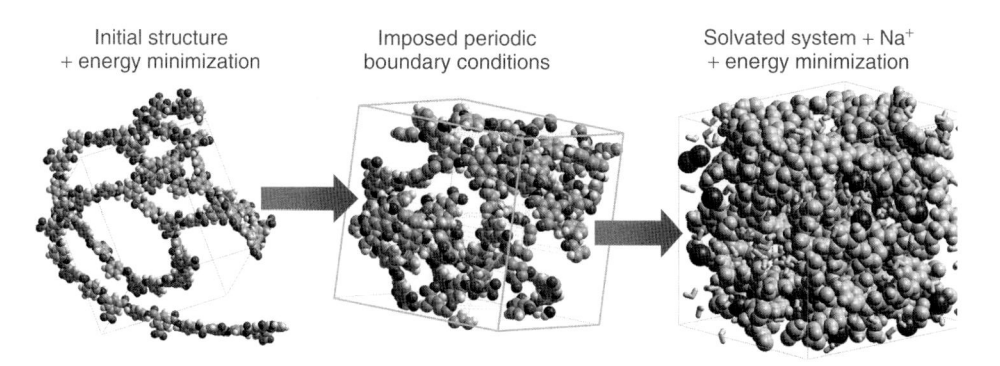

Figure 10.8 Folding of a partially cross-linked energy-minimized PA model into a periodic cell followed by solvation and introduction of Na+ ions and other solutes to complete the model system (Ridgway, unpublished).

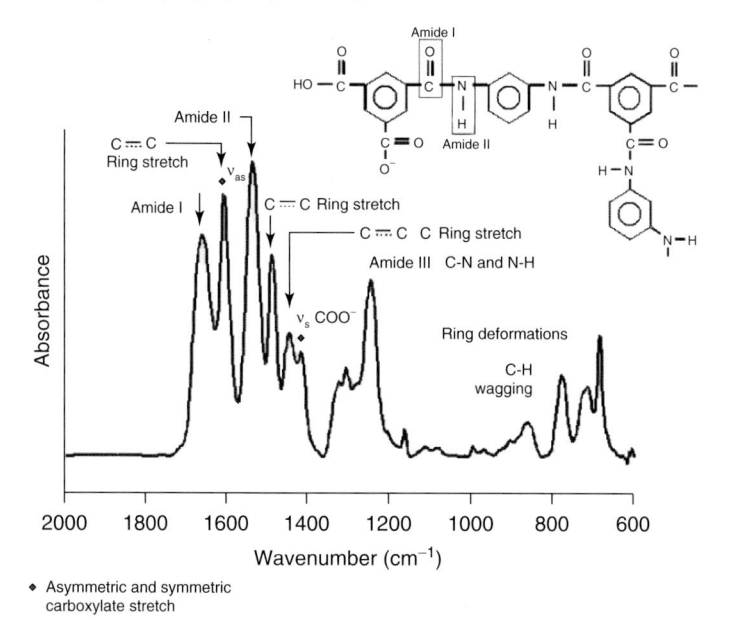

Figure 10.9 ATR-FTIR spectra of an isolated PA active layer with absorption band identification (K. Ishida, Orange County Water District, Fountain Valley, CA, unpublished)

the antisymmetric (~1600 cm^{-1}) and symmetric (~1400 cm^{-1}) carboxylate bands decreases. The ATR-FTIR spectra thus provide a semiquantitative measurement of the relative degree of cross-linking and carboxylate functional groups in the membrane. In a membrane with no cross-linking, the COO–/Am$_{II}$ ratio has a theoretical value of ~0.5 (~2 amide bonds per COO– group), whereas in membranes that approach complete cross-linking, the ratio approaches zero. In practice, the COO–/Am$_{II}$ ratio is observed to vary from about 0.25 to 0.42 depending on the type of commercial PA membrane material examined (K. Ishida, Orange County water

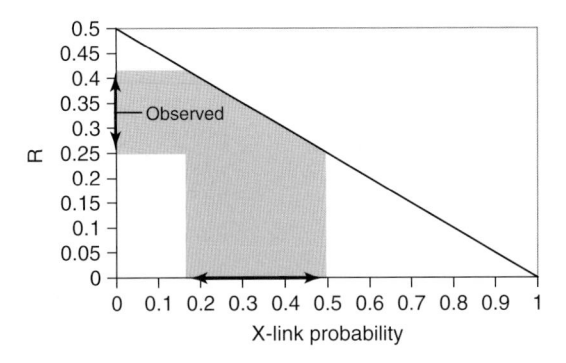

Figure 10.10 Relationship between the cross-link frequency in a PA membrane (x-axis) and the COO–/Am_{II} ratio (R) as determined by ATR-FTIR spectrometry (Ridgway and Ishida, unpublished data)

District, Fountain Valley, CA, unpublished). As shown in Figure 10.10, the observed range corresponds to cross-link probabilities between ∼0.18 and ∼0.5, suggesting that the cross-linking is largely incomplete. In contrast, Coronell et al. [14] report cross-link efficiencies of about 94% as determined from the concentration of Ag^+ ions (quantified by RBS spectrometry) associated with free carboxylate groups in PA active layers. The degree of polymer cross-linking helps determine the size and distribution of nanopores that allow passage of water and solutes through the membrane interior. Each cross-link formed potentially represents creation (or perhaps stabilization) of a new cavity (i.e., nanopore) inside the membrane. Interestingly, in MD simulations, Kotelyanskii et al. [27] found little difference in the water transport behaviors of cross-linked and non-cross-linked versions of aromatic PA membrane models in terms of their water self-diffusion coefficients and jump frequencies.

10.2.3
Simulation of Water and Solute Transport Behaviors

Molecular simulations have been used in recent years to predict the water and solute transport behaviors of various PA RO membrane materials [29, 34]. Collectively, these studies have clearly demonstrated the utility of this approach and molecular simulations are now being utilized more frequently in the fundamental design and engineering of novel membrane materials [34]. Molecular simulations also have the added benefit of being able to elucidate transport or rejection mechanisms that would otherwise be extremely challenging to decipher using current experimental techniques alone [27].

Prediction of the water and solute transport properties of synthetic permselective RO or NF membrane materials is typically based on the widely recognized classical solution-diffusion theory [9, 35, 36]. The two fundamental relationships that govern the semipermeable characteristics of most RO membranes are given by the equations in Figure 10.11. The solvent and solute diffusivities, D_{bm} and D_{am}, respectively, may be computed from the root mean squared displacements of these

10.2 Molecular Simulations of Polymeric Membrane Materials for Water Treatment Applications | 263

$$\text{Water flu } (J_b) = \frac{-D_{bm} \, C_{bm} \, V_{bm}(\Delta_P - \Delta_\pi)}{RT\sigma}$$

Where: D_{bm} = solvent diffusion coefficient (in membrane)
C_{bm} = solvent concentration in membrane
V_{bm} = solvent moral volume
RT = gas constant × temperature
σ = solvent flow resistance (membrane thickness)
$\Delta P - \Delta \pi$ = applied pressure − osmotic pressure

$$\text{Solute flux } (J_a) = \frac{-D_{am} \, K_a \, V_{bm}(\Delta C_a)}{\sigma}$$

Where: D_{am} = solvent diffusion coefficient (in membrane)
K_a = solvent parititon coefficient ([feed]/[membrane])
ΔC_a = solute concentration gradient ($C_{feed} - C_{perm}$)
σ = solute transport resistance (membrane thickness)

Figure 10.11 Solvent and solute transport equations that describe the solution-diffusion theory for permselective RO membranes.

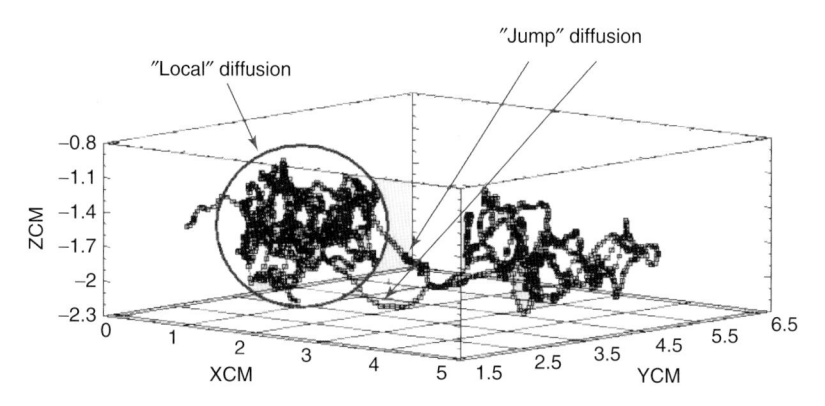

Figure 10.12 Center-of-mass diffusion trajectory of NDMA in a solvated PA membrane model $t = 15-35$ ps from an NVT simulation @300 °K (Ridgway, unpublished).

molecular species as a function of time in MD simulations of sufficient duration ($d <\Delta r2>/dt$). Such simulations must be long enough to account for a statistically significant sampling of all possible water motions (water phase space) within the membrane interior, including normal Fickian self-(local) diffusion and infrequent jumps. As a general rule, $d <\Delta r2>/dt$ must approach an asymptotic value over long timescales (in practice, >0.3 ns as a minimum). Kotelyanskii et al. [27] have noted that the mean squared water displacements may not be proportional to time (i.e., nonlinear with time) and that this behavior is expected for a jump diffusion process when the observation timescale is short compared to the jump frequency, τ. The characteristic jump frequency of water is about 1–2 ns, and the mean jump length is ~ 3 Å [27, 28]. Jump diffusion may also be exhibited by larger solutes, such as low-molecular-weight organics incorporated into the solvated membrane interior. As shown in Figure 10.12, nitrosodimethylamine (NDMA) exhibited two

Table 10.2 Model and experimental values for membrane water and Na$^+$ diffusion (D_{mem}), as well as model predictions of membrane water flux.

Water D_{mem} (simulation: NPT cross-linked model; cm^2 s^{-1})	Water D_{mem} (computed from measured water flux; cm^2 s^{-1})	Salt D_{mem} (NPT cross-linked model; cm^2 s^{-1})	Membrane water flux (simulations; m s^{-1})	Membrane water lux (experiment; m s^{-1})	References
~2.0×10^{-6}	~1.0×10^{-6}	~4.0×10^{-4}	NC	NC	Kotelyanskii et al. [24]
~5.0×10^{-6}	NC	NC	1.4×10^{-6}	7.7×10^{-6}	Harder et al. [26]

NC, not computed in this study.

jumps (out and back) in the early (15–35 ps) unstable portion of an NVT simulation of a fully solvated (23 wt%) PA membrane system (Figure 10.8).

The equilibrium concentration of the solvent inside the membrane (C_{bm}) and the membrane solute partition coefficient (K_a) can be similarly computed directly from MD simulations in which the bulk (55 M) water and solute are allowed to equilibrate against the membrane model, as described by Harder et al. [29]. The membrane diffusivity and solubility values estimated from the MD simulations may then be used to directly compute the theoretical water and solute fluxes from the equations presented in Figure 10.11. This type of analysis has only been performed in a few instances (that have been published), and these data are summarized in Table 10.2. Inspection of these data reveals approximate agreement between model predictions and experimental determinations for membrane water flux and salt (sodium + chloride) rejection. Presumably, future refinements to the membrane models will result in greater agreement between predictions from molecular simulations and experimental determinations.

It should be noted that MD simulations predict that the water and solute diffusivities inside the membrane are not constant but rather vary significantly as a function of overall hydrated system density, as was originally reported by Kotelyanskii et al. [27]. This dependence presumably results from electrostatic and van der Waals interactions between polar groups of the PA membrane (e.g., amine nitrogen and carbonyl oxygen) and water and solute molecules.

Several recent publications have addressed the role of the fractional free volume (FFV) on the performance (water and salt transport) of various PA membranes as determined from MD simulations [34, 37, 38]. The FFV is proportional to $V - V_0/V$, where V is the total system (model) volume, V_0 is the specific volume CV_w, where V_w is the van der Waals space of the polymer and C is a proportionality constant whose value depends on the type of hard-surface probe rendering implemented (typically a Connolly surface with a probe radius of 1.3–1.6 Å; Figure 10.13). FFVs may be directly computed from dissection of membrane molecular models, as illustrated in Figure 10.14. Apart from the FFV, the contents of each slice may be independently analyzed as a function of simulation time for mass, charge,

Figure 10.13 Connolly probe surface rendering of a hydrated 23 wt% aromatic cross-linked PA membrane model (probe radius = 1.4 Å). Nanopores constitute the fractional free volume of the membrane.

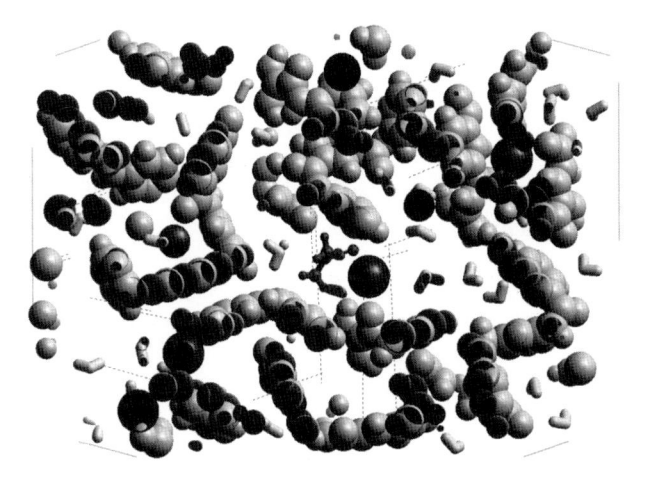

Figure 10.14 A 3.5 Å thick slice through the midpoint of a hydrated PA membrane model. The FFV for each such slice is represented by the open space. Water molecules = tubes; membrane = gray vdw spheres; Na^+ = black vdw spheres; and NDMA (center) = ball and stick (Ridgway, unpublished).

elemental composition, molecule identity, and other information to quantify how system dynamics influence the distributions of properties of interest and also to better delineate the locations and sizes of nanopores (see example in Figure 10.15).

Kao *et al.* [37] found that for aromatic PA pervaporation membranes used for ethanol separations, the FFV could be increased by direct incorporation of bulky monomer groups (e.g., 1,4-bis[4-aminophenoxy]2,5-di-*tert*-butylbenzene or arylene ether groups) into the polymer backbone. MD simulations of the various novel polymer systems were able to predict FFVs that closely correlated with experimental values determined from PALS. The higher FFV membranes were correlated with improved alcohol separation behaviors.

A similar use of MD simulations to evaluate the FFVs of PA RO membranes prepared from a series of methyl-substituted diamines was recently reported by

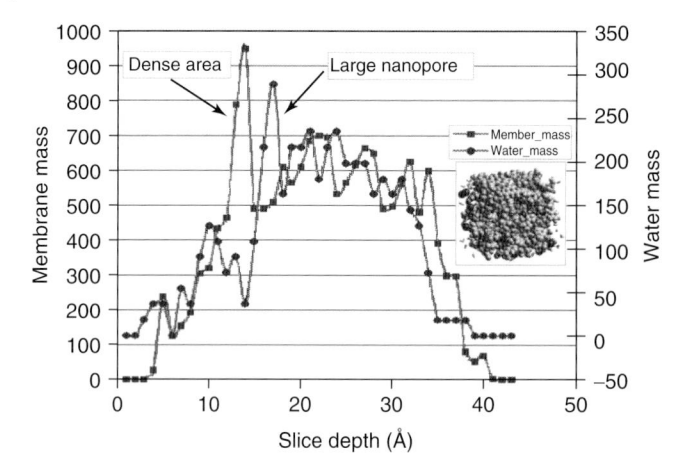

Figure 10.15 Mass distributions for membrane and water atoms at $t = 0$ ps of MD as a function of slice location over a solvated PA membrane model (inset). Regions denoting high membrane density and a large water-filled nanopore are marked (arrows).

Shintani et al. [38]. Methyl substitution increased the FFVs and the interstitial space of the PA chains as predicted from MD simulations and confirmed by PALS (Table 10.1). Changes in the FFVs were correlated with enhanced chlorine resistance and decreased salt rejection of the synthetic membranes.

Recently, Peng et al. [34] used FFVs calculated from MD simulations to estimate the water flux and salt rejection behaviors of a series of ultrathin cross-linked poly(vinyl alcohol)-polysulfone interfacial composite membranes. The free volume measurements of the nanofiltration-like membrane materials derived directly from the MD simulations strongly agreed with composite membrane permeability, including the pure water flux and rejections of sodium, chloride, and sulfate.

10.2.4
Concluding Remarks

Over the past decade, the use of molecular simulations has rapidly evolved from a largely theoretical exercise for the elucidation of RO/NF membrane structure and function to practical applications in the a priori design and development of membrane materials with improved performance and antifouling characteristics. Despite such rapid progress in modeling cross-linked PA membranes and related polymer materials used in water purification processes, a number of important membrane features and properties have yet to be adequately accounted for in modeled systems. Perhaps, the main area of concern is that the PA membranes are not morphologically uniform polymer films or dense layers, but rather they exhibit considerable nanoscale heterogeneity that has not yet been fully interpreted in terms of where water/solute separations take place. Freger [12] and Pacheco et al. [20] have clearly demonstrated that the ultrastructural heterogeneity of PA membranes is reflected in an equally nonuniform chemical makeup for the various structural regions that can be delineated by electron microscopy. In addition, the ionization potentials of carboxylate and amine functional groups in the PA membrane vary significantly depending on their local physicochemical environment, as indicated by studies recently reported by Coronell and coworkers [14, 15, 21, 22]. These different membrane nanoenvironments could be expected to affect the partitioning of water, salts, and organic solutes into the PA membrane separation layer, which in turn determines if and how they are rejected by the membranes [13, 39–41]. Future efforts to simulate transport processes in these complex membrane materials may require multiscale and/or continuum-scale modeling, or possibly other more novel approaches that can better account for the kinds of chemical and structural diversity observed in these materials.

10.3
Molecular Simulation of Inorganic Desalination Membranes

Desalination by RO involves the use of a membrane to remove salt ions from water. In commercial desalination plants, these membranes have always been organic materials, initially cellulose acetate, and now more often some type of PA material. There is considerable interest in developing new types of materials for RO, capable of reducing the energy costs of desalination by having greater selectivity for salt rejection, a faster flow rate for a given applied pressure, or a greater resistance to fouling than current membranes.

In addition, to investigate modified versions of the current PA materials and new organic materials, a number of inorganic materials have also been suggested as possible RO membranes [42]. These materials include silica and clays, but the most popular inorganic materials for desalination are zeolites. Zeolites are naturally occurring or synthetic nanoporous crystalline materials, consisting of an aluminosilicate framework of corner sharing SiO_4 or AlO_4 tetrahedral with cations present in the channels in order to charge compensate for the presence of aluminum when necessary [43]. Normally, the incorporation of aluminum within the zeolite framework leads to such structures being hydrophilic. However, there has been considerable recent interest in preparing purely siliceous materials, either directly or by dealumination [44–46]. In this form, the zeolitic silica becomes hydrophobic, and thus the flow rate of water through the zeolite is increased because of weaker interactions with the channel walls, something observed both experimentally [47] and in simulation [48]. It is possible that these siliceous materials could be tailored for the purpose of desalination, by selecting pore dimensions that are too small to allow the easy diffusion of solvated sodium or chloride ions through the membrane. There has already been experimental work carried out investigating zeolites as desalination membranes, both on their own [49, 50] and incorporating zeolite nanoparticles within traditional TFC membranes to make a composite material of improved efficiency [51]. While good salt rejection appears to be achievable with such zeolitic membranes, the exchange of metal ions can alter the silicate structure, leading to changes in the overall performance [52].

Much of the interest shown in zeolites as desalination membranes was sparked by the results of a paper describing the simulation of a zeolitic membrane separating water from NaCl solution [53]. In this work, a region of pure water was separated from the salt solution by two zeolite membranes. Over the course of these simulations, water molecules were able to permeate through the membranes, but the salt ions were not. While this is a very promising result, there are a number of factors to be considered. First, the time frame of these simulations (100 ps) is very short, making it difficult to extrapolate the behavior of the system over longer timescales. Second, the force field used to describe the membrane must be viewed with some caution. The zeolite atoms were tethered to their initial positions by a harmonic force, effectively making the membrane more rigid than it probably should be (see the following section for more details regarding this problem). Moreover, the zeolite atoms were electrostatically neutral, something that seems highly anomalous, as most zeolite force fields assign significant charges to the silicon, aluminum, and oxygen atoms that make up the zeolite, something supported by first-principles QM calculations [54]. Finally, the zeolite membranes seem to have been terminated by simply cleaving the surface, whereas in reality there would almost certainly be hydroxyl groups compensating for broken Si–O linkages in an aqueous environment.

Beyond this study, there has been remarkably little published work directly investigating the potential of zeolites as desalination membranes. However, there have been a number of simulations investigating the behavior of water within siliceous zeolites [55–62]. The most popular zeolites for simulation are silicalite or dealuminated zeolite Y (DAY). Very little work has looked at zeolites with other pore topologies, but one recent paper [63] investigated a series of zeolites with one-dimensional channel structures that may lead to more directed diffusion of water, useful in desalination. There are also a number of studies that have considered the behavior of ions within zeolite frameworks. Before discussing these results in detail, it may be useful to discuss the most important factors that need to be considered when simulating these systems.

10.3.1
Modeling of Zeolites

When modeling any material, the first thing to decide on is the appropriate modeling/simulation technique for investigating the properties that are most relevant to the problem of interest. First-principle techniques, based on the density functional theory (DFT) [64–67], can provide valuable information about the adsorption of molecules within zeolites or the reactivity at acidic sites within aluminosilicate zeolites, especially, where questions of proton transfer arise [64]. These methods are best suited to problems involving strong interactions, such as those found under hydrophilic conditions. Although there is progress being made in the treatment of weaker interactions during physisorption with the advent of van der Waal's functional [68], empirical dispersion corrections [69], and the use of post-Hartree–Fock methods for periodic solids [70], the use of QM techniques remains challenging for water in siliceous materials. In contrast, force field methods are readily applicable to such systems, although careful calibration of the parameterized interactions is necessary.

In order to determine the diffusion behavior of water within materials, the most appropriate simulation technique is MD. This classical approach propagates the positions of nuclei by solving Newton's equations of motion to capture the time evolution of the system. The numerical precision of the method depends on the magnitude of the finite time step used. To accurately sample the fastest nuclear motions, such as the vibration of O–H bonds, this requires time steps of the order of femtoseconds. Because of this limit on the time step, the amount of real time that can be sampled is typically restricted to several picoseconds when using QM evaluation of the forces, whereas force field methods routinely allow simulations of many nanoseconds or more. Consequently, when considering the diffusion of water within zeolites, the use of QM techniques is not appropriate because of the limited timescale that is accessible. Although unable to provide information about dynamic properties, the Monte Carlo (MC) techniques, especially the Grand Canonical MC, can provide valuable information that is difficult to obtain through other methods, such as adsorption isotherms [59–61, 71].

Both classical MD and MC describe the potential energy of the system through a force field and because of the considerable interest shown in zeolites, a wide range of force fields have been developed for investigating the behavior of water within zeolites [55–58, 61, 62, 72–74]. As with the choice of simulation technique, it is important to ensure that the force field used is appropriate for the investigation of the particular properties that are of interest. To comprehensively review all the possible force fields is beyond the scope of this chapter; instead, we will highlight some of the most important factors to consider when choosing a force field.

Whether the zeolite framework will be treated as rigid or flexible is one of the first factors that require consideration. A zeolite framework that is flexible will be more computationally expensive but is likely to describe the system more accurately. The effect of approximating the system as being rigid has been investigated by Demontis and Suffritti [75], who compared the results of diffusion within flexible and rigid silicalite against each other. The overall conclusion was that the lattice vibrations of the zeolite play an important role in the diffusion of species within the framework, and thus, a flexible zeolite model was preferred.

Whether or not to use a force field that accounts for the polarizability of atoms in some manner (and if so by what means) is another factor in the choice of model. Including polarizability in the force field allows the species to adapt to their local environment to a greater degree, but again is accompanied by an increase in the computational cost of the simulations. Methods that have been used to treat polarization in zeolites include using an electric field dependent potential [56, 61], the electronegativity equalization method [58] for determining geometry-dependent charges, and use of a shell model [72]. Rigid-ion models are unable to capture the low-temperature monoclinic distortion of silicalite [76] and are also less likely to predict the order of phase stability of silica correctly [77]. We note that a recent report [62] of a modified partial charge rigid-ion force field reproducing the monoclinic distortion of silicalite was as a result of discontinuities in the energy surface. Accurate summation of the attractive C6 terms in this model, through the use of an Ewald-like technique [78], results in the monoclinic form optimizing to the higher symmetry orthorhombic structure, as would be expected for a rigid-ion model.

The zeolite force field may have been designed with a particular water model in mind, in which case using a different water model should only be done after careful thought. Alternatively, the zeolite force field might be developed with no specific water model in mind; in such a case, both the water–water and the zeolite–water interactions need to be determined. Some models may describe one of these interactions well but fail to fully capture the other. As with the zeolite, the computational expense of the model is likely to be a factor in the final choice of water model. Rigid water models will allow a larger time step to be used and there are a number of rigid water models that describe the properties of bulk water very well, for example, SPC/E, TIP4P, TIP4P-Ew, and SWM4-NDP. However, the algorithm required to propagate rigid bodies is more complex and so a flexible water model can be almost as efficient, depending on the method of implementation.

If using a zeolite force field that has not had a water model developed specifically for it, or changing the water model in a previous force field, then it is highly likely that the water–zeolite interaction will need to be refitted. What data the force field should be refitted against depends on the properties that are currently under investigation. One of the challenges in this field is the difficulty in obtaining accurate data for the straightforward calibration of force fields, as much of the available experimental information pertains to water in aluminosilicates.

10.3.2
Behavior of Water within Zeolites

Now that some of the factors that need to be considered before deciding on a simulation strategy have been discussed, the behavior of water within the zeolitic systems can be looked at in more detail. As previously mentioned, silicalite is one of the most studied siliceous zeolites and there have been a number of MD simulations of water diffusing within this material [48, 55–58, 62, 79]. Silicalite is an all-silica zeolite having two different interconnecting channel systems, with both channels consisting of 10-membered rings [80]. Parallel to the y-axis, based on the standard crystallographic settings, are elliptical straight channels $\sim 5.7 \times 5.2$ Å in diameter, while parallel to the x-axis are sinusoidal channels, circular in cross section with a diameter of ~ 5.4 Å (Figure 10.16).

The diffusion coefficients of water within silicalite, as computed by a number of simulation studies, as well as experimentally determined from pulsed field gradient (PFG) NMR measurements [55], are given in Table 10.3. This data is for a system under standard conditions and at a loading of eight water molecules per unit cell. In addition to the three-dimensional diffusion coefficient, the component along each of the three axes is also given. More insight regarding the nature of the diffusion of water within microporous materials [81], such as silicalite, can be determined

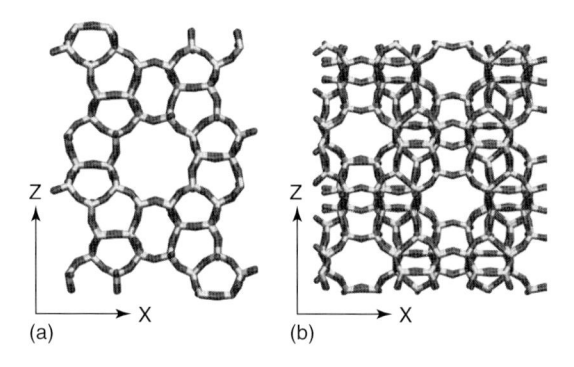

Figure 10.16 Representation of the zeolite silicalite viewed down (a) the y-axis showing the straight channels and (b) the x-axis showing the sinusoidal channels. The silicon and oxygen atoms of the zeolite are colored light gray and medium gray, respectively.

Table 10.3 The diffusion coefficients, β and δ, of water within the zeolite silicalite at a loading of eight waters per unit cell at 300 K and 1 atm, as determined from a number of simulations.

Study/model	Diffusion coefficients (10^{-9} m^2 s^{-1})					
	D	D_x	D_y	D_z	β	δ
Bussai et al. [55]	3.3	2.6	6.5	0.19	1.06	5.76
Demontis et al. [56]	8.6	7.9	15.9	2.0	1.19	5.95
Smirnov and Bougeard [57]	2.3/4.1	—	—	—	—	—
Fleys et al. [58]	8.83	2.33	6.17	0.67	1.14	6.34
Ari et al. [48]	1.94	0.69	1.24	0.00	—	—
Bordat et al. [62]	—	3.5	6.35	0.90	1.11	5.47
Hughes et al. [63] (polarizable force field)	1.65	2.16	2.36	0.43	1.18	5.24
Hughes et al. [63] (rigid-ion force field)	1.18	1.35	1.89	0.29	1.23	5.62
Experimental PG-NMR [55]	1.7	—	—	—	—	—

The overall experimental diffusion coefficient is also included for comparison.

from the calculation of two parameters, β and δ. β measures the discrepancy from a simple random walk model and is calculated from

$$\beta = \frac{c^2/D_z}{a^2/D_x + b^2/D_y} \quad (10.1)$$

where a, b, and c are the dimensions of the silicalite unit cell and D_x, D_y, and D_z are the components of the diffusion coefficient along the x-, y-, and z-axes, respectively. The random walk model applies when $\beta = 1$. If $\beta < 1$, then the water molecules prefer to move between the different channel types, whereas if $\beta > 1$, (which is the case for water in silicalite under these conditions), then the water molecules prefer to diffuse along the same channel type after passing an intersection. The diffusion anisotropy parameter, δ, is given by

$$\delta = \frac{1/2(D_x + D_y)}{D_z} \tag{10.2}$$

and in silicalite should be $\sim$4.4 for a random walk model. The values of these two parameters determined from the results of the different MD simulations are also given in Table 10.3. The diffusion parameters show a spread of values; this is not particularly surprising as even small changes in the simulation parameters can result in quite a large change in the diffusion coefficient. For instance, the studies of Ari et al. [48] and Fleys et al. [58] used the same force field but different methods to evaluate the electrostatic forces and there is significant difference in the results obtained. Equally, there are a number of factors (such as defects within the zeolite or a small amount of Al impurities in the structure) that make the experimental measurement of this quantity difficult [55, 56]. However, while the value of the diffusion coefficient does vary between the simulations, a number of factors are consistent. First, in all these simulations, the diffusion of the water is anisotropic, with the greatest rate of migration parallel to the y-axis with which the straight channels are aligned, and diffusion parallel to the z-axis being the lowest owing to the fact that the water molecules have to diffuse along a combination of the straight and sinusoidal channels. Second, all the simulations exhibit $\beta > 1$, indicating that the water molecules prefer to diffuse along a single channel, rather than to move between channels. Finally, for those simulations where δ was calculated, $\delta > 4.4$. Thus, while there are quantitative differences between the different simulations, there is good overall qualitative agreement.

The standard conditions for the simulation of silicalite are those described earlier (i.e., 0% Al, 300 K, 1 atm, and eight water molecules per unit cell), but the effects of temperature [52, 54], loading [54, 59], and aluminum content [45] on diffusion of water have all been investigated. The general findings were along the lines that one might expect; the diffusion of water within the zeolite increases with rising temperature and decreases as the aluminum content and the amount of water within the zeolite increase. One point worth noting is that the diffusion rate of the water within the zeolite, even under unfavorable conditions, is appreciably greater than the diffusion rate of water within the PA membranes currently used in RO, which have a diffusion rate of $\approx 0.2 \times 10^{-9}$ m^2 s^{-1} [27, 82].

The behavior of water molecules within silicalite differs significantly from molecules in the bulk liquid, with vaporlike characteristics being observed [56, 58] (at 300 K, 1 atm, and eight waters per unit cell). This difference in behavior has also been shown in the Grand Canonical MC simulations [59, 61] where the dipole moment was calculated as being $\sim$10% smaller than that of bulk water for the confined water molecules. Given that water force fields are parameterized to describe the bulk liquid, this tendency is likely to be even greater in the real system.

While silicalite is the most studied siliceous zeolite, other systems have also been examined. For example, Fleys et al. [58] compared the diffusion of water within silicalite and DAY. DAY is a synthetic material that has the same structure as the mineral faujasite, but with all the aluminum atoms replaced by silicon. The pore structure in DAY is larger than in silicalite, with 12-membered rings forming channels with a diameter of ~7.5 Å. For a system at 300 K, 1 atm, and a loading of eight water molecules per unit cell, $D = 5.72 \times 10^{-9}$ m^2 s^{-1} in DAY compared with $D = 8.83 \times 10^{-9}$ m^2 s^{-1} in silicalite. They attributed the lower rate of diffusion in DAY to the fact that in silicalite the narrower channels of the zeolite mean that the effect of the lattice vibrations of the zeolite framework had a greater positive effect on the diffusion of water molecules. In addition, they found that silicalite is more disruptive to hydrogen bonding of water molecules, meaning that the water in DAY behaves in a manner that is closer to that of bulk water than the water in silicalite. The different pore dimensions of the two systems thus cause significant changes in the behavior and structure of the water within the zeolite. As in the case of silicalite, increasing the amount of water molecules per unit cell for DAY reduces the rate of diffusion, whereas increasing the temperature has the converse effect.

Silicalite has a two-dimensional channel system and DAY a three-dimensional one, but there are also zeolites with one-dimensional channel systems. These zeolites, when purely siliceous, resemble carbon nanotubes (CNTs) in that both systems have nanometer-wide hydrophobic pores. The diffusion of water within CNTs has attracted a great deal of interest [78–80,83] and calculations of the rate of diffusion of water predict it to be high. It is surprising, therefore, that little work has been performed to study the diffusion of water within 1D zeolite systems; as far as the authors are aware, the only paper investigating these systems is that of Hughes et al. [63]. They studied the diffusion of water within four different one-dimensional zeolite systems of varying channel size and shape, namely, MTF, SFF, VET, and GON, as given by their standard three-letter framework codes. MTF, VET, and SFF all have nearly circular channels of different sizes, whereas GON has an elliptical channel structure. The channels of MTF are smaller than those of silicalite, consisting of an eight-ring channel (diameter 3.75 Å); those of SFF are almost the same size (10-membered ring, diameter 5.5 Å), whereas in VET the channels are a little larger (12-ring structure, 5.9 Å diameter). GON also has channels larger than silicalite, consisting of 12-membered rings with dimensions of ~5.9 × −5.0 Å. The systems were simulated using two different force fields, one consisting of a polarizable shell model and the other a partial charge rigid-ion model and at a variety of different loadings. The diffusion rates of water within these zeolites are given in Table 10.4, while Figure 10.17 shows snapshots from simulations of the SFF zeolite at a variety of different water loadings. At the lower loadings, the water molecules often form hydrogen-bonded chains.

The values of the diffusion coefficients calculated for the two force fields do differ, especially at low loadings. However, as for silicalite, there is significant qualitative agreement. At the highest loadings, the diffusion coefficients of SFF, VET, and GON are very similar, whereas the rate of diffusion in MTF is somewhat lower. This indicates that once a channel has reached a certain size, ~5.5 Å in diameter, the influence of this dimension on the rate of water diffusion is minor. The agreement between the two force fields is relatively good for the two zeolites with smaller channels, MTF and SFF; only at a very low loading of a single water molecule per channel is there serious disagreement. However, for VET and GON,

Table 10.4 Diffusion coefficients of water within the four zeolites with 1D channel structures at different water loadings (all at 300 K and 1 atm).

Zeolite	Water loading per channel	Diffusion coefficients (10^{-9} m^2 s^{-1})	
		Polarizable	Rigid ion
MTF	1	2.26 ± 0.92	5.96 ± 1.43
	3	0.62 ± 0.11	0.44 ± 0.14
	5	0.37 ± 0.07	0.40 ± 0.08
	17	0.24 ± 0.02	0.15 ± 0.02
SFF	1	1.91 ± 0.61	4.21 ± 0.64
	3	1.00 ± 0.40	1.96 ± 0.56
	5	0.73 ± 0.16	0.49 ± 0.09
	10	0.53 ± 0.12	0.45 ± 0.03
	32	0.49 ± 0.02	0.38 ± 0.02
VET	1	2.39 ± 1.39	22.09 ± 5.11
	3	2.06 ± 0.47	9.43 ± 2.98
	6	1.49 ± 0.29	5.23 ± 1.3
	32	0.41 ± 0.01	0.34 ± 0.02
GON	1	8.01 ± 3.68	25.43 ± 6.26
	3	7.16 ± 0.89	9.57 ± 3.15
	6	2.36 ± 0.55	5.11 ± 1.10
	32	0.51 ± 0.03	0.32 ± 0.02

Data taken from Hughes et al. [63].

the diffusion coefficients obtained using the TIP4P-Ew model were appreciably larger than those given by the shell model (except at the highest loading). In addition, despite the fact that the channels of the two zeolites are approximately the same size, the shell model gives a diffusion coefficient for GON, which is considerably larger than that of VET, whereas the rates of diffusion calculated using the rigid-ion model were very similar. The adsorption energy profile of a water molecule within the zeolite as a function of position along the channel has been calculated and this explains the reason for the above differences. There is a large energy barrier to the diffusion of water within VET for the polarizable force field, whereas the nonpolarizable rigid-ion model fails to capture all the details of the zeolite–water interaction and the potential energy landscape is much flatter. These differences are most important at low water concentrations where the zeolite–water interaction dominates the behavior of the system. However, as the number of water molecules present in each channel increases, the water–water interactions become the dominant force in determining behavior. This means that water molecules are unable to get "trapped" at points along the zeolite and the differences in the rates of diffusion for the two zeolites become less marked. The results of this study underline the importance of choosing the correct force field for investigating the properties that one is interested in. At low loadings, polarizability is an important effect and probably should be incorporated in the force field.

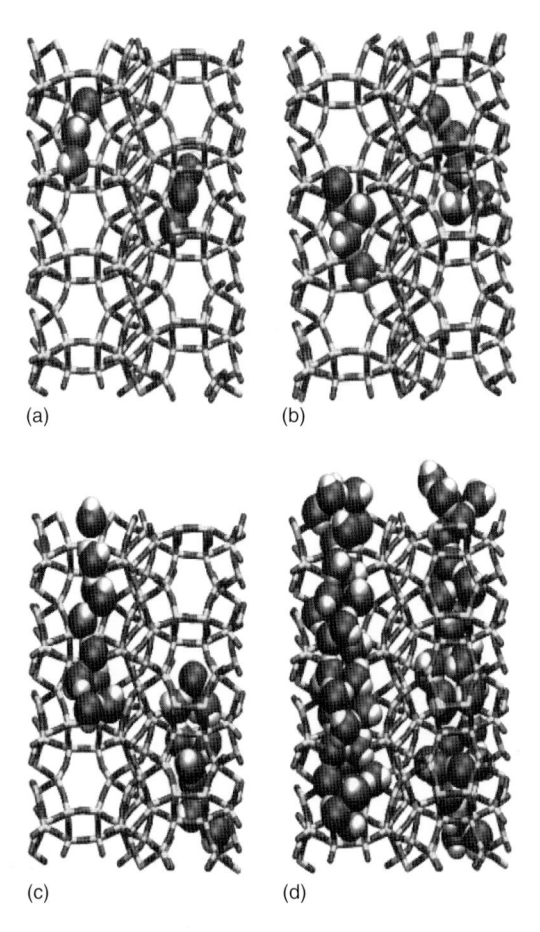

Figure 10.17 Snapshots of water molecules diffusing within the 1D zeolite SFF, at 300 K, 1 atm, and water loadings of (a) 3, (b) 5, (c) 10, and (d) 32 water molecules per channel. The system is viewed down the x-axis and the silicon and oxygen within the zeolite structure are colored light gray and medium gray, respectively, and the oxygen and hydrogen of the water molecules are colored gray and white, respectively.

As for silicalite, the diffusion rates of water within both DAY and the one-dimensional zeolites compare well with that of water within the PA membranes. However, one-dimensional zeolites are unable to match the predicted rapid diffusion of water within CNTs. The rate of diffusion of a water molecule in short, defect-free CNTs of a similar pore radius (an internal diameter of 4.7 Å) was calculated as being half that of a water molecule in the aqueous phase [84], whereas the diffusion rates of water at the highest loadings in the one-dimensional zeolites were in the range of a sixth to an eighth of the diffusion rate of bulk water, depending on the force field used. Even so, all simulations studying the diffusion of water within siliceous zeolites indicate that they should be competitive for practical application in terms of flow rate.

10.3.3
Zeolites and Salt Ions

There have been a number of simulations of zeolites systems where Na^+ was acting as a counterion for aluminum within the zeolite framework in the presence of water [48, 85], but there has not been a great deal of literature systematically investigating the behavior of salt ions within fully siliceous zeolites. One of the difficulties of simulating salt ions is the increasing complexity of the force field; not only do the water–water and water–zeolite interactions need to be balanced but the salt–water and salt–zeolite interactions must also be taken into account. The interaction of many salt ions in bulk water has been investigated in some detail, and while there is still some uncertainty in a number of properties, such as the coordination number of waters in the first solvation shell [86], a number of models, believed to model the bulk water–salt ion interactions well, have been developed. How transferable these parameters are to water molecules confined within zeolites is an open question. In addition, there have been relatively few potentials developed for the interaction of the salt ions with the zeolite framework, and this is especially true in the case of anions.

More important than simulating the behavior of salt ions already incorporated within the zeolite is determining whether or not the salt rejection properties of zeolites are sufficient for commercial use. There have been a few papers that have investigated the relative preference of salt ions for the two different environments, bulk solution versus zeolite. As previously mentioned, the study of Lin and Murad [53] showed that there was no diffusion of salt ions across a zeolite membrane in the presence of a concentration gradient, at least on the picosecond timescale. This result seems to be supported by the study of Gren et al. [87], where a thin slice (~25 Å) of zeolite A (LTA) was placed in water. Here, sodium ions, initially within the zeolite framework, were found to leach into the bulk solution, indicating a preference for that environment. Unfortunately, the statistics were inadequate for any quantitative analysis of this preference, through the calculation of a partition function.

One method of determining the thermodynamic preference of salt ions for bulk solution relative to the zeolite channel is to use free energy techniques. In particular, the free energy perturbation (FEP) method [88] allows the free energy difference between two states, labeled 0 and 1, to be calculated from the following equation

$$\Delta G = -\frac{1}{\beta} \ln \langle \exp[-\beta \Delta U] \rangle_0 \tag{10.3}$$

where ΔU is the difference in potential energy between the two states and $\langle \cdots \rangle_0$ is the ensemble average of configurations sampled in state 0. FEP has been used successfully to calculate the hydration/dehydration free energy of water [89, 90] and a number of different ions [91, 92]. The application of this method was recently extended [63] to the calculation of the free energy needed to insert/remove salt ions from the one-dimensional zeolites MTF, SFF, VET, and GON. In state 1, the salt

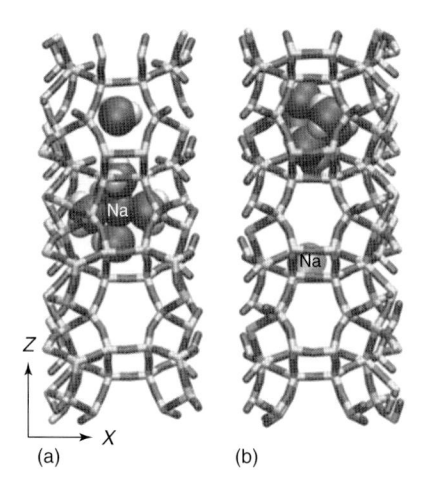

Figure 10.18 A sodium ion and its first solvation shell in the SFF zeolite: in state 1, (a) the interactions of the ions with the other species are present, in state 0, (b) the interaction of the ion with the other species has been scaled to zero. From the difference in the potential energy of the two states, it is possible to calculate the free energy needed to insert/remove a sodium ion from the SFF zeolite. Silicon and oxygen within the zeolite structure are colored light gray and medium gray, respectively. Oxygen and hydrogen in the water molecules are colored medium gray and white, respectively. Sodium is colored gray and labeled Na.

Table 10.5 Calculated free energies for a variety of different reactions.

Zeolite	ΔG (kJ mol^{-1})				Overall free
	Ion from zeolite	Water from zeolite	Ion from solution	Water from solution	Energy difference (kJ mol^{-1})
MTF	502.2 ± 3.7	46.0 ± 3.3	371.1	174.0	−3.1
SFF	479.3 ± 2.2	53.6 ± 2.5	371.1	174.0	12.2
VET	503.7 ± 5.1	50.5 ± 3.0	371.1	174.0	−9.1
GON	498.0 ± 2.9	46.5 ± 3.4	371.1	174.0	0.6

From left to right, the free energy needed to remove a Na$^+$ ion from the hydrated zeolite, the free energy needed to remove six water molecules from the zeolite, the free energy of dehydration of a Na$^+$ ion, the solvation free energy of a water molecule in aqueous solution, and the overall free energy needed to remove a Na$^+$ ion and six water molecules from bulk solution and place them inside the zeolite systems [63].

ion interacts with the other species, whereas in state 0, the interaction of the ion with the zeolite and water molecules has been removed, as shown in Figure 10.18.

The free energies required to remove a sodium ion and the six water molecules in its first solvation from the zeolite, the bulk solution, and the overall free energy difference, are shown in Table 10.5. The free energy differences for all four zeolites are close to zero, and so this result that would seem to indicate that a zeolitic RO membrane may encounter problems with the rejection of sodium ions. However, unless a counterion (such as Cl^- or OH^-) enters the zeolite simultaneously, there will be an extra free energy penalty due to charge separation. As the dielectric constant of siliceous zeolites is low (for the four 1D zeolites used in this study the dielectric constant was calculated to lie in the range of 4–5) when compared with water (78.4), this penalty will be significant. As the same study showed that the removal of a Cl^- ion from a hydrated zeolite takes only a few tens of kJ mol^{-1}, as opposed to the hundreds of kilojoule per mole needed to remove a chloride ion from bulk water, it seems highly unlikely that chloride ions will diffuse into the zeolite membrane. Overall, while Na^+ ions might enter the zeolite channels, they are unlikely to diffuse into the bulk of the membrane, instead staying within a few amperes from the membrane interface. It is also important to note that this study only considers the limiting thermodynamics of the bulk system. The diffusion of ions across the zeolite–water interface is likely to exhibit an activation barrier that would further enhance the level of salt rejection.

10.3.4
Concluding Remarks

Zeolites are the most promising of all inorganic RO membranes; however, despite this there have been relatively few simulation studies performed on such systems with specific regard to desalination. In contrast, there is a significant amount of literature on the behavior of water within zeolites, as well as a lesser amount investigating systems consisting of salt ions, water molecules, and a zeolite. All the studies of water within zeolites indicate that the diffusion rate of water within a zeolite membrane is likely to exceed the diffusion rate of water within the PA membranes that are currently used in commercial desalination plants. Thus, zeolites satisfy the need for high flow rates that are required for potential RO membranes. The second criteria, a high salt rejection rate, has not been investigated as thoroughly and more work needs to be carried out in this area. The rejection of Cl^- by zeolite membranes can be very promising, but there are some indications that Na^+ ions may be more problematic with a possible build up of ions near the membrane interface. In the aforementioned work, it is assumed that diffusion occurs only through the bulk zeolite crystallites. Extended defects, such as grain boundaries and intergranular interfaces, represent potential sources of failure for the membrane by allowing salt diffusion to occur through a facile pathway. The final requirement of any RO membrane is a resistance to fouling and this factor is the least investigated of all, perhaps in part because most foulant molecules in RO are organic, and the parameterization of a force field able to model organic–inorganic interactions is more challenging.

In terms of further investigation, there are a number of directions that would seem to be profitable. The first area is to investigate the diffusion of water molecules and salt ions across zeolite interfaces; this would provide useful information regarding the flow rate of any zeolite membrane. Even more importantly, some of the questions remaining about the salt rejection properties of zeolite membranes could be answered. A second area for future investigation is the behavior of zeolite membranes in the presence of foulants commonly found in waters.

The interest in zeolites as potential desalination membranes is only likely to grow in the near future and computer simulation will provide an important tool in guiding the development of such systems.

10.4
Molecular Simulation of Membrane Fouling

A major operational issue and a limiting factor associated with RO and related membrane technologies is membrane fouling [4]. The effects of membrane fouling are detected as a reduction in the efficiency of the RO process, as reflected in higher operating pressures and lower permeate fluxes. This is due to the fact that membrane fouling is generally caused by the adsorption of solutes to the membrane surface, as well as pore clogging (in porous membranes) [93]. Such fouling mechanisms effectively reduce the membrane surface area.

Depending on the nature of the fouling species, four broad classifications of membrane fouling may be defined: (i) colloidal/particulate fouling, (ii) microbial/biofouling, (iii) dissolved organic fouling, and (iv) inorganic fouling (also commonly referred to as *scaling*) [4, 94, 95]. It should be noted that these fouling mechanisms are rarely independent of each other. Indeed, autopsy studies have found that combinations of these foulants are frequently found on the surface of fouled membranes [96]. However, each class of foulant has its own amelioration strategy and shall be treated independently here.

Of the classes of foulants mentioned, dissolved organic fouling is more difficult to predict because of the wide range of organic compounds found in natural waters and wastewaters. Amelioration of organic fouling for MF/UF is achieved by chemical coagulation using aluminum- or ferric-based coagulants and chemically enhanced backwashing using chlorine-based compounds for membranes that are tolerant of oxidants. Colloidal fouling can be effectively controlled by implementing a UF or a microfiltration pretreatment step to the RO process. Microbial fouling is caused by microbes forming a biofilm on the surface of the membrane. This can be reduced through the addition of antimicrobial oxidants to the feed water – usually chloramines for desalination processes as RO membranes are degraded in the presence of free chlorine. For inorganic fouling, scaling is caused by the precipitation of inorganic salts on the membrane surface, especially in waters that are high in calcium, silica, phosphate, and carbonate, as well as other ions. This can be reduced operationally by adding antiscalant compounds, decreasing the pH of the feed, reducing the recovery rate, and/or altering the pH.

Fouling by dissolved organic compounds is a poorly understood phenomenon [94], and effective pretreatment strategies can be difficult to standardize because of the variations in composition and concentration of organic compounds at specific times and locations. One research approach is to focus on the nature of the membrane surface itself. Several studies have found that membrane fouling by dissolved organics is significantly reduced when the surface is smooth and/or hydrophilic [97–103]. However, some flux is lost when the membrane has a smooth surface, and recent work [104] has shown that a rough surface increases flux – at least initially, before fouling shows any noticeable effect on the operating conditions. These results suggest that the most likely mechanism involved in the organic fouling of membrane materials involves hydrophobic interactions between the solute and the surface. Such fouling is enhanced by surface roughness by providing pockets for mechanically trapping foulants, as well as providing an increased surface area for potential hydrophobic interactions at the molecular level.

While the empirical study of membrane fouling is well defined in the laboratory via the use of methods such as spectroscopy, spectrometry, and microscopy [96], the same cannot be said with respect to simulating fouling mechanisms at the molecular level. Indeed, to the best of our knowledge, there has only been one such publication to date that explicitly explores the molecular interactions between a model organic foulant molecule and a membrane surface, where a model humic acid compound fouling a polyethersulfone (PES) membrane was considered [105]. This will be discussed in Section 10.4.2.

10.4.1
Molecular Modeling of Potential Organic Foulants

Before considering the molecular modeling of the interactions of organic foulant candidates with membranes, it is pertinent to examine what types of organic species are present in the dissolved organic fraction found in natural raw waters or in wastewaters that are to be recycled. These dissolved organic compounds are generally referred to as *natural organic matter* (NOM) in raw waters and *effluent organic matter* (EfOM) in wastewaters. NOM and EfOM consist of a mixture of plant and animal products in various stages of decomposition, as well as compounds synthesized biologically and chemically [106].

Usually, the species that contribute to NOM and EfOM are divided into two groups; humic substances and non-humic substances [106]. Humic substances are generally described as large, amorphous polyelectrolyte particles of varying size and structure, formed via the natural degradation of nonhumic substances [107]. Nonhumic substances include compounds such as carbohydrates, proteins, peptides, amino acids, fats, waxes, resins, pigments, and various low-molecular-weight species [106]. Both the humic and nonhumic compounds can lead to membrane fouling, although the location of fouling appears to be within the pores for humic substances and via surface coverage or cake layer formation for nonhumic substances [108]. As the initial molecular dynamic modeling of membrane fouling has considered fouling by humic substances, we shall concentrate on this system initially.

Humic materials vary greatly with respect to their elemental composition, size, functional groups, structure, chelating ability, and age [107]. There is even discussion [109–113] in the literature as to whether humic substances are true macromolecules or supramolecular structures formed from smaller subunits and

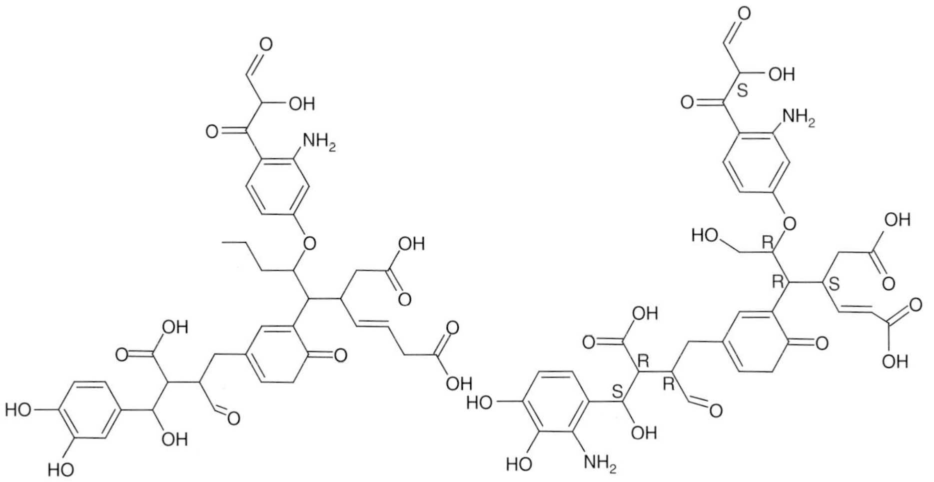

Figure 10.19 Structural representation of the Steelink model (top) and the Temple-Northeastern-Birmingham (TNB) model (bottom). The most stable TNB stereoisomer, SRRSRRS, is also defined here.

held together via relatively weak noncovalent interactions, with soil scientists modeling humic acids via the latter scenario [114–116], while water chemists consider the former model [112]. In this regard, there is no singular identifiable "humic acid molecule"; rather, several structural models have been developed, which contain well-defined functional groups as determined by experimental analysis.

The primary functional groups and structures of importance in humic acid are the presence of both aliphatic and aromatic hydrocarbons, with attached amide, carboxyl, hydroxyl, and ketone groups. Two of the most popular models in the literature are the Steelink model [117] and the Temple-Northeastern-Birmingham (TNB) model [118] (Figure 10.19). As can be seen from Figure 10.19, there is little difference between these two structures, with the TNB model building on and refining the earlier published Steelink model. Owing to this similarity, as well as the fact that the TNB model is more recent, further discussion shall be focused on this model.

The TNB model was first proposed as an improvement on the Steelink model, with the TNB model deemed to be more consistent with the chemical composition and functional group analysis obtained via analytical measurements [119]. This model molecule, with seven chiral centers, may exist in one of 64 chemically distinct stereoisomers [119]. Therefore, the first modeling study carried out for the TNB model molecule was a study to determine the lowest energy stereo isomer.

This investigation, by Jansen *et al.* [119], was carried out in order to determine the dominant stable species. All 64 unique stereoisomers were subjected to MM-based geometry optimization, and the 5 lowest energy structures were then subjected to random conformational searching, whereby the lowest energy conformers of these stereoisomers were further analyzed via MD and geometry optimization [119]. The methods used in these procedures may be considered archaic by current standards; the MM2 force field and Powell conjugate-gradient minimization algorithms were used initially, with the conformational searching carried out using either a random incremental pulse sequence (RIPS) or random structure search (RSS) algorithm to sample the computational space [119]. The result of this extensive conformational searching was that the RSSRSSR isomer was the significantly lower energy stereoisomer, as shown in Figure 10.19. It was also noted that common features of the lowest energy structures generated in this study were stacking interactions between two of the aromatic rings, as well as the amine and one of the carboxyl groups being sterically unhindered [119]. This latter point is important, as it was believed by the authors that this feature "may be important in humic acid polymerization" [119]. This statement reflected the thinking at the time of this study that humic acids were macromolecular biopolymers – indeed, the fact that this model was constructed with an amide group on one side of the molecule and a carboxylic acid group at the other belies this thinking. This idea was formally postulated by further work from the members of this group [118].

Davies et al. [118] further investigated the TNB model using molecular modeling, this time as a biopolymer. Several monomers were joined together via peptide bonds between the unhindered amide and carboxyl groups identified in the previous study. This biopolymer was then subjected to a similar modeling regime, using the MM2 force field as utilized in Sybyl, although no further detail was given. The result of this modeling was a helical structure with interesting properties – hydrophobic "patches" were observable on the exterior of the modeled structure, punctuated by carboxylic acidic groups which are the primary metal-binding sites available to humic substances. The elliptical interior is also hydrophobic [118]. Unfortunately, this study did not thoroughly investigate, via the molecular modeling techniques, metal ions interacting with this proposed helical biopolymeric structure, even though the main focus of this work was the identification of tight binding sites in isolated humic substances, especially with regard to the solubility of trace elements in natural aqueous environments [120, 121]. Instead, this modeling seemed to be restricted only to provide qualitative, or illustrative, information via MD, rather than a rigorous investigation of metal binding.

To more accurately and thoroughly investigate the interactions of metal ions and the TNB model, MD is required, rather than just the geometry optimization used in the previously examined studies. The difference between these methods is the addition of thermal energy, which is propagated to the appropriate ensemble through the equations of motion, as discussed in Section 10.3.1.

Keeping with the theme of modeling trace elements and NOM, an MD investigation was carried out to model the complexation of Cs^+ and Cl^- ions with the TNB NOM model [122]. The conditions of the MD simulations run as part of this study by Xu et al. [122] included the carboxyl groups of the TNB model being completely deprotonated to replicate the pH neutral conditions of the equivalent experiments that were also carried out. With a -3 charge on the TNB (because of the presence of three carboxylic acid groups), this was charge balanced for the simulation via the addition of three Cs^+ ions [122]. This system was then hydrated with 556 water molecules in a cubic MD box with periodic boundary conditions [122]. Further simulations were also run with more ions added, both Cs$^+$ and Cl$^-$, in place of some of the water molecules. Also, an equivalent set of "neat" simulations were run at the equivalent aqueous CsCl concentration for comparison [122]. In all, 16 MD simulations were collected, with relative concentrations of CsCl between 0.1 and 4 M.

All simulations were run using the consistent valence force field [123] (CVFF) to describe the aqueous NOM interactions and the simple point charge [124] (SPC) potentials were applied to the bulk water molecular interactions [122] as well as the Cs$^+$-H$_2$O and Cl$^-$-H$_2$O, as previously described in the literature [125, 126]. As thermodynamic equilibrium was required by the authors of this study, constant-volume (NVT-ensemble) pre-equilibrium runs were performed for all simulated systems, which included an initial run at 600 K for 20 ps using a 1 fs time step [122]. From this elevated temperature, each system was reduced in 50 K steps until 300 K was reached, with an equivalent simulation carried out at each temperature step. A further 20 ps pre-equilibration MD run was carried out at a constant pressure (1 bar) to optimize the models. These optimized structures were used as input for the production MD runs, which consisted of 20 ps to allow each system to equilibrate, followed by 100 ps where the equilibrium dynamic trajectory was recorded at 10 ps intervals for statistical significance [122]. From these production runs, a range of statistical functions could be applied, which described the interactions of Cs$^+$ and Cl$^-$ ions with the TNB model.

The results of these MD simulations corresponded well with experimental ^{133}Cs and ^{35}Cl NMR data also collected during this study [122]. The MD simulations indicated that the Cs$^+$ ions interacted relatively weakly with the NOM model via outer-shell interactions and that the Cl$^-$ ions did not interact with the NOM model at all [122]. This reconciliation of the experimental and theoretical models indicates that (i) the TNB model, even when simulated as a single monomer unit, rather than the polymer that it was designed for, is indicative of humic acid isolated from natural surface waters and studied experimentally with respect to Cs$^+$ interactions and (ii) all parameters for the atoms involved in these simulations were appropriate to describe laboratory results.

Following this work, Kalinichev and Kirkpatrick [114] expanded on these previous findings by extending their study to Na$^+$, Mg^{2+}, and Ca^{2+} using a molecular modeling approach exactly the same as those described previously for the Cs$^+$ study. The results of this study found that Na$^+$ weakly interacts, Ca^{2+} ions strongly interact, and Mg^{2+} showed no interaction [114]. With a number of cations of various sizes and charges being modeled with the TNB model of NOM, it could be concluded that larger ions of the same charge interacted with the negative carboxylate groups in a stronger manner than smaller ions and that a higher charge results in a stronger interaction [114]. From this work, even though the authors recognized the deficiencies of the model compared to "real-world" NOM assemblies, some insights could be gleaned regarding the behavior of NOM in surface waters, particularly the impact of calcium ions that were estimated to interact with ∼33% of all carboxylate groups present in NOMs [114]. However, a comparison with the experiment with respect to these interactions with the alkali and alkali earth metals was not present in this work.

Further work by this team [127] was completed using a different modeling scheme, reflecting the evolution of more realistic force fields and more precise water descriptions. This study used the TIP3P water model, which is an evolution of the SPC model used previously. Owing to this change in the water model, a force field change was also required to describe all other atoms within the simulation, as the force field parameters are dependent on the water model. In this work, the AMBER FF03 force field was selected and the partial charges were described using a utility in AMBER tools [128], based on the AM1-BCC charge generation method [127].

The simulation conditions were also markedly different from the preceding studies, which is reflective of the exponential increase in computing power – the model systems now included eight TNB molecules with the equivalent number of cations to balance the charge of the system [127]. These systems were also solvated with 4215 water molecules in a considerably larger periodic box [127]. Further, starting structures for the production run simulations were carried out by varying the pressure of the system rather than the temperature, before 10 ns production runs were collected and analyzed [127]. The results of this study provided excellent insights into the potential aggregation of NOM monomers into supramolecular assemblies mediated by the presence of Ca^{2+} ions, as bridging between TNB molecules by Ca^{2+}, which is known experimentally [129], was observed for the first time computationally [127]. This finding not only provided more evidence for the effectiveness of the TNB model when computationally modeling NOM but also highlighted the importance of metal ions in NOM aggregation and the potential impact on membrane fouling. A study investigating the interactions between the TNB NOM model and a polyethersulfone membrane in the presence of cations has been carried out and reported in the literature [105]; this is discussed in Section 10.4.2. It is pertinent at this point to discuss other NOM model compounds for use in molecular modeling.

There are a number of alternative NOM models described in the literature, one example of which is poly(acrylic acid) (PAA) [107]. This compound (Figure 10.20) was selected for study as a humic acid analog because of the simplicity of this polymer – it is essentially a one-dimensional string of carboxylic acid moieties, which, as mentioned previously, are responsible for the metal-binding characteristics of NOM [107]. This simple model lacks other features and functional groups of NOMs as defined experimentally, but this shortcoming is overcome by the fact that the PAA model was used to computationally examine aspects of metal ion

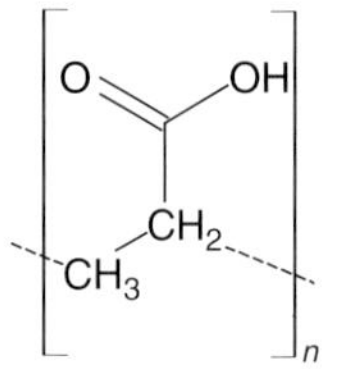

Figure 10.20 Structural representation of poly(acrylic acid).

complexation only [107]. Also, these PAA oligomers were studied individually via quantum chemical methods, so secondary features such as aggregation were not modeled as per the MD studies previously discussed. Indeed, owing to the fact that these PAA oligomers and their metal–ion interactions were studied via DFT, a simplification of the model (i.e., PAA over the TNB model) may have been required for a less computationally expensive study.

All calculations carried out on these PAA oligomers (where $n = 2-5$) were at the DFT level of theory, using the B3LYP hybrid functional with the SVP + sp basis set (split valence polarization basis set with additional s functions on all atoms and additional p shell functions on the heavy atoms). All four polymer lengths were investigated with respect to hydrogen bonding between the PAA carboxyl groups and water, acetic acid, and MCPA ((4-chloro-2-methylphenoxy)acetic acid) – a common herbicide), in turn, with full geometry optimizations performed for each structure in both the gas phase, as well as aqueous phase, with use of the COSMO model [107]. Also, cation bridges using a hydrated Ca^{2+} ion between the PAA and the MCPA were also examined [107]. This comprehensive study, while providing excellent detail on the thermodynamics of calcium–carboxylate binding and hydrogen-bonding interactions, provides little to the NOM investigations, owing to the narrow focus demanded by the use of quantum chemical calculations.

A more appropriate computational model to investigate interactions between ions and humic acid analogs was presented by Guardado et al. [130]. This work used a combination of MM and semi-empirical quantum chemical calculations to examine a range of ionic interactions [130]. MM, using both the MM + and AMBER force fields, were used initially to produce energy-minimized structures of a humic acid analog with either a metal ion or a phosphate ion [130]. Where an interaction between specific regions of the molecules have achieved a bond distance equivalent, these structures were then subjected to the semiempirical PM3 calculation (which, as it is a quantum chemical method and specifically accounts for electrons, is able to create new bonds) to quantify this binding interaction [130]. However, in order to compare these computational results with experimental results, salicylic acid was used as the NOM model [130]. While salicylic acid (IUPAC name 2-hydroxybenzoic acid; see Figure 10.21a) contains some of the functional groups of NOM, specifically an aromatic group, a carboxyl group, and a hydroxyl group, other aspects such as an aliphatic character and ketone groups are absent [130].

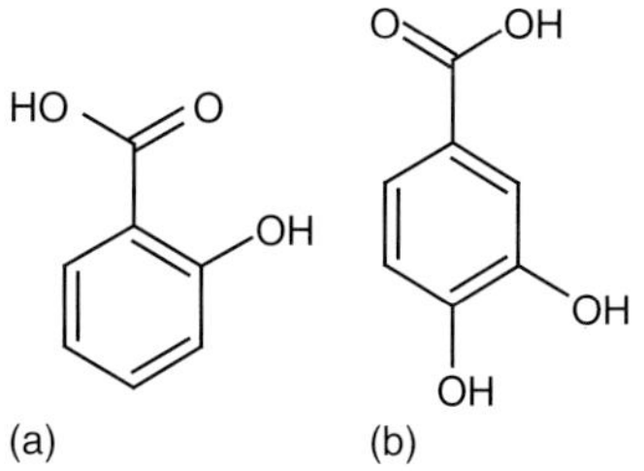

Figure 10.21 Structural representations of (a) salicylic acid and (b) 3,4-dihydroxybenzoic acid.

Also, the inherent "floppiness" of NOM, as well as the ability to aggregate into supramolecular aggregates, are absent from this model. As per the study discussed earlier, the NOM model for this study was selected purely on the metal-binding characteristics and the ease of study rather than being truly representative of NOM.

A further example of this is the use of 3,4-dihydroxybenzoic acid (Figure 10.21b) as a humic acid model [131]. This was selected as a NOM analog to model transient radicals in humic substances mainly due to the ease of experimental investigation [131], coupled with the discussion provided earlier for the salicylic acid study.

Perhaps the most ambitious attempt to computationally investigate alternative models for NOMs was published by Zhang and LeBoeuf [132]. The models investigated in this study included a softwood lignin building block [133], a hardwood lignin building block [134], a modified kerogen building block [135], and an n-hexane soot model [136]. This study was designed to examine a range of aqueous thermodynamic properties of these models and compare these results to experiment, in order to possibly suggest modifications to NOM models to more accurately describe these aspects [132]. MD was used to computationally describe these model compounds, using a number of different force fields [132]. The results were generally comparable with experiments; however, some aspects, such as the glass-transition temperature, were affected by the size of the periodic cell [132].

Therefore, with respect to the different models available to model the humic fraction of NOM, it would appear that the TNB model is the most powerful candidate to approximate NOM computationally. The model contains all of the functional groups expected in NOM, and yet is small enough to easily model multiple molecules in the presence of water and ions using MD. It also demonstrates good metal-binding characteristics as well as self-aggregating behavior – also aspects of experimentally studied NOM. Indeed, the TNB model was used in the one study published at the time of writing to investigate the effects of NOM and metal ions on a UF membrane.

10.4.2
Modeling of Membrane Fouling

In what is, surprisingly, the only available example of membrane fouling studied via molecular modeling in the literature, the work by Ahn *et al.* [105] looked at the effects of metal ions on a NOM model in the presence of a polymeric membrane material. This work specifically aimed to provide insights into the problematic area of adsorption of macromolecules onto the membrane surface, as this potential mechanism of fouling is poorly understood when compared with macroscale effects such as the formation of cake layers [105]. Owing to the inherent nanoscale of molecular interactions, MD provides the ideal tool to examine this issue in detail.

Although there are a number of studies [24, 137–140] in the literature, whereby RO and zeolite membranes have been examined computationally to provide further insights into the behavior of water and ions with these membranes, these studies have not previously been extended to examine fouling – an equally important issue in membrane technologies. The approach by Ahn *et al.* [105] utilized MD to probe

10.4 Molecular Simulation of Membrane Fouling

interactions between that most commonly used NOM model, the TNB structure, with a PES membrane in the presence of metal ions. This particular combination of compounds was chosen because of the hypothesis of the researchers that the TNB NOM model would interact with, and therefore adsorb to, the surface of the membrane via a cation bridge between the partial negative charge that resides on the sulfonyl groups of the membrane surface and the deprotonated carboxyl groups of the TNB molecule [105]. Owing to the sulfonyl groups on the surface of the membrane, and hence, a residual partial charge, the modeled PES membrane was hydrophilic in nature. In practice, while these membranes are often hydrophilic, the surface properties of the membranes may be altered by the addition of additives such as polyvinylpyrrolidone (PVP), although this is not always disclosed. As a result, modeling the molecular characteristics of membranes with a nominal composition becomes a key concern, and may be one of the reasons that MD modeling of membrane fouling mechanisms has not been used more frequently.

The conditions of the simulations used in Ahn *et al.* [105] were similar to those discussed previously – the hydrated system was modeled under periodic boundary conditions using the SPC water model to simulate the bulk water. All other atomic interactions were described by a combination of the CVFF and the CLAYFF force field [105]. While this combination of force fields is an interesting approach, the specific details of precisely how this combination is achieved are not explicitly provided. While assumptions could be made on the basis of the parameterization of both the CVFF and CLAYFF force fields, the exact combination of these two force fields was not adequately defined. Therefore, no further comment can be made regarding the validity of this approach.

The systems for simulation were constructed to reflect laboratory conditions – the polymer membrane system was compressed stepwise from a very low starting density of 0.0115 g cm^{-3} (the raw result of the polymer building program) to 1.30 g cm^{-3}, which is close to the experimental value of 1.37 g cm^{-3} [105]. It must be noted at this point that no attempt was made to model pores in this membrane; only interactions between the metal ions, the NOM model, and the polymer surface were simulated [105]. The 60 residue monomer was relaxed dynamically *in vacuo* to ensure a low-energy state, before being hydrated and simulated further until the now hydrated polymer system was at equilibrium [105]. Once an appropriate model of the aqueous polymer was achieved, one TNB model molecule was added along with ions (either two divalent cations (Ca^{2+} or Mg^{2+}) plus one Cl$^-$ ion, or four Na$^+$ ions plus one Cl$^-$) were added in place of some of the water molecules [105]. Initially, the metal ions were placed between the membrane surface and the NOM model, specifically at distances of ∼5 Å from a sulfonyl group and ∼6 Å from a carboxyl group [105]. All systems were initially modeled for approximately 100 ps to reach equilibrium, after which the coordinates and trajectories were recorded at 10 fs intervals over 300 ps of simulation time for statistical analysis [105]. A control run of only the hydrated PES system was also simulated [105].

This control simulation showed that the parameterization of the PES polymer was appropriate; hydrogen bonds were observed between water molecules and the sulfonyl surface groups, which resulted in a slowing of the diffusive mobility of the water molecules, the diffusion constant changing from 2.33×10^{-5} cm^2 s^{-1} in the bulk water to 1.33×10^{-5} cm^2 s^{-1} at the surface [105]. The diffusion constants from the simulation, calculated from their mean square displacement, are described in Equation 10.4).

$$D_i = \frac{\text{MSD}_i(t)}{6t} = \frac{|\mathbf{r}_i(t) - \mathbf{r}_i(0)|^2}{6t} \quad (10.4)$$

where D_i is the diffusion coefficient of the species i; $\mathbf{r}(t)$ is the vector of atomic coordinates at time t; and the angular brackets denote the averaging over all species of type i present in the system and over all possible time origins along the MD trajectory, $t = 0$ [105].

With evidence that the model membrane surface was interacting electrostatically with the water molecules, as would be expected of a hydrophilic surface, the effects of this surface on metal ions and the NOM model could be analyzed. As with previous computational experiments reported in the literature, the Ca^{2+} ions interacted strongly with the carboxylic acid moieties of the TNB model and the Mg^{2+} interacted weakly with these groups [105]. Interestingly, none of the cations investigated were observed to significantly interact with the partially charged sulfonyl groups of the PES surface during either the 100 ps pre-equilibration step or the 300 ps production simulation [105]. In fact, the average distance of the divalent cations from the oxygen atom of the nearest sulfonyl group was ~7 Å, with the Na$^+$ ions diffusing out of the membrane – NOM interface completely [105]. On further investigation, it was revealed that the partial charges on the atoms as used in the simulations were the source of the non-interaction between the sulfonyl groups and the divalent cations [105]. The sulfonyl oxygen atoms carried a partial charge of -0.3368 e in these simulations, which is relatively smaller than both the -0.82 e of the water oxygen atom and the -0.57 e on the deprotonated carboxyl oxygen atoms [105]. Therefore, the sulfonyl oxygen atoms could not compete with the stronger charges of the other atoms.

These findings were inconsistent with the proposed hypothesis of fouling in this system, as the relatively weak negative partial charge of this hydrophilic membrane was shielded from interacting with the metal ions via hydration [105]. While it would appear that cationic bridging is not a source of fouling in hydrophilic membrane systems, there is still evidence in this paper for the aggregation of NOM species mediated by Ca^{2+}, because of the strong binding interactions known for molecules containing carboxyl groups.

Owing to the design of this experiment, other aspects of these systems that may result in fouling were not able to be investigated, such as fouling due to direct hydrophobic interactions (because of the hydropobic nature of the TNB model compound) or the fact that membranes are typically slightly negative when measured experimentally [32, 141].

Another type of molecule naturally found in NOM and EfOM – polyuronides – are also a target for modeling studies. Polyuronide compounds, which are hydrophilic carboxylate-containing polysaccharide chains, are believed to be responsible for

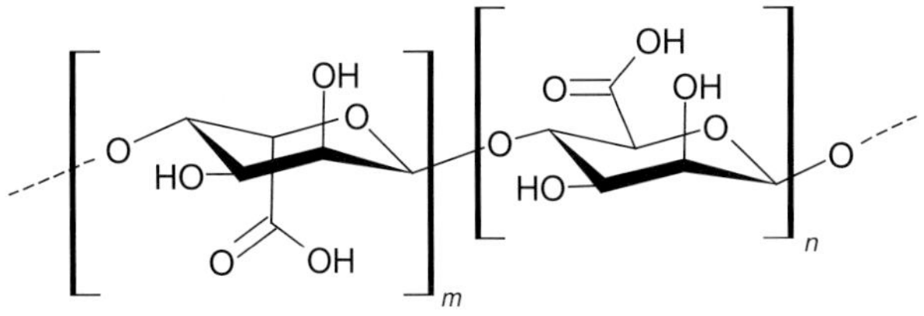

Figure 10.22 Schematic diagram of the structure of alginates, showing α-L-guluronic acid (L) and β-D-mannuronic acid (R).

aggregation of soil particles [142]. Perhaps the most widely studied of this class of molecules is the relatively simple structure of alginic acid (also referred to as *alginate*), which is a binary copolymer of 1,4-linked α-L-guluronic acid and β-D-mannuronic acid (Figure 10.22).

Alginates are currently under investigation as a model for membrane fouling, as such compounds are contained in NOM and EfOM. Flemming and Ridgway [31] and Reinhard et al. [143] have described MD simulations indicating that natural bacterial alginates containing acetyl substitutions on β-D-mannuronic acid residues undergo rapid sorption to solvated PA membrane surfaces. These compounds contain carboxylate groups and have the ability to chelate metal ions, forming supramolecular aggregates in the presence of Ca^{2+}. Further to this, alginates are relatively easy to study experimentally, as they are water soluble and are able to be purchased cheaply and with good purity.

From a theoretical point of view, short-chain alginate chains are also relatively easy to study, as many of the parameters required for MD investigations (i.e., bond lengths, bond angles, and bond rigidity) of these alginate monomer units can be easily adapted from the equivalent saccharide parameters already defined for most force fields. However, it is recommended that the electrostatic charges of all atoms be recalculated using a quantum chemical method, as the addition of a carboxylate group to a saccharide molecule has a significant electronic effect.

In initial experimental investigations, alginate (especially in the presence of Ca^{2+} ions) has been shown to significantly foul PA [143] and hydrophobic membranes [144]. To assist in determining if such fouling is due to molecular interactions or a colloidal mechanism, molecular modeling experiments involving short chains of alginate and a hydrophobic polypropylene surface, which approximated the membrane surface, have been carried out by the authors (Orbell, Stewart, and Gray: unpublished results). These simulations showed that the short individual alginate chains, in the presence of Ca^{2+} ions, have the ability to interact with the hydrophobic polymer surface on a molecular level (Figure 10.23a), through hydrophobic "patches" present on the β-D-mannuronic acid monomer units (Figure 10.23b). This work is continuing at present; however, these initial results suggest that this could be a good model for understanding organic fouling of hydrophobic membranes by hydrophilic compounds.

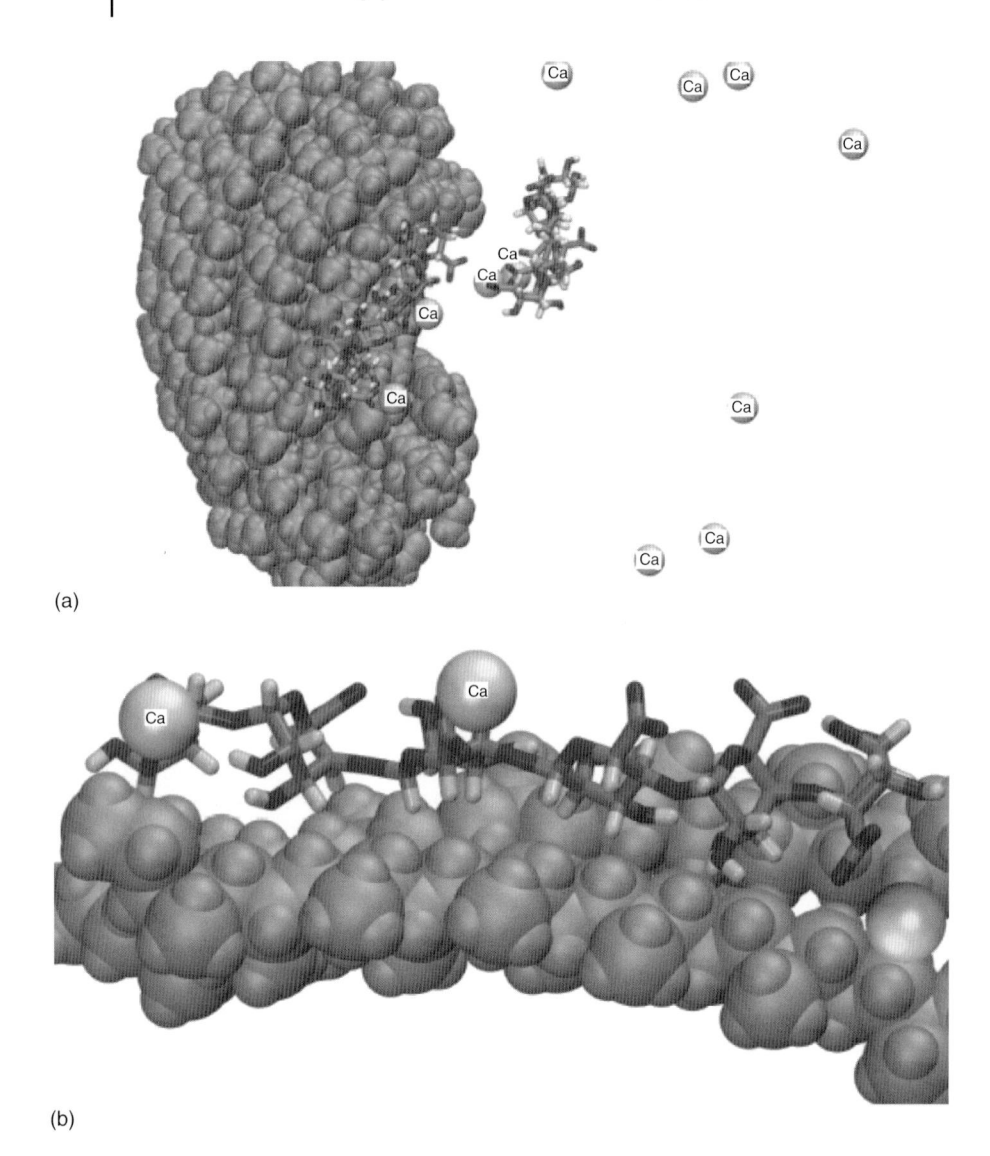

Figure 10.23 (a) Molecular dynamic simulation frame showing alginate chain (multishaded stick structure) interacting with both calcium ions (light gray spheres labeled) and polypropylene surface (dark gray atoms). Water has been removed for clarity. (b) Close-up view of the alginate-calcium ions-polypropylene interaction for (a). Noninteraction polypropylene chains and water have been removed for clarity (calcium atoms are the light gray spheres labeled Ca, alginate the multishaded stick structure and polypropylene the dark gray atoms).

10.4.3
Future Directions

While some initial modeling has been carried out previously with respect to the interactions between polymer membrane surfaces and foulants in the presence of metal ions (especially Ca^{2+}), there is a need for more of these types of studies. Intermolecular interactions between membrane surfaces and organic foulants are poorly understood because they are generally more difficult to study experimentally than other fouling modes, such as colloidal fouling and scaling. However, intermolecular fouling mechanisms are just as important as these "bulk" modes, especially given that membranes fouled in this way are more difficult to clean than cake layers.

Further work in this area must include the thorough experimental characterization of the membrane surfaces, with respect to charges (including partial charges as measured by the zeta potential), functional groups, and topology. If any of these aspects in a model are significantly different from the experimental conditions, then any outcomes elucidated from these computational experiments will be inaccurate. Such physicochemical investigations would need to be carried out for each type of membrane to be studied computationally in order to obtain the most realistic model possible. This will allow a more accurate representation of the important electrostatic interactions that occurs between the organic foulant and the membrane.

The development of a different NOM model compound or a suite of compounds that may be modeled together – especially with respect to hydrophobic interactions – may also be an area of future investigation. NOM that is isolated from raw water is known to contain aliphatic and aromatic groups; however, current models either hide these important groups (by attaching hydroxyl and carboxyl groups, which mask these hydrophobic areas) or the model compounds are too large to effectively simulate in the presence of a membrane surface and metal ions. However, any newly developed NOM model would need to be rigorously tested against experimental data to ensure that the theoretical behaviors are similar to actual NOM, especially with regard to the ability to chelate metal ions, and particularly to display self-aggregation in the presence of Ca^{2+}.

10.4.4
Concluding Remarks

Fouling in membrane-based technologies is an important issue, as fouled membranes are less efficient, resulting in higher operating pressures or lower fluxes. Membrane fouling has a number of different causes, namely, colloidal/particulate fouling, microbial/biofouling, dissolved organic fouling, and scaling. While some aspects of these fouling modes are well understood and can be mitigated via alteration of operating conditions, fouling by NOM is less understood because of the complex nature of NOM and the variety of membrane surfaces employed.

NOM are found in all surface waters and in many wastewater streams. While there are a number of components that make up NOM, the most prevalent is the humic substances, which is the result of natural decomposition and degradation of biomolecules. Humic substances are referred to as such because there is not one definitive structure for humic acid – the composition of these substances depends on the location and local environment of the sample site, and fluctuates with time.

While it is well established that NOM fouls polymeric membranes, the exact mechanism has been difficult to prove. However, it is hypothesized that the humic substances found in NOM interact on a molecular scale and adsorb onto the surface of the membrane. The mechanism of this adsorption is still under investigation; however, studies on the molecular scale can be difficult in the laboratory. In this regard, computational chemistry, in particular MD, may provide some insights into this mechanism.

MD is suitable for studying the intermolecular interactions, particularly hydrogen bonding, and other electrostatic interactions in aqueous environments. In this regard, NOM models have previously been investigated with respect to their interactions with metal ions, with these investigations comparing favorably with experimental results. Equally, membranes have also been successfully modeled computationally and studied with respect to their water transport and ion rejection ability. The next step is to model foulants, in the presence of divalent cations, with well-defined membrane surfaces over relatively larger time periods than is currently reported in the literature. Simulations of nanosecond length, containing several NOM structures, an appropriate number of ions, and a large membrane surface area, would improve the understanding of intermolecular interactions at the membrane surface.

References

1. Zhu, F., Tajkhorshid, E., and Schulten, K. (2004) Theory and simulation of water permeation in aquaporin-1. *Biophys. J.*, **86**, 50–57.
2. Crozier, P.S., Stevens, M.J., Forrest, L.R., and Woolf, T.B. (2003) Molecular dynamics simulation of dark-adapted rhodopsin in an explicit membrane bilayer: coupling between local retinal and larger scale conformational change. *J. Mol. Biol.*, **333**, 493–514.
3. Pasenkiewicz-Gierula, M., Murzyn, K., Róg, T., and Czaplewski, C. (2000) Molecular dynamics simulation studies of lipid bilayer systems. *Acta Biochim. Pol.*, **47**, 601–611.
4. Malaeb, L. and Ayoub, G.M. (2011) Reverse osmosis technology for water treatment: state of the art review. *Desalination*, **267**, 1–8.
5. Radjenovic, J., Petrovic, M., Venturac, F., and Barcelo, D. (2008) Rejection of pharmaceuticals in nanofiltration and reverse osmosis membrane drinking water treatment. *Water Res.*, **42**, 3601–3610.
6. Bellona, C. and Drewes, J. (2005) The role of membrane surface charge and solute physico-chemical properties in the rejection of organic acids by NF membranes. *J. Membr. Sci.*, **249** (1–2), 227–234.
7. Cadotte, J.E. (1981) Interfacially synthesized reverse osmosis membrane. US Patent 4,277,344, filed Feb. 22, 1979 and issued Jul. 7, 1981.
8. Larson, R.E., Cadotte, J.E., and Petersen, R.J. (1981) The FT30 seawater reverse osmosis membrane-element test results. *Desalination*, **38**, 473–483.
9. Baker, R.W. (2006) *Membrane Technology and Applications*, 2nd edn, John Wiley & Sons, Inc., Hoboken, NJ.
10. Morgan, P.W. (1965) *Condensation Polymers by Interfacial and Solution Methods*, Interscience, New York.

11. Freger, V. (2005) Kinetics of film formation by interfacial polycondensation. *Langmuir*, **21**, 1884–1894.
12. Freger, V. (2003) Nanoscale heterogeneity of polyamide membranes formed by interfacial polymerization. *Langmuir*, **19**, 4791–4797.
13. Mi, B., Cahill, D., and Marinas, B.J. (2007) Physico-chemical integrity of nanofiltration/reverse osmosis membranes during characterization by Rutherford backscattering spectrometry. *J. Membr. Sci.*, **291**, 77–85.
14. Coronell, O., Gonzalez, M.I., Marinas, B.J., and Cahill, D. (2010) Ionization behavior, stoichiometry of association, and accessibility of functional groups in the active layers of reverse osmosis and nanofiltration membranes. *Environ. Sci. Technol.*, **44**, 6808–6814.
15. Coronell, O., Marinas, B.J., Zhang, X., and Cahill, D. (2008) Quantification of functional groups and modeling of their ionization behavior in the active layer of FT30 reverse osmosis membrane. *Environ. Sci. Technol.*, **42**, 5260–5266.
16. Ishida, K.P., Bold, R.M., and Phipps, D.W. (2005) Identification and Evaluation of Unique Chemicals for Optimum Membrane Compatibility and Improved Cleaning Efficiency. Final Report Research Contract No. 41812, California Department of Water Resources, Sacramento, CA.
17. Kim, S.H., Kwak, S.Y., and Suzuki, T. (2005) Positron annihilation spectroscopic evidence to demonstrate the flux-enhancement mechanism in morphology-controlled thin-film composite (TFC) membrane. *Environ. Sci. Technol.*, **39**, 1764–1770.
18. Sharma, V.K., Singh, P.S., Gautam, S., Maheshwari, P., Dutta, D., and Mukhopadhyay, R. (2009) Dynamics of water sorbed in reverse osmosis polyamide membrane. *J. Membr. Sci.*, **326**, 667–671.
19. Lipp, P., Gimbel, R., and Himmel, F.H. (1994) Parameters influencing the properties of FT30 membranes. *J. Membr. Sci.*, **95**, 185–197.
20. Pacheco, F.A., Pinnau, I., Reinhard, M., and Leckie, J.O. (2010) Characterization of isolated polyamide thin films of RO and NF membranes using novel TEM techniques. *J. Membr. Sci.*, **358**, 51–59.
21. Coronell, O., Marinas, B.J., and Cahill, D. (2009) Accessibility and ion exchange stoichiometry of ionized carboxylic groups in the active layer of FT30 reverse osmosis membrane. *Environ. Sci. Technol.*, **43**, 5042–5048.
22. Coronell, O., Marinas, B.J., and Cahill, D. (2011) Depth heterogeneity of fully aromatic polyamide active layers in reverse osmosis and nanofiltration membranes. *Environ. Sci. Technol.*, **45**, 4513–4520.
23. Tang, C.Y., Kwon, Y.-N., and Leckie, J.O. (2007) Probing the nano- and micro-scales of reverse osmosis membranes – a comprehensive characterization of physicochemical properties of uncoated and coated membranes by XPS, TEM, ATR-FTIR, and streaming potential measurements. *J. Membr. Sci.*, **287**, 146–156.
24. De Baerdemaeker, J., Boussu, K., Djourelov, N., Van der Bruggen, B., Dauwe, C., Weber, M., and Lynn, K.G. (2007) Investigation of nanopores in nanofiltration membranes using slow positron beam techniques. *Phys. Status Solidi C*, **4**, 3804–3809.
25. He, Y., Tang, Y., and Wang, X. (2011) Dissipative particle dynamics simulation on the membrane formation of polymer-diluent system via thermally induced phase separation. *J. Membr. Sci.*, **368**, 78–85.
26. Kotelyanskii, M.J., Wagner, N.J., and Paulaitis, M.E. (1996) Building large amorphous polymer structures: atomistic simulation of glassy polystyrene. *Macromolecules*, **29**, 8497–8506.
27. Kotelyanskii, M.J., Wagner, N.J., and Paulaitis, M.E. (1998) Atomistic simulation of water transport in the reverse osmosis membrane FT30. *J. Membr. Sci.*, **139**, 1–16.
28. Kotelyanskii, M.J., Wagner, N.J., and Paulaitis, M.E. (1999) Molecular dynamics simulation study of the

mechanisms of water diffusion in a hydrated, amorphous polyamide. *Comput. Theor. Polym. Sci.*, **9**, 301–306.
29. Harder, E., Walters, D.E., Bodnar, Y.D., Faibish, R.S., and Roux, B. (2009) Molecular dynamics study of a polymeric reverse osmosis membrane. *J. Phys. Chem. B*, **113**, 10177–10182.
30. Riley, R.L., Ishida, K.P., Lin, S.W., Murphy, A., and Ridgway, H.F. (2001) Development of Improved Membranes for ROWPU Spiral-wound Elements. Final Report Contract DAAD 19-99-C-0003, Department of the Army, Army Research Office, Research Triangle Park, NC.
31. Flemming, H.-C. and Ridgway, H. (2009) in *Marine and Industrial Biofouling*, Springer Series on Biofilms (eds H.-C. Flemming, P.S. Murthy, R. Venkatesen, and K.E. Cooksey), Springer-Verlag, Berlin, pp. 103–117.
32. Childress, A.E. and Elimelech, M. (1996) Effect of solution chemistry on the surface charge of polymeric reverse osmosis and nanofiltration membranes. *J. Membr. Sci.*, **119**, 253–268.
33. Rodriguez, G., Buonora, S., Knoell, T., Phipps, D., and Ridgway, H. (2004) Rejection of Pharmaceuticals by Reverse Osmosis Membranes: Quantitative Structure Activity Relationship (QSAR) Analysis. NWRI Project No. 01-EC-002, National Water Research Institute, Fountain Valley, CA.
34. Peng, F., Jiang, Z., and Hoek, E.M.V. (2011) Tuning the molecular structure, separation performance and interfacial properties of poly(vinyl alcohol)-polysulfone interfacial composite membranes. *J. Membr. Sci.*, **368**, 26–33.
35. Lonsdale, H.K., Merten, U., and Riley, R.L. (1965) Transport properties of cellulose acetate osmotic membranes. *J. Appl. Polym. Sci.*, **9**, 1341–1362.
36. Merten, U. (1966) in *Desalination by Reverse Osmosis* (ed. U. Merten), MIT Press, Cambridge, MA, pp. 15–54.
37. Kao, S., Huang, Y., Liao, K., Hung, W., Chang, K., Guzman, M., Huang, S., Wang, D., Tung, K., Lee, K., and Lai, J. (2010) Applications of positron annihilation spectroscopy and molecular dynamics simulation to aromatic polyamide pervaporation membranes. *J. Membr. Sci.*, **348**, 117–123.
38. Shintani, T., Shimazu, A., Yahagi, S., and Matsuyama, H. (2009) Characterization of methyl-substituted polyamides used for reverse osmosis membranes by positron annihilation lifetime spectroscopy and MD simulation. *J. Appl. Polym. Sci.*, **113**, 1757–1762.
39. Mi, B., Coronell, O., Marinas, B.J., Watanabe, F., Cahill, D., and Petrov, I. (2006) Physico-chemical characterization of NF/RO membrane active layers by Rutherford backscattering spectrometry. *J. Membr. Sci.*, **282**, 71–81.
40. Zhang, X., Cahill, D., Coronell, O., and Marinas, B. (2007) Partitioning of salt ions in FT30 reverse osmosis membranes. *Appl. Phys. Lett.*, **91**, 181904.
41. Hughes, Z.E. and Gale, J.D. (2010) A computational investigation of the properties of a reverse osmosis membrane. *J. Mater. Chem.*, **20**, 7788–7799.
42. Li, D. and Wang, H. (2010) Recent developments in reverse osmosis desalination membranes. *J. Mater. Chem.*, **20**, 4551–4556.
43. Breck, D.W. (1974) *Zeolite Molecular Sieves: Structure, Chemistry and Use*, John Wiley & Sons, Inc., New York.
44. Muller, M., Harvey, G., and Prins, R. (2000) Comparison of the dealumination of zeolites beta, mordenite, ZSM-5 and ferrierite by thermal treatment, leaching with oxalic acid and treatment with $SiCl_4$ by ^1H, ^{29}Si and ^{27}Al MAS NMR. *Microporous Mesoporous Mater.*, **34**, 135–147.
45. Morales-Pacheco, P., Alvarez-Ramirez, F., del Angel, P., Bucio, L., and Dominguez, J.M. (2007) Synthesis and structural properties of zeolytic nanocrystals I. MFI type zeolites. *J. Phys. Chem. C*, **111**, 2368–2378.
46. Apelian, M.R., Fung, A.S., Kennedy, G.J., and Degnan, T.F. (1996) Dealumination of zeolite beta via dicarboxylic acid treatment. *J. Phys. Chem.*, **100**, 16577–16583.
47. Caro, J., Hocevar, S., Karger, J., and Riekert, L. (1986) Intracrystalline

self-diffusion of H_2O and CH_4 in ZSM-5 zeolites. *Zeolites*, **6**, 213–216.
48. Ari, M.U., Ahunbay, M.G., Yurtsever, M., and Erdem-Senatalar, A. (2009) Molecular dynamics simulation of water diffusion in MFI-type zeolites. *J. Phys. Chem. B*, **113**, 8073–8079.
49. Li, L.X., Dong, J.H., Nenoff, T.M., and Lee, R. (2004) Desalination by reverse osmosis using MFI zeolite membranes. *J. Membr. Sci.*, **243**, 401–404.
50. Duke, M.C., O'Brien-Abraham, J., Milne, N., Zhu, B., Lin, J.Y.S., and da Costa, J.C.D. (2009) Seawater desalination performance of MFI type membranes made by secondary growth. *Sep. Purif. Technol.*, **68**, 343–350.
51. Jeong, B., Hoek, E., Yan, Y., Subramani, A., Huang, X., Hurwitz, G., Ghosh, A., and Jawor, A. (2007) Interfacial polymerization of thin film nanocomposites: a new concept for reverse osmosis membranes. *J. Membr. Sci.*, **294**, 1–7.
52. Zhu, B., Zou, L., Doherty, C.M., Hill, A.J., Lin, Y.S., Hue, X., Wang, H., and Duke, M. (2010) Investigation into the effects of ion and water interaction on structure and chemistry of silicalite MFI type zeolite for its potential use as a seawater desalination membrane. *J. Mater. Chem.*, **20**, 4675–4683.
53. Lin, J. and Murad, S. (2001) A computer simulation study of the separation of aqueous solutions using thin zeolite membranes. *Mol. Phys.*, **99**, 1175–1181.
54. Gonze, X., Allan, D.C., and Teter, M.P. (1992) Dielectric tensor, effective charges, and phonons in alpha-quartz by variational density-functional perturbation-theory. *Phys. Rev. Lett.*, **68**, 3603–3606.
55. Bussai, C., Vasenkov, S., Liu, H., Bohlmann, W., Fritzsche, S., Hannongbua, S., Haberlandt, R., and Karger, J. (2002) On the diffusion of water in silicalite-1: MD simulations using ab initio fitted potential and PFG NMR measurements. *Appl. Catal. Gen.*, **232**, 59–66.
56. Demontis, P., Stara, G., and Suffritti, G.B. (2003) Behavior of water in the hydrophobic zeolite silicalite at different temperatures. A molecular dynamics study. *J. Phys. Chem. B*, **107**, 4426–4436.
57. Smirnov, K. and Bougeard, D. (2003) Including the polarization in simulations of hydrated aluminosilicates. Model and application to water in silicalite. *Chem. Phys.*, **292**, 53–70.
58. Fleys, M., Thompson, R.W., and MacDonald, J.C. (2004) Comparison of the behavior of water in silicalite and dealuminated zeolite Y at different temperatures by molecular dynamic simulations. *J. Phys. Chem. B*, **108**, 12197–12203.
59. Desbiens, N., Boutin, A., and Demachy, I. (2005) Water condensation in hydrophobic silicalite-1 zeolite: a molecular simulation study. *J. Phys. Chem. B*, **109**, 24071–24076.
60. Fleys, M. and Thompson, R. (2005) Monte Carlo simulations of water adsorption isotherms in silicalite and dealuminated zeolite Y. *J. Chem. Theory Comput.*, **1**, 453–458.
61. Puibasset, J. and Pellenq, R.J.M. (2008) Grand canonical Monte Carlo simulation study of water adsorption in silicalite. *J. Phys. Chem. B*, **112**, 6390–6397.
62. Bordat, P., Cazade, P.-A., Baraille, I., and Brown, R. (2010) Host and adsorbate dynamics in silicates with flexible frameworks: empirical force field simulation of water in silicalite. *J. Chem. Phys.*, **132**, 094501.
63. Hughes, Z.E., Carrington, L.A., Raiteri, P., and Gale, J.D. (2011) A computational investigation into the suitability of purely siliceous zeolites as reverse osmosis membranes. *J. Phys. Chem. C*, **115**, 4063–4075.
64. Shah, R., Payne, M.C., Lee, M.H., and Gale, J.D. (1996) Understanding the catalytic behavior of zeolites: a first-principles study of the adsorption of methanol. *Science*, **271**, 1395–1397.
65. Haase, F. and Sauer, J. (2000) Ab initio molecular dynamics simulation of methanol interacting with acidic zeolites of different framework structure. *Microporous Mesoporous Mater.*, **35–6**, 379–385.

66. Benco, L., Demuth, T., Hafner, J., Hutschka, F., and Toulhoat, H. (2001) Adsorption of linear hydrocarbons in zeolites: a density-functional investigation. *J. Chem. Phys.*, **114**, 6327–6334.
67. Hytha, M., Stich, I., Gale, J.D., Terakura, K., and Payne, M.C. (2001) Thermodynamics of catalytic formation of dimethyl ether from methanol in acidic zeolites. *Chem.—Eur. J.*, **7**, 2521–2527.
68. Dion, M., Rydberg, H., Schroder, E., Langreth, D.C., and Lundqvist, B.I. (2004) Van der Waals density functional for general geometries. *Phys. Rev. Lett.*, **92**, 246401.
69. Grimme, S., Antony, J., Ehrlich, S., and Krieg, H. (2010) A consistent and accurate ab initio parameterization of density functional dispersion correction (DFT-D) for the 94 elements H-Pu. *J. Chem. Phys.*, **132**, 154104.
70. Maschio, L., Usvyat, D., Manby, F.R., Casassa, S., Pisani, C., and Schutz, M. (2007) Fast local-MP2 method with density-fitting for crystals. I. Theory and algorithms. *Phys. Rev. B*, **76**, 075101.
71. Adams, D.L. (1974) Grand canonical ensemble Monte-Carlo for a leonnard-jones fluid. *Mol. Phys.*, **29**, 307–311.
72. Sanders, M.J., Leslie, M., and Catlow, C.R.A. (1984) Interatomic potentials for SiO_2. *J. Chem. Soc., Chem. Commun.*, 1271–1273.
73. Faux, D.A., Smith, W., and Forester, T. (1997) Molecular dynamics studies of hydrated and dehydrated Na^+-zeolite-4A. *J. Phys. Chem. B*, **101**, 1762–1768.
74. Ockwig, N.W., Cygan, R.T., Criscenti, L.J., and Nenoff, T.M. (2008) Molecular dynamics studies of nanoconfined water in clinoptilolite and heulandite zeolites. *Phys. Chem. Chem. Phys.*, **10**, 800–807.
75. Demontis, P. and Suffritti, G.B. (2009) A comment on the flexibility of framework in molecular dynamics simulations of zeolites. *Microporous Mesoporous Mater.*, **125**, 160–168.
76. Bell, R.G., Jackson, R.A., and Catlow, C.R.A. (1990) Computer-simulation of the monoclinic distortion in silicalite. *J. Chem. Soc., Chem. Commun.*, **10**, 782–783.
77. Martonak, R., Donadio, D., Oganov, A.R., and Parrinello, M. (2007) Predicting crystal structures: the parrinello-rahman method revisited. *Phys. Rev. B*, **76**, 014120.
78. Williams, D.E. (1989) Accelerated convergence treatment of Rn lattice sums. *Crystallogr. Rev.*, **2**, 3–23.
79. Bussai, C., Hannongbua, S., Fritzsche, S., and Haberlandt, R. (2002) Ab initio potential energy surface and molecular dynamics simulations for the determination of the diffusion coefficient of water in silicalite-1. *Chem. Phys. Lett.*, **354**, 310–315.
80. von Koningsveld, H., Bekkum, H.V., and Jansen, J.C. (1987) On the location and disorder of the tetrapropylammonium (TPA) ion in zeolite ZSM-5 with improved framework accuracy. *Acta Crystallogr., Sect. B*, **43**, 127–132.
81. Karger, J. and Ruthven, D.M. (1992) *Diffusion in Zeolites and Other Microporous Solids*, John Wiley & Sons, Inc., New York.
82. Hummer, G., Garde, S., Garcia, A.E., and Pratt, L.R. (2000) New perspectives on hydrophobic effects. *Chem. Phys.*, **258**, 349–370.
83. Corry, B. (2008) Designing carbon nanotube membranes for efficient water desalination. *J. Phys. Chem. B*, **112**, 1427–1434.
84. Kalra, A., Garde, S., and Hummer, G. (2003) Osmotic water transport through carbon nanotube membranes. *Proc. Natl. Acad. Sci. U.S.A.*, **100**, 10175–10180.
85. Demontis, P., Jobic, H., Gonzalez, M.A., and Suffritti, G.B. (2009) Diffusion of water in zeolites NaX and NaY studied by quasi-elastic neutron scattering and computer simulation. *J. Phys. Chem. C*, **113**, 12373–12379.
86. Patra, M. and Karttunen, M. (2004) Systematic comparison of force fields for microscopic simulations of NaCl in aqueous solutions: diffusion, free energy of hydration, and structural properties. *J. Comput. Chem.*, **25**, 678–689.

87. Gren, W., Parker, S.C., Slater, B., and Lewis, D.W. (2010) Structure of zeolite A (LTA) surfaces and the zeolite A/water interface. *J. Phys. Chem. C*, **114**, 9739–9747.
88. Chipot, C. and Pohorille, A. (2007) *Free Energy Calculations*, Springer, New York.
89. Horn, H., Swope, W., and Pitera, J. (2005) Characterization of the TIP4P-Ew water model: vapor pressure and boiling point. *J. Chem. Phys.*, **123**, 194504.
90. Karamertzanis, P.G., Raiteri, P., and Galindo, A. (2010) The use of anisotropic potentials in modeling water and free energies of hydration. *J. Chem. Theory Comput.*, **6**, 1590–1607.
91. Joung, I.S. and Cheatham, T.E. (2008) Determination of alkali and halide monovalent ion parameters for use in explicitly solvated biomolecular simulations. *J. Phys. Chem. B*, **112**, 9020–9041.
92. Raiteri, P., Gale, J.D., Quigley, D., and Rodger, P.M. (2010) Derivation of an accurate force-field for simulating the growth of calcium carbonate from aqueous solution: a new model for the calcite-water interface. *J. Phys. Chem. C*, **114**, 5997–6010.
93. Bartels, C., Hirose, M., Rybar, S., and Franks, R. (eds) (2008) *Optimum RO System Design with High Area Spiral Wound Elements*. EDS/EuroMed Conference, 2008, Dead Sea, Jordan.
94. Amy, G. (2008) Fundamental understanding of organic matter fouling of membranes. *Desalination*, **231**, 44–51.
95. Speth, T., Gusses, A., and Summers, R. (2000) Evaluation of nanofiltration pretreatments for flux loss control. *Desalination*, **130**, 31–44.
96. Tran, T., Bolto, B., Gray, S., Hoang, M., and Ostarcevic, E. (2007) An autopsy study of a fouled reverse osmosis element used in a brackish water treatment plant. *Water Res.*, **41**, 3915–3923.
97. Khulbe, K.C. and Matsuura, T. (2000) Characterization of synthetic membranes by Raman spectroscopy, electron spin resonance, and atomic force microscopy: a review. *Polymer*, **41**, 1917–1935.
98. Riedl, K., Girard, B., and Lencki, R.W. (1998) Influence of membrane structure on fouling layer morphology during apple juice clarification. *J. Membr. Sci.*, **139**, 155–166.
99. Zhu, X. and Elimelech, M. (1997) Colloidal fouling of reverse osmosis membranes: measurements and fouling mechanisms. *Environ. Sci. Technol.*, **31**, 3654–3662.
100. Elimelech, M., Zhu, X., Childress, A.E., and Hong, S. (1997) Role of membrane surface morphology in colloidal fouling of cellulose acetate and composite aromatic polyamide reverse osmosis membranes. *J. Membr. Sci.*, **127**, 101–109.
101. Hirose, M., Ito, H., and Kamiyama, Y. (1996) Effect of skin layer surface structures on the flux behaviour of RO membranes. *J. Membr. Sci.*, **121**, 209–215.
102. Mondal, S. and Wickramasinghe, S.R. (2008) Produced water treatment by nanofiltration and reverse osmosis membranes. *J. Membr. Sci.*, **322**, 162–170.
103. Knoell, T., Safarik, J., Cormack, T., Riley, R., Lin, S.W., and Ridgway, H. (1999) Biofouling potentials of microporous polysulfone membranes containing a sulfonated polyetherethersulfone/polyethersulfone block co-polymer: correlation of membrane surface properties with bacterial attachment. *J. Membr. Sci.*, **157**, 117–138.
104. Kwak, S.-Y. and Ihm, D.W. (1999) Use of atomic force microscopy and solid-state NMR spectroscopy to characterize structure–property-performance correlation in high-flux reverse osmosis (RO) membranes. *J. Membr. Sci.*, **158**, 143–153.
105. Ahn, W.-Y., Kalinichev, A.G., and Clark, M.M. (2008) Effects of background cations on the fouling of polyethersulfone membranes by natural organic matter: experimental and molecular modeling study. *J. Membr. Sci.*, **309**, 128–140.
106. Sutzkover-Gutman, I., Hasson, D., and Semiat, R. (2010) Humic substances fouling in ultrafiltration processes. *Desalination*, **261**, 218–231.

107. Aquino, A.J.A., Tunega, D., Pašalić, H., Haberhauer, G., Gerzabek, M.H., and Lischka, H. (2008) The thermodynamic stability of hydrogen bonded and cation bridged complexes of humic acid models – a theoretical study. *Chem. Phys.*, **349**, 69–79.
108. Gray, S.R., Ritchie, C.B., Tran, T., and Bolto, B.A. (2007) Effect of NOM characteristics and membrane type on microfiltration performance. *Water Res.*, **41**, 3833–3841.
109. Swift, R.S. (1999) Macromolecular properties of soil humic substances: fact, fiction, and opinion. *Soil Sci.*, **164**, 790–802.
110. Hayes, M.H.B. and Clapp, C.E. (2001) Humic substances: considerations of compositions, aspects of structure, and environmental influences. *Soil Sci.*, **166**, 723–737.
111. Piccolo, A. (2002) The supramolecular structure of humic substances: a novel understanding of humus chemistry and implications in soil science. *Adv. Agron.*, **75**, 57–134.
112. Leenheer, J.A. and Croué, J.P. (2003) Characterizing aquatic dissolved organic matter. *Environ. Sci. Technol.*, **37**, 18A–26A.
113. Sutton, R. and Sposito, G. (2005) Molecular structure in soil humic substances: the new view. *Environ. Sci. Technol.*, **39**, 9009–9015.
114. Kalinichev, A.G. and Kirkpatrick, R.J. (2007) Molecular dynamics simulation of cationic complexation with natural organic matter. *Eur. J. Soil Sci.*, **58**, 909–917.
115. Simpson, A.J., Kingery, W.L., Hayes, M.H.B., Spraul, M., Humpfer, E., and Dvortsak, P. (2002) Molecular structure and associations of humic substances in the terrestrial environmental. *Naturwissenschaften*, **89**, 84–88.
116. Peña-Méndez, E.M., Gajdošová, D., Novotná, K., Prošek, P., and Havel, J. (2005) Mass spectroscopy of humic substances of different origin including those from Antarctica – a comparative study. *Talanta*, **67**, 880–890.
117. Steelink, C. (1985) *Implications of Elemental Characteristics of Humic Substances in Humic Substances in Soil, Sediment, and Water: Geochemistry, Isolation, and Characterization*, John Wiley & Sons, Inc., New York.
118. Davies, G., Fataftah, A., Cherkasskiy, A., Ghabbour, E.A., Radwan, A., Jansen, S.A., Kolla, S., Paciolla, M.D., Sein, L.T. Jr.,, Buermann, W., Balasubramanian, M., Budnick, J., and Xing, B. (1997) Tight metal binding by humic acids and its role in biomineralization. *J. Chem. Soc., Dalton Trans.*, 4047–4060.
119. Jansen, S.A., Malaty, M., Nwabara, S., Johnson, E., Ghabbour, E.A., Davies, G., and Varnum, J.M. (1996) Structural modeling in humic acids. *Mater. Sci. Eng. C*, **4**, 175–179.
120. Buffle, J. (1988) *Complexation Reactions in Aquatic Systems: An Analytical Approach*, Ellis Horwood Ltd, Chicester.
121. Tipping, E. (2002) *Cation Binding by Humic Substances*, Cambridge University Press, Cambridge.
122. Xu, X., Kalinichev, A.G., and Kirkpatrick, R.J. (2006) ^{133}Cs and ^{35}Cl NMR spectroscopy and molecular dynamics modeling of Cs$^+$ and Cl$^-$ complexation with natural organic matter. *Geochim. Cosmochim. Acta*, **70**, 4319–4331.
123. Kitson, D.H. and Hagler, A.T. (1988) Theoretical studies of the structure and molecular dynamics of a peptide crystal. *Biochemistry*, **27**, 5246–5257.
124. Berendsen, H.J.C., Postma, J.P.M., van Gunsteren, W.F., and Hermans, J. (1981) in *Intermolecular Forces* (ed. B. Pullman), Riedel, Dordrecht, p. 331.
125. Dang, L.X. and Smith, D.E. (1993) Molecular dynamics simulations of aqueous ionic clusters using polarizable water. *J. Chem. Phys.*, **99**, 6950–6956.
126. Dang, L.X. (1995) Mechanism and thermodynamics of ion selectivity in aqueous solutions of 18-crown-6 ether – a molecular dynamics study. *J. Am. Chem. Soc.*, **117**, 6954–6960.
127. Iskrenova-Tchoukova, E., Kalinichev, A.G., and Kirkpatrick, R.J. (2010) Metal cation complexation with natural organic matter in aqueous solutions: molecular dynamics simulations and

128. Case, D.A., Darden, T.A., Cheatham, T.E. III,, Simmerling, C.L., Wang, J., Duke, R.E., Luo, R., Merz, K.M., Pearlman, D.A., Crowley, M., Walker, R.C., Zhang, W., Wang, B., Hayik, S., Roitberg, A., Seabra, G., Wong, K.F., Paesani, F., Wu, X., Brozell, S., Tsui, V., Gohlke, H., Yang, L., Tan, C., Mongan, J., Hornak, V., Cui, G., Beroza, P., Mathews, D.H., Schafmeister, C., Ross, W.S., and Kollman, P.A. (2006) *AMBER9*, University of California, San Fransisco, CA.

127. potentials of mean force. *Langmuir*, **26**, 15909–15919.

129. Li, Q.L. and Elimelech, M. (2004) Organic fouling and chemical cleaning of nanofiltration membranes: measurements and mechanisms. *Environ. Sci. Technol.*, **38**, 4683–4693.

130. Guardado, I., Urrutia, O., and Garcia-Mina, J.M. (2008) Some structural and electronic features of the interaction of phosphate with metal-humic complexes. *J. Agric. Food Chem.*, **56**, 1035–1042.

131. Witwicki, M., Jezierska, J., and Ozarowski, A. (2009) Solvent effect on EPR, molecular and electronic properties of semiquinone radical derived from 3,4-dihydroxybenzoic acid as model for humic acid transient radicals: highfield EPR and DFT studies. *Chem. Phys. Lett.*, **473**, 160–166.

132. Zhang, L. and LeBoeuf, E.J. (2009) A molecular dynamics study of natural organic matter: 1. Lignin, kerogen and soot. *Org. Geochem.*, **40**, 1132–1142.

133. del Rio, J.C. and Hatcher, P.G. (1996) Structure characterization of humic substances using thermochemolysis with tetramethylammonium hydroxide, in *Humic and Fulvic Acids: Isolation, Structure, and Environmental Role* (eds J.S. Gaffney, N.A. Marley, and S.B. Clark), American Chemical Society, Washington, DC.

134. Nimz, H. (1974) Beech lignin – proposal of a constitutional scheme. *Angew. Chem.*, **86**, 336–344.

135. Behar, F. and Vandenbroucke, M. (1987) Chemical modeling of kerogens. *Org. Geochem.*, **11**, 15–24.

136. Ahkter, M.S., Chughtai, A.R., and Smith, D.M. (1985) The structure of hexane soot – I – spectroscopic studies. *Appl. Spectrosc.*, **39**, 143–153.

137. Tamai, Y., Tanaka, H., and Nakanishi, K. (1995) Molecular design of polymer membranes using molecular simulation technique. *Fluid Phase Equilib.*, **104**, 363–374.

138. Hofman, D., Fritz, L., Ulbrich, J., Schepers, C., and Bohning, M. (2000) Detailed atomistic molecular modelling of small molecule diffusion and solution processes in polymeric membrane materials. *Macromol. Theory Simul.*, **9**, 293–327.

139. Jia, W. and Murad, S. (2006) Molecular dynamics simulations of pervaporation in zeolite membranes. *Mol. Phys.*, **104**, 3033–3043.

140. Murad, S. and Nitsche, L.C. (2004) The effect of thickness, pore size and structure of a nanomembrane on the flux and selectivity in reverse osmosis separations: a molecular dynamics study. *Chem. Phys. Lett.*, **397**, 211–215.

141. Jucker, C. and Clark, M.M. (1994) Adsorption of aquatic humic substances on hydrophobic ultrafiltration membranes. *J. Membr. Sci.*, **97**, 37–52.

142. DeRamos, C.M., Irwin, A.E., Nauss, J.L., and Stout, B.E. (1997) ^{13}C NMR and molecular modeling studies of alginic acid binding with alkaline earth and lanthanide metal ions. *Inorg. Chim. Acta*, **256**, 69–75.

143. Reinhard, M., Ridgway, H., and Steinle-Darling, E. (2006) Molecular Basis of Reverse Osmosis Membrane Fouling: Analysis of the Membrane Adsorption Behaviour of Bacterial Alginate, A Model of Extracellular Substance. Final Report Grant # 04-Fc-81-1045A, Bureau of Reclamation, Denver, CO.

144. Gray, S.R., Dow, N., Orbell, J.D., Tran, T., and Bolto, B.A. (2011) The significance of interactions between organic compounds on low pressure membrane fouling. *Water Sci. Technol.*, **64**, 632–639.

11
Conclusions: Some Potential Future Nanotechnologies for Water Treatment

Mikel Duke

This book has provided an overview of the status of some of the more studied nanotechnologies from key researchers in the field. However, the purpose of this final chapter is to shed some light on several emerging nanotechnologies that are only just now being investigated in regard to water treatment. Although the practical aspects are not fully explored, some promising nanomaterials emerging for water treatment use are nanotubes, graphene, aquaporins, and metal–organic frameworks/polymer organic frameworks (MOFs/POFs). Despite their early stage in development, they still offer new thinking and functionalities that might be harnessed for real water treatment issues. Owing to the more fundamental nature of the research, these new materials have the potential to become disruptive technologies if an application is found. Therefore, they yield both new ways of thinking and a chance to make significant changes.

11.1
Nanotubes

Nanotubes, principally carbon nanotubes (CNTs), are a relatively recent material that is basically a tube made of graphite. Being a new material, researchers have been fascinated by its properties. A review was recently conducted on the developments in CNTs for water purification [1], and their key features that have motivated researchers are shown in Figure 11.1.

Possible applications of nanotubes in water treatment have harnessed the following properties:

- fast molecular flow inside tubes;
- high-strength nanometer dimension fibers;
- high aspect ratio; and
- electrical conductivity.

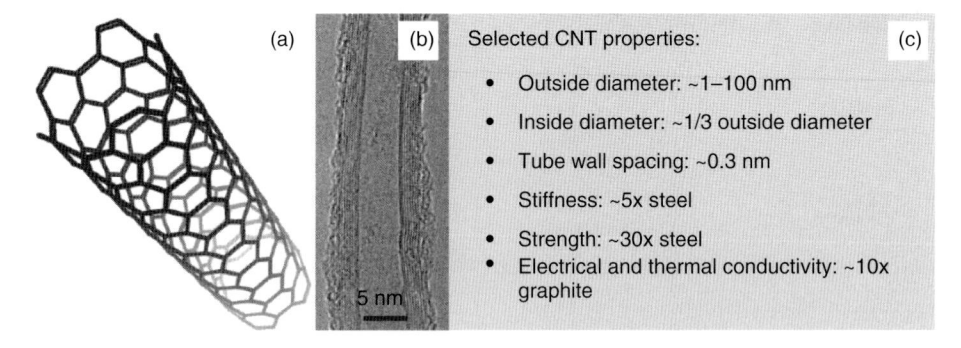

Figure 11.1 Schematic of a CNT (a), including transmission electron micrograph of a CNT showing (b) a number of concentric graphite walls and (c) the list of beneficial properties [1]. (Source: Reprinted with permission from the corresponding author.)

Selected CNT properties:
- Outside diameter: ~1–100 nm
- Inside diameter: ~1/3 outside diameter
- Tube wall spacing: ~0.3 nm
- Stiffness: ~5x steel
- Strength: ~30x steel
- Electrical and thermal conductivity: ~10x graphite

11.1.1
Fast Molecular Flow

CNTs possess an interesting feature in their ability to transport molecules inside the tubes orders of magnitude faster than other materials [2–7]. CNTs uniquely possess "slippery" or frictionless surfaces and could thus diffuse water with little resistance. This led to a drive by researchers to explore this in more detail, both experimentally and theoretically. To date, the progress has gone as far as modeling and laboratory testing, but on introduction of species such as salt and dissolved gases, the fast transport may be compromised. Also, nanotubes are not easily fabricated with small enough pore sizes to reject salt. So research is ongoing to steer the unique fast transport to practical benefits for water treatment.

11.1.2
CNTs as High Strength Fibers

Another useful feature of CNTs is their rigid fiber structure that is nanometers in diameter. This gives the fiber a high surface area to volume ratio and ability to produce fine fibrous materials that can act as electrodes or membrane barriers (Figure 11.2). Recently emerged research shows that CNTs can now be used to provide a nonwoven structure, which may be organized with very small pores around the nanotubes. Being hydrophobic in their native form, their water vapor diffusing properties have been tried in a laboratory-scale membrane distillation setup. The CNTs were also used as a convenient surface to deposit functional compounds including hydrophobic polytetrafluoroethylene (PTFE) and tested also for membrane distillation [8–10]. Despite the test using membrane distillation to validate performance, the virtues of the material as a nanoporous filter are yet unexplored.

The unique small-diameter fiber has also been utilized for capacitive deionization shown in Figure 11.3 [11], promising to offer high adsorptive capacity of

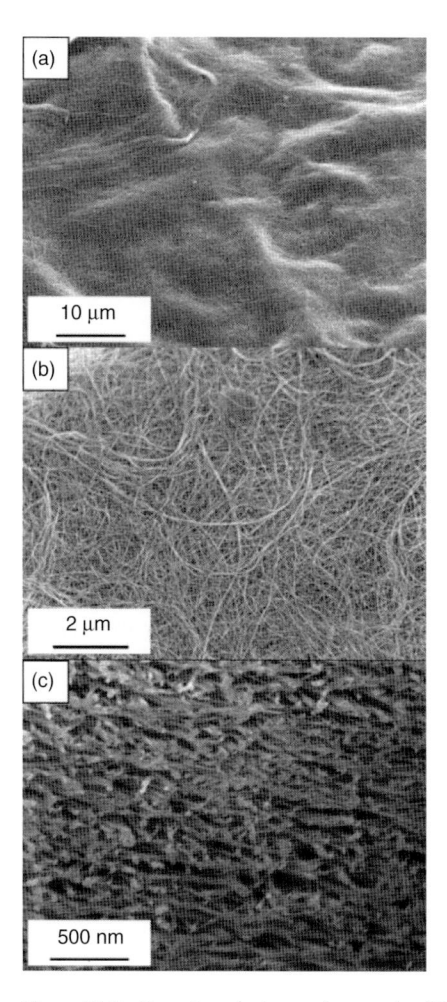

Figure 11.2 Scanning electron micrograph of CNT "buckypapers" showing surface (a) and (b) and cross-section (c). (Source: Reprinted with permission from [10].)

ions in tandem with excellent electrical conductivity [12, 13]. Despite potential advantages, the challenge to fabricate CNTs as electrodes is the subject of current research [14]. Further challenges include developing the technology into a cost-effective system with acceptable performance after multiple adsorption–desorption cycles.

11.1.3
High Aspect Ratio

One of the more practical applications of CNTs stems from their high aspect ratio feature. This has been harnessed for water treatment by incorporating the

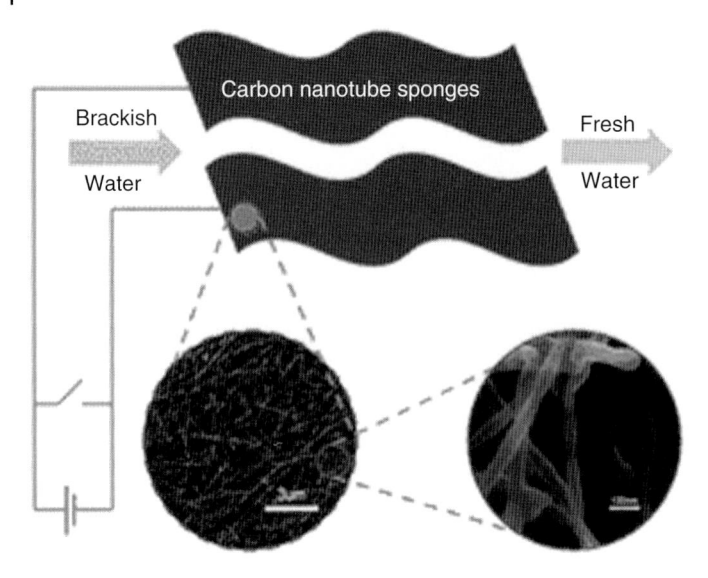

Figure 11.3 The representative use of CNTs in spongelike material to capture the superior surface area and electrical conductivity for capacitive deionization. (Source: Reproduced from [11] with permission from The Royal Society of Chemistry.)

nanotubes into membranes, promising to improve water transport and chemical resistance compared to conventional commercial materials. CNT composites for desalination were explored in Chapter 7, but the concept has been extended to filtration membranes, which is briefly mentioned here. For example, CNTs were added to polysulfone ultrafiltration membranes after first functionalizing the surface with 5-isocyanato-isophthaloyl chloride to compatibalize them with the host polymer [15]. In general, CNTs added up to 0.5 wt% lead to smoother surfaces compared to polysulfone without them. Pure water flux peaked at 0.19 wt% because of increased hydrophilicity of the membrane. Addition of CNTs reduced rejection of PEG-2000 because of increased pore size, whereas bovine serum albumin adsorption decreased, indicating reduced fouling potential. Similar increases in hydrophilicity and permeability have been observed for the same polymer in other work [16], but recent focus has been to add other novel functions including antibacterial nanosilver coatings on top of the CNT composite polysulfone membrane [17].

11.1.4
Electrical Conductivity

High electrical conductivity of CNTs has been harnessed for increasing catalytic activity of TiO_2 [18]. With a small amount of CNT added to TiO_2, greater photocatalytic activity was found but too much lead to blocking of the UV light.

11.2
Graphene

Graphene has been identified as an emerging material for materials research in 2007 [19]. The unique material possesses a chemical feature similar to CNTs but can be considered as a planar version of its tubular counterpart. Being only one atom thick, it can be easily mass produced in graphene oxide (GO) form. Besides its nanodimension pores, it also possesses great mechanical strength, flexibility, electrical conductivity, and high specific surface area. Therefore, GO has compelled researchers to explore its applications in water treatment as films, electrodes, adsorption, and catalysis. GO can be produced safely using the "Hummer's method," which involves the modification of natural graphite flakes with potassium permanganate and sodium nitrate in concentrated sulfuric acid [20].

11.2.1
Graphene Barrier Material

GO's unique barrier property has been achieved by making GO laminates formed by layered and interlocked micrometer-sized GO crystallites. Using the Hummer's method to produce GO suspension in water, GO laminates were made by spray or spin coating on to Cu foil (Figure 11.4) [21]. A 1 cm hole was etched into the foil

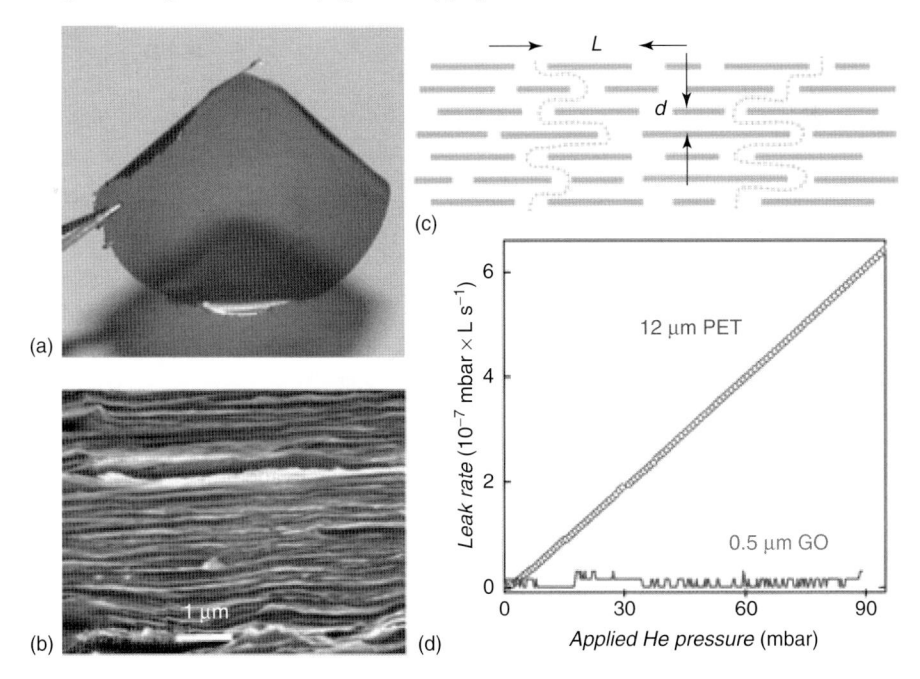

Figure 11.4 The GO membrane film with the thickness of ∼1 μm (a), electron micrograph of section (b), molecular permeation pathway (c), and He permeance measurement compared against polyethylene terephthalate (PET) (d). (Source: From [21] Reprinted with permission from AAAS.)

to leave freestanding GO films varying in thickness from 0.1 to 10 µm consisting of GO crystals of several micrometers (dimension "L" in Figure 11.4c), with space of around 1 nm between layers (dimension "d" in Figure 11.4c). These films were able to withstand differential pressure of 10 kPa. Permeation testing of these films was found to diffuse water rapidly but exhibited no measurable permeation of ethanol, hexane, acetone, decane, and propanol after several days. GO appeared to provide this unique permeation characteristic by forming a water monolayer within the material. Clearly, an interesting next step would be to observe what happens when salt is included to observe if a desalination effect is possible.

11.2.2
Desalination and Heavy Metal Adsorption

GO also has advantages similar to CNTs in that it has a more accessible porous structure, giving it the ability to adsorb more compounds, for example, ions. One use of this feature is in capacitive deionization [22]. In this application, graphene was prepared as graphene nanoflakes (GNFs) using a modified Hummer's method. GNF electrodes were prepared by pressing a mixture of GNFs, activated carbon, and PTFE as a binder, which led to increased deionization performance compared to activated carbon. GO materials have also been explored to remove heavy metals for remediation in water and waste water [23].

11.2.3
Catalytic Assistance

Other features of graphene include its electrical conductivity, which, as in CNTs, has been utilized in water treatment by mixing with photocatalytic TiO_2 to improve reaction efficiency for removing organic contaminants from waste water [24]. It has been found that TiO_2–GO composite materials yield enhanced removal of water pollutants, alkylphenol, compared to commercial TiO_2, explained as being possibly due to (i) coadsorption of alkylphenol molecules aiding contact with photocatalysis or (ii) GO acted to inhibit the recombination of electron and hole pairs of TiO_2, thus assisting the oxidation of the organic molecule [25].

11.3
Aquaporins

Aquaporins are a unique biological structure that potentially brings biomaterials to water treatment. They comprise a biological lipid membrane, spanning proteins common in living cells. These proteins utilize a distinct fast water transport feature, transporting water at rates far greater than synthetic membranes. Recently, these proteins were embedded into a synthetic film, leading to a functional biomimetic membrane as shown in Figure 11.5 [26]. The research of this membrane was directed for forward osmosis.

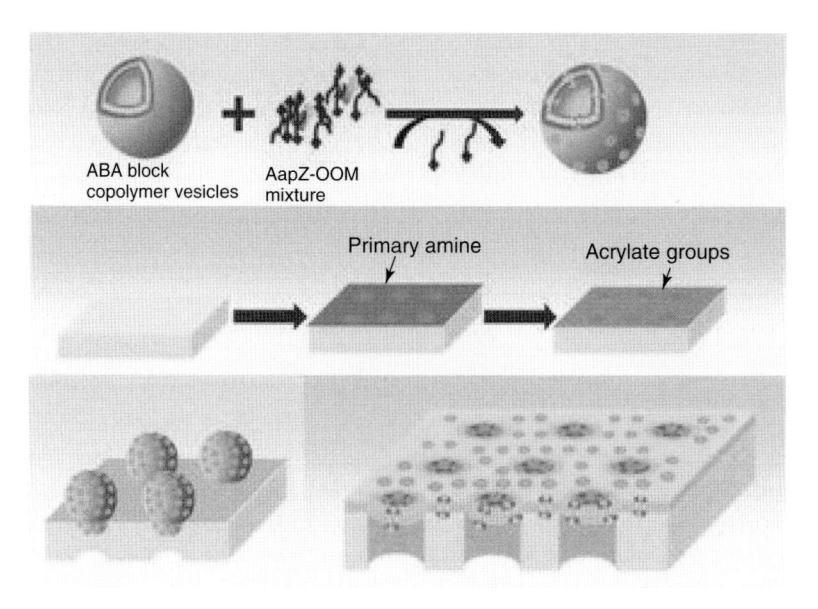

Figure 11.5 Synthesis of a biomimetic membrane based on aquaporins [26]. (Source: Reproduced with permission.)

Desalination was also explored with biomimetic membranes with aquaporins, tested in a membrane setup using a transmembrane pressure of 5 bar [27]. Over a membrane area of 0.2 cm^2 fed with 200 mg l^{-1} of NaCl solution, flux was measured at 16 l m^{-2} h^{-1} bar. This value is comparable with commercial polymer reverse osmosis membranes operating at low pressure but is hard to compare exactly as high salt concentration (and thus high pressure) conditions would need to be tested to fully justify any benefits. Further, salt rejection reached 45%, being over half of the commercial types (typically >99% rejection), leading to the conclusion that this initial work demonstrated a nanofiltration effect. Further work is clearly needed to prove the selective water permeating effect of aquaporin-based membranes.

11.4
Metal–Organic, Zeolitic Imidazolate, and Polymer Organic Frameworks

A new class of porous crystalline materials known as *MOFs* and *POFs* have made great advances to the membrane science field [28, 29]. These materials possess a wide selection of nanometer-sized pores, high surface areas, chemically tailorable cavities, and flexible structure. Such properties have great potential for water treatment, including size exclusion of ions (desalination) and tailored chemical interactions. Considerable attention has been focused on MOF materials, especially membranes made from these materials. Zeolitic imidazolate frameworks (ZIFs), a subfamily of MOFs based on transition metals and imidazolates as linkers, with exceptional thermal and chemical stability, are very promising for the fabrication of MOF membranes [30, 31]. ZIF materials are considered nontoxic, require little energy to produce, and are chemically and structurally configurable to achieve desired functionality. While most MOFs are found to be unstable in water, certain ZIFs, such as ZIF-8 and ZIF-11 (shown in Figure 11.6), are however known to be stable under aqueous and even acidic conditions [32] because of their tetrahedral structures similar to that of aluminosilicate zeolitic networks.

The ability to add functional groups has some practical use for water treatment, for example, targeted pollutant removal or chemical sensing in water treatment. One of the most robust techniques for adding desired functionalities to MOF is by post synthesis modification (PSM) of the organic linking group [33]. Specifically for ZIF, Yaghi and coworkers [34] modified a ZIF-90 structure by covalent modification of the free aldehyde functionality using ethanalamine via imine condensation. A review by Wang and Cohen on PSM of MOFs indicated that numerous avenues are available to add desired functionalities to these unique structures unlike other materials (i.e., pure zeolites) owing to their modifiable organic linking groups [33]. More recently, in 2010, Yaghi and coworkers further extended the field of tailored modifications for multivariate MOF functionalities utilizing the organic linker's orientation, number, relative position, and ratio along the backbone. There are many options to vary the functionalities of MOFs (and likewise ZIFs) to achieve the desired selective performance [35]. The same work showed improved selective uptakes of H_2 and CO_2 for gas application, so this work extending to water

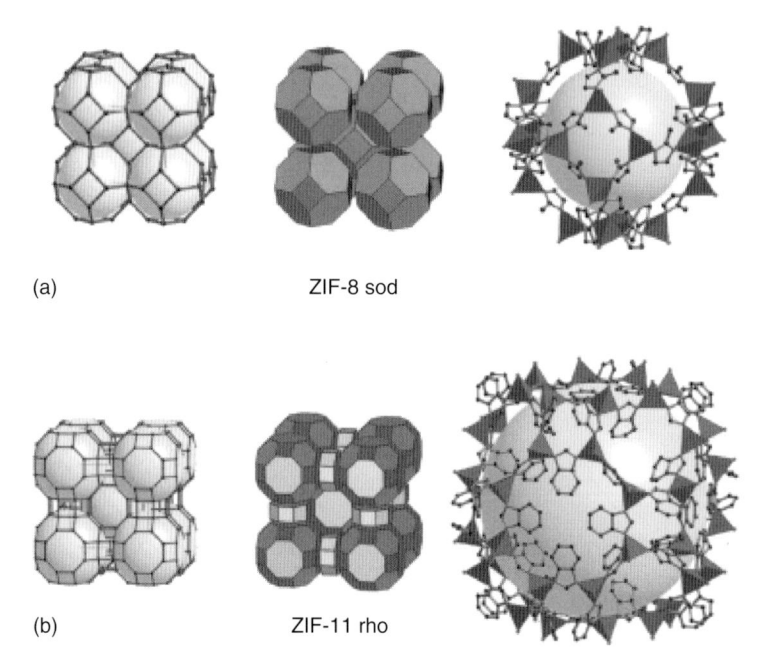

(a)   ZIF-8 sod

(b)   ZIF-11 rho

Figure 11.6 Single-crystal X-ray structures of ZIF-8 (a) and ZIF-11 (b), shown as stick diagram (left) and tiling (center). Expanded view of largest cage (right). (Source: Reproduced with permission. Copyright (2006) National Academy of Sciences, U.S.A.)

treatment is scientifically novel and a key element in discovering new avenues for functionalized ZIF materials.

More recently, POFs have been developed as the metal-free equivalent of MOFs and offer further advantages of being lightweight and purely organic [28]. The highly cross-linked nature of these materials allows them to possess higher thermal stabilities compared to simple organic polymers. Similar to MOFs, POFs also allow for facile functionalization by incorporating various functional groups into the microporous material, allowing for tailored interaction with guest species. Simple one-pot synthesis methods can produce POF materials with surface areas exceeding 1500 $m^2 g^{-1}$.

In terms of applications in water treatment, high surface area MOF, ZIF, or POF materials can be used as specialist sorbents or purely for their nanoporous cage structures. Already, ZIF materials have been proposed as sensors of chemical vapors and gases [36]. As mentioned earlier, facile PSM of the organic linking groups has already been explored to achieve desired functionalities but has not been explored for water treatment applications including targeted adsorption, reaction, or chemical sensing. A work recently published on ZIF application in water treatment involved a molecular dynamics simulation, which demonstrated that ZIF-8 possesses the essential structure for water diffusion and ion rejection and could thus be applied for desalination [37]. Therefore, despite the small amount of documented work specifically in water treatment, these novel materials offer great potential for researchers.

11.5
Conclusions

Nanotechnologies for water treatment are an exciting field for researchers and industry, with new and exciting properties still being found. Nanotubes, graphene, aquaporins, and MOFs/POFs have been studied in some detail for water treatment, offering unique features that have never been seen before on any other material. Although most of the work remains under academic investigation, with time the new phenomena will find virtue in treating water, but not necessarily in the ways they are being currently applied. Therefore, there is great scope for further creativity to not only discover new nanomaterials and their properties but also to find ways in which the new properties can be practically harnessed to make a positive impact on solving our water issues.

References

1. Sears, K., Dumée, L., Schütz, J., She, M., Huynh, C., Hawkins, S., Duke, M., and Gray, S. (2010) Recent developments in carbon nanotube membranes for water purification and Gas separation. *Materials*, **3** (1), 127–149.

2. Chen, H. and Sholl, D.S. (2006) Predictions of selectivity and flux for CH4/H2 separations using single walled carbon

nanotubes as membranes. *J. Membr. Sci.*, **269** (1–2), 152–160.

3. Corry, B. (2008) Designing carbon nanotube membranes for efficient water desalination. *J. Phys. Chem. B*, **112**, 1427–1434.

4. Hummer, G., Rasalah, J.C., and Noworyta, J.P. (2001) Water conduction through the hydrophobic channel of a carbon nanotube. *Nature*, **414**, 188–190.

5. Skoulidas, A., Ackerman, D.M., Johnson, K., and Sholl, D.S. (2002) Rapid transport of gases in carbon nanotubes. *Phys. Rev. Lett.*, **89**, 185901:1–185901:4.

6. Striolo, A. (2006) The mechanism of water diffusion in narrow carbon nanotubes. *Nano Lett.*, **6**, 633–639.

7. Waghe, A., Rasaiah, J.C., and Hummer, G. (2002) Filling and emptying kinetics of carbon nanotubes in water. *J. Chem. Phys.*, **117**, 10789–10795.

8. Dumée, L., Campbell, J.L., Sears, K., Schütz, J., Finn, N., Duke, M., and Gray, S. (2011) The impact of hydrophobic coating on the performance of carbon nanotube bucky-paper membranes in membrane distillation. *Desalination*, **283** (0), 64–67.

9. Dumée, L., Germain, V., Sears, K., Schütz, J., Finn, N., Duke, M., Cerneaux, S., Cornu, D., and Gray, S. (2011) Enhanced durability and hydrophobicity of carbon nanotube bucky paper membranes in membrane distillation. *J. Membr. Sci.*, **376** (1–2), 241–246.

10. Dumée, L.F., Sears, K., Schütz, J., Finn, N., Huynh, C., Hawkins, S., Duke, M., and Gray, S. (2010) Characterisation and evaluation of carbon nanotube bucky-paper membranes for direct contact membrane distillation. *J. Membr. Sci.*, **351** (1–2), 36–43.

11. Wang, L., Wang, M., Huang, Z.-H., Cui, T., Gui, X., Kang, F., Wang, K., and Wu, D. (2011) Capacitive deionization of NaCl solutions using carbon nanotube sponge electrodes. *J. Mater. Chem.*, **21** (45), 18295–18299.

12. Dai, K., Shi, L., Fang, J., Zhang, D., and Yu, B. (2005) NaCl adsorption in multi-walled carbon nanotubes. *Mater. Lett.*, **59** (16), 1989–1992.

13. Dai, K., Shi, L., Zhang, D., and Fang, J. (2006) NaCl adsorption in multi-walled carbon nanotube/active carbon combination electrode. *Chem. Eng. Sci.*, **61** (2), 428–433.

14. Nie, C., Pan, L., Li, H., Chen, T., Lu, T., and Sun, Z. (2012) Electrophoretic deposition of carbon nanotubes film electrodes for capacitive deionization. *J. Electroanal. Chem.*, **666** (0), 85–88.

15. Qiu, S., Wu, L., Pan, X., Zhang, L., Chen, H., and Gao, C. (2009) Preparation and properties of functionalized carbon nanotube/PSF blend ultrafiltration membranes. *J. Membr. Sci.*, **342** (1–2), 165–172.

16. Wu, H., Tang, B., and Wu, P. (2010) Novel ultrafiltration membranes prepared from a multi-walled carbon nanotubes/polymer composite. *J. Membr. Sci.*, **362** (1–2), 374–383.

17. Kim, E.-S., Hwang, G., Gamal El-Din, M., and Liu, Y. (2012) Development of nanosilver and multi-walled carbon nanotubes thin-film nanocomposite membrane for enhanced water treatment. *J. Membr. Sci.*, **394–395**, 37–48.

18. Yu, Y., Yu, J.C., Yu, J.-G., Kwok, Y.-C., Che, Y.-K., Zhao, J.-C., Ding, L., Ge, W.-K., and Wong, P.-K. (2005) Enhancement of photocatalytic activity of mesoporous TiO_2 by using carbon nanotubes. *Appl. Catal., A*, **289** (2), 186–196.

19. Geim, A.K. and Novoselov, K.S. (2007) The rise of graphene. *Nat. Mater.*, **6** (3), 183–191.

20. Hummers, W.S. and Offeman, R.E. (1958) Preparation of graphitic oxide. *J. Am. Chem. Soc.*, **80** (6), 1339–1339.

21. Nair, R.R., Wu, H.A., Jayaram, P.N., Grigorieva, I.V., and Geim, A.K. (2012) Unimpeded permeation of water through helium-leak-tight graphene-based membranes. *Science*, **335** (6067), 442–444.

22. Li, H., Zou, L., Pan, L., and Sun, Z. (2010) Novel graphene-like electrodes for capacitive deionization. *Environ. Sci. Technol.*, **44** (22), 8692–8697.

23. Luo, L.L., Gu, X.X., Wu, J., Zhong, S.X., and Chen, J.R. (2012) Advances in graphene for adsorption of heavy

23. metals in wastewater. *Adv. Mater. Res.*, **550–553**, 2121–2124.
24. Aglietto, I. (2010) Advanced Photocatalytic Oxidation with Graphene for Wastewater Treatment. ENT Magazine (Mar.–Apr. 33–36).
25. Basheer, C. (2012) Application of titanium dioxide-graphene composite material for photocatalytic degradation of alkylphenols. *J. Chem.*, **2013**, 1–10 (Article ID 456586).
26. Wang, H., Chung, T.-S., Tong, Y.W., Jeyaseelan, K., Armugam, A., Chen, Z., Hong, M., and Meier, W. (2012) Highly permeable and selective pore-spanning biomimetic membrane embedded with aquaporin Z. *Small*, **8** (8), 1185–1190.
27. Duong, P.H.H., Chung, T.-S., Jeyaseelan, K., Armugam, A., Chen, Z., Yang, J., and Hong, M. (2012) Planar biomimetic aquaporin-incorporated triblock copolymer membranes on porous alumina supports for nanofiltration. *J. Membr. Sci.*, **409–410** (0), 34–43.
28. Pandey, P., Katsoulidis, A.P., Eryazici, I., Wu, Y., Kanatzidis, M.G., and Nguyen, S.T. (2010) Imine-linked microporous polymer organic frameworks. *Chem. Mater.*, **22** (17), 4974–4979.
29. Zacher, D., Shekhah, O., Woll, C., and Fischer, R.A. (2009) Thin films of metal-organic frameworks. *Chem. Soc. Rev.*, **38**, 1418–1429.
30. Huang, A.S., Bux, H., Steinbach, F., and Caro, J. (2010) Molecular-sieve membrane with hydrogen permselectivity: ZIF-22 in LTA topology prepared with 3-aminopropyltriethoxysilane as covalent linker. *Angew. Chem. Int. Ed.*, **49**, 4958–4961.
31. Li, Y.-S., Liang, F.-Y., Bux, H., Feldhoff, A., Yang, W.-S., and Caro, J. (2009) Molecular sieve membrane: supported metal–organic framework with high hydrogen selectivity. *Angew. Chem. Int. Ed.*, **49** (3), 548–551.
32. Park, K.S., Ni, Z., Cote, A.P., Choi, J.Y., Huang, R., Uribe-Romo, F.J., Chae, H.K., O'Keeffe, M., and Yaghi, O.M. (2006) Exceptional chemical and thermal stability of zeolitic imidazolate frameworks. *Proc. Natl. Acad. Sci. U.S.A.*, **103** (27), 10186–10191.
33. Wang, Z. and Cohen, S.M. (2008) Post-synthetic modification of metal–organic frameworks. *Chem. Soc. Rev.*, **38**, 1315–1329.
34. Morris, W., Doonan, C.J., Furukawa, H., Banerjee, R., and Yaghi, O.M. (2008) Crystals as molecules: postsynthesis covalent functionalization of zeolitic imidazolate frameworks. *J. Am. Chem. Soc.*, **130**, 12626–12627.
35. Deng, H., Doonan, C.J., Furukawa, H., Ferreira, R.B., Towne, J., Knobler, C.B., Wang, B., and Yaghi, O.M. (2010) Multiple functional groups of varying ratios in metal-organic frameworks. *Science*, **327** (5967), 846–850.
36. Lu, G. and Hupp, J.T. (2010) Metal-organic frameworks as sensors: a ZIF-8 based fabry-Pérot device as a selective sensor for chemical vapors and gases. *J. Am. Chem. Soc.*, **132**, 7832–7833.
37. Hu, Z., Chen, Y., and Jiang, J. (2011) Zeolitic imidazolate framework-8 as a reverse osmosis membrane for water desalination: insight from molecular simulation. *J. Chem. Phys.*, **134**, 134705.

Index

a

adsorption 10
advanced oxidation processes (AOPs) 7, 10–11
– Fenton-like reactions 12–16
– nanocatalytic wet oxidation 17–18
– photo-Fenton reactions 16–17
alginates 289
Analyticon Instruments Corporation 43
anatase 39
anion adsorption 73–74
anion exchange membranes (AEMs) 126, 128, 137, 142, 148
antifouling property, improving 149–150
aquaporins 306–307
aromatic polyamides. See polyamides (PAs)
atomic force microscopy (AFM) 178
attenuated total reflectance (ATR) technique 179
– Fourier transform infrared (ATR-FTIR) spectrometry 260, 261
2,2′-azobisisobutyramidine dihydrochloride (AIBA) 96

b

benzophenone (BPO) 100
bimetallic nanoparticles (BNPs) 7, 27–29
biofouling. See membrane fouling
biopolymer 282
boron-doped diamond (BDD) 38
bovine serum albumin (BSA) 99, 100, 103
brookite 39

c

calcination process 39
carbon nanotubes (CNTs) 40–41, 175–176, 182, 183, 185, 273, 275, 301
– properties
– – electrical conductivity 304
– – fast molecular flow 302
– – high aspect ratio 303–304
– – as high strength fibers 302–303
catalytic wet air oxidation (CWAO), nanocatalyst-based 17
catalytic wet oxidation (CWO) 11, 17–18
cation exchange membranes (CEMs) 126, 128, 139, 142, 148
cellulose acetate 163
cellulose diacetate (CDA) 163
ceramic membranes 195
– application in microfiltration and ultrafiltration of wastewater 202–204
– – cost 212–213
– – membrane microstructure 204–205
– – surface charge properties 206–208
– – technical process 208–212
– – wettability 205–206
– membrane preparation 196
– – extrusion 196
– – sol–gel process 196–198
– surface water and seawater clarification 198
– – membrane microfiltration, of surface water 199–201
– – seawater RO pretreatment 201–202
cetyltrimethylammonium bromide (CTAB) 72, 228
charge induced on film membranes 142–143
chemical initiators 95–97
chemical liquid deposition (CLD) 228

chemical oxygen demand (COD)   37, 41, 43
– exhaustive degradation mode   50–53
– partial oxidation mode   53–54
– photocatalytic determination   43–46
– probe-type TiO_2 photoanode   46–50
chemical surface modifications   90
chemical vapor deposition (CVD)   228
chronopotentiogram   131, 133
chronopotentiometry   131
colloidal fouling   279
colloidal nanoparticles   9
composite ion exchange membranes   143–147
composite seeding process   227
concentration polarization   86
– and limiting current density   128
– – electroconvection   130
– – gravitational convection   130
– – overlimiting current density   128–129
– – water dissociation   129–130
conducting polymers, structures of   147–148
conjugate-gradient method   256
consistent valence force field (CVFF)   283, 287
contact angle measurement   178
convection   127
cross-flow velocity (CFV)   86, 199, 201–202
cross-linking   68, 98, 101, 139, 141, 147, 173, 250, 251, 255, 256, 260–262, 266, 309

d

dealuminated zeolite Y (DAY)   268, 272–273
dense nonaqueous-phase liquid (DNAPL)   21, 22
density functional theory (DFT)   268
deposition-precipitation method   10
desalination   2, 4
differential scanning calorimetry   254
differential staining techniques   251
diffusion   127, 132, 263
3,4-dihydroxybenzoic acid   286
direct *in situ* crystallization, for zeolite membrane preparation   220–221
direct polymerization from monomer units   139, 141
direct potable reuse   2
dodecyltrimethoxysilane (DMS)   228
Donnan exclusion   126, 131
Dow-FilmTec Corporation   256
dynamic light scattering (DLS)   171

e

effluent organic matter (EfOM)   280
electroconvection   130
electrodialysis (ED)   125, 126, 128, 149, 150, 151, 152
electrodiffusion   132
electron–hole recombination reaction   35
electron microscopy   177
electroosmosis   130
emulsified zero-valent iron (EZVI)   21–22
emulsions   202–203
energy dispersive X-ray (EDX) spectroscopy   177–178
extrusion process   196

f

Fenton-like reactions   12
– iron oxide as heterogeneous nanocatalyst   12–16
Fenton process   11
fillers, in membrane materials   165
flame ionization detector (FID)   55
flocculation/coagulation, pretreatment with   199
flow injection analysis (FIA)   43, 54, 55, 56
fouling resistance and chlorine stability   184–185
Fourier transform infrared spectroscopy (FTIR) spectrometry   179
fractional free volume (FFV)   264–266
free energy perturbation (FEP) method   276
free radical graft polymerization (FRGP)   91, 92
fuel cells   126

g

gamma-induced graft polymerization   97–99
gas separation (GS)   195, 196, 197
– inorganic-membrane-based   198
glancing angle deposition   58
graft polymerization   90, 91–92
– irradiation-induced   97–101
– plasma-initiated   101–104
– reaction schemes   92–94
– surface activation with chemical initiators   95–97
– surface activation with vinyl monomers   94–95
graphene   305
– aquaporins   306–307
– barrier material   305–306
– catalytic assistance   306
– desalination and heavy metal adsorption   306
gravitational convection   130

h

heavy metal ions adsorption   68–72
hemolytic fission   97
heterogeneous membranes   133–134
high-performance liquid chromatography (HPLC)   41, 55, 56
homogeneous membranes   133
humic substances   280
Hummer's method   305
hybrid ozonation and ceramic ultrafiltration   201
hybrid system, for desalination cost reduction   150–152
hydrogen peroxide   13, 15
hydrothermal synthesis   168, 170
hyperchem modeling   257

i

impregnated catalysts   9–10
infrared (IR) spectroscopy   179
Inopor Corporation, Germany   197
inorganic desalination molecular simulation   267–268
– water behavior within zeolites   270–275
– zeolites and salt ions   276–278
– zeolites modeling   268–270
inorganic fillers fabrication and characterization   168–172
inorganic fouling   279
inorganic membranes   164
inorganic–organic composite materials   143
interfacial polymerization   172–175
– with inorganic fillers   175–177
internal reflection spectroscopy (IRS)   179
intrachain cross-linking   256
ion exchange capacity (IEC)   127, 137, 138, 139, 147
ion exchange membranes (IEMs)   125
– commercial, properties   136–137
– fundamentals, and transport phenomena   125–127
– – concentration polarization and limiting current density   128–130
– – ion transport   127
– – structure and surface heterogeneity   130–134
– future perspectives   150
– – hybrid system   150–152
– – small-scale seawater desalination   152
– material development and characterization tools   135
– – composite ion exchange membranes   143–147
– – membranes with specific properties   147–150
– – polymer-based IEMs development   135–143
ion transport   127
iron oxide, as heterogeneous nanocatalyst   12–16
irradiation-induced graft polymerization   97
– gamma-induced   97–99
– UV-induced   99–101
i–v characteristic   132

j

jump diffusion   263

k

keto-molinate isomer   24
Kotelyanskii model   256

l

limiting current density. See under concentration polarization
long-range electrostatics   254

m

maghemite nanocrystals   78
mass spectroscopy (MS) detectors   55
membrane fouling   86–89, 199
– molecular simulation   279–280
– – modeling   286–291
– – potential organic foulants molecular modeling   280–286
– strategies for mitigation   89–91
membrane surface nanostructuring   85–86
– graft polymerization   91–92
– – irradiation-induced   97–101
– – plasma-initiated   101–104
– – reaction schemes   92–94
– – surface activation with chemical initiators   95–97
– – surface activation with vinyl monomers   94–95
– membrane fouling   86–89
– – strategies for mitigation   89–91
3-mercaptopropyltrimethoxysilane (MPTMS)   68
mesoporous materials   67
– anion adsorption   73–74
– heavy metal ions adsorption   68–72
– organic pollutant
– – adsorption   74–77
– – photocatalytic degradation   79–82

mesoporous materials (*contd.*)
- sorbent multifunctional modification 77–79
- titania 80–82
mesoporous silica–carbon (MPSC) 76
metallic nanoparticles 7
metal–organic framework (MOFs) 301, 307–308–309
meta-phenylenediamine (MPD) 251, 255, 256, 257, 259–260
methylene blue (MB) 74, 75, 81
microbial fouling 279
microcystins 77–78
microfiltration 85, 249, 2479
- applications for industrial-scale waterworks 201
- and ceramic membrane application in ultrafiltration of wastewater 202–204
- – cost 212–213
- – membrane microstructure 204–205
- – surface charge properties 206–208
- – technical process 208–212
- – wettability 205–206
- hybrid ozonation and ceramic ultrafiltration 201
- pretreatment with flocculation/coagulation 199
- transmembrane pressure (TMP) effect and cross-flow velocity (CFV) 199–200
- ultrasound cleaning 200
microheterogeneous model 131, 134
micrometer-scale particles 23
microwave synthesis, for zeolite membrane preparation 223–227
migration 127
mixed matrix membranes (MMMs) 165
molecular dynamics. See molecular scale modeling
molecular scale modeling 249
- inorganic desalination molecular simulation 267–268
- – water behavior within zeolites 270–275
- – zeolites and salt ions 276–278
- – zeolites modeling 268–270
- membrane fouling molecular simulation 279–280
- – membrane fouling modeling 286–291
- – potential organic foulants molecular modeling 280–286
- polymeric membrane materials molecular simulations 249–250, 266–67
- – modeling strategies 255–262
- – RO membranes 250–254
- – water and solute transport behavior simulation 262–266
Monte Carlo (MC) technique 269
multiwalled carbon nanotubes (MWCNTs) 175–176, 182, 183, 185
mutual irradiation 98

n

N_2 adsorption 234
NANOCAT® 12
nanocatalysts 8. See also individual entries
- colloidal nanoparticles 9
- supported nanoparticles 9–10
nanofiltration 3, 85, 195, 197, 213, 249, 262, 266
nanoiron particles, organic contaminant transformation by 23
nanoscale zero-valent iron (nZVI) 7, 18–20
- bimetallic particles 27–29
- degradation mechanism 22–25
- field application 25–26
- synthesizing methods 20–22
nanotubes. See carbon nanotubes
narrow linear range 44
natural organic matter (NOM) 199, 201, 229, 280, 283–284, 286–288, 291–292
Nernst–Planck flux equation 127
neutral red (NR) 75
nitrosodimethylamine (NDMA) 263
noble metals 27, 81
nondispersive infrared (NDIR) detector 42
nonhumic substances 280
nuclear magnetic resonance (NMR) 68
nucleation 170

o

Ohmic region 133
oily wastewater 202
ordered mesoporous carbons (OMCs) 72, 75, 78
organic fouling 279
organic pollutant
- adsorption 74–77
- photocatalytic degradation 79–82
organics destruction in water, via iron nanoparticles 7
- advanced oxidation processes (AOPs) 10–12
- – Fenton-like reactions 12–16
- – nanocatalytic wet oxidation 17–18
- – photo-Fenton reactions 16–17
- nanocatalysts 8

– – colloidal nanoparticles 9
– – supported nanoparticles 9–10
– nano zero-valent iron (nZVI) 18–20
– – bimetallic particles 27–29
– – degradation mechanism 22–25
   field application 25–26
– – synthesizing methods 20–22
overlimiting current density 128–129
oxidation–reduction potential (ORP) 19
ozone

p

periodic mesoporous organosilicas (PMOs) 72
permselectivity 147
peroxidation 98
persulfate initiators 96
pervaporization (PV) 195, 196, 197
pH 29
photocatalysis 11
photoelectrocatalysis, at TiO_2 nanomaterials 36–37
– analytical signal generation 57–58
– sensing application 46–56
photoelectrochemical detector (PECD) 53, 56
photoelectrochemical probe for rapid determination of COD (PeCOD) 47, 55
photoelectrochemical universal detector, for organic compounds 55
photo-Fenton reactions 16–17
photohole 33, 34
photo-induced polymerization. See UV-induced polymerization
photometric method 46
physical surface treatment 89–90
piezoelectric gas sensors 58
plasma-initiated graft polymerization 101–104
plasma treatment, of surface 90
poly(acrylic acid) (PAA) 284–285
poly(aryl ether ketone) 137–139
poly (ether ether ketone). See poly(aryl ether ketone)
polyamides (PAs) 96, 163, 173, 180, 184–186, 252, 253
polyaniline (PANI) 148
polyarylene polymers 135
polyether sulfone (PES) 85, 86, 95, 96–97, 99–100, 103, 135–136, 139, 141, 287
– polymers, sulfonation reaction 138

polymer backbone direct modification 135–139
polymer-based IEMs development 135
– charge induced on film membranes 142–143
– direct polymerization from monomer units 139, 141
– polymer backbone direct modification 135–139
polymer grafting 90
polymeric membrane materials molecular simulations 249–250, 266–67
– modeling strategies 255–262
– RO membranes 250–254
– water and solute transport behavior simulation 262–266
polymer organic frameworks (POFs) 307, 309
polyphenylene oxides (PPO) 139
polypyrrole 148
polysulfonate (PS) 164, 165, 173
polytetrafluoroethylene (PTFE) 302
polyuronide compounds 288
polyvinylpyrrolidone (PVP) 287
positron annihilation lifetime spectroscopy (PALS) 234, 254
post synthesis modification (PSM) 308
postsynthetic treatment, for zeolite membrane preparation 228–229
preirradiation 98

q

QuantumFlux (Qfx) seawater RO element 185
quantum mechanical (QM) calculations 250, 268, 269
quasielectric neutron scattering (QENS) 254

r

radiation-induced grafting 142
random walk model 271
rapid thermal processing (RTP) 222
redox initiators 96
reflective index detectors (RIDs) 55
relative standard deviation (RSD) 44
reverse osmosis (RO) 3, 85, 163, 168, 187, 229, 279. See also seawater RO (SWRO) pretreatment
Ridgway algorithm 259
rubbing and leveling method 223
Rutherford backscattering spectrometry (RBS) 179, 251, 253
rutile 39

s

salt ions 276–278
seawater RO (SWRO) pretreatment
– ceramic membrane application for industrial-scale plant 202
– operational parameters effect 201–202
seeded secondary growth, for zeolite membrane preparation 221–223
self-cleaning, TiO_2 58
SGE International Pty. Ltd 42
silicalite 268, 270, 272–273, 275
silica membranes 198
single-crystal X-ray structures 308
single-electron transfers 23
single-walled carbon nanotubes (SWCNTs) 183
small-scale seawater desalination, for desalination 152
sol–gel method 9, 168, 196–198
– colloidal 196
– dip-coating 37
sonolysis 11
sorbent multifunctional modification 77–79
sputtering deposition 37
Steelink model 281
Stroke radius 130
structure-directing agents (SDAs) 170
sulfonated poly (ether ether ketone) (sPEEK) 138–139
sulfonated polyethersulfone (sPES) 139
surface activation
– with chemical initiators 95–97
– with vinyl monomers 94–95
surface charge 178
surface coating 228
surface hydrophilicity 178
surface modification, of membrane 89–90
surface water and seawater clarification, using ceramic membranes 198
– microfiltration 199
– – applications for industrial-scale waterworks 201
– – hybrid ozonation and ceramic ultrafiltration 201
– – pretreatment with flocculation/coagulation 199
– – transmembrane pressure (TMP) effect and cross-flow velocity (CFV) 199–200
– – ultrasound cleaning 200
surfactants 90
synchrotron X-ray powder diffraction (the Australian Synchrotron) 234

t

target areas, for water treatment and desalination nanotechnology development 1–6
– future 1–2
– practical considerations 2–3
– water treatment market 3–4
Temple-Northeastern-Birmingham (TNB) model 281–284, 286–288
tetraethoxy orthosilicate (TEOS) 68, 76, 198, 228
tetramethylammonium hydroxide (TMAOH) 171
tetrapropyl ammonium (TPA) 220
thin film composite and nanocomposite membranes 86, 163–168, 177–179, 251
– commercialization and future developments 185–187
– fabrication and characterization
– – interfacial polymerization 172–177
– – inorganic fillers fabrication and characterization 168–172
– properties tailored by fillers addition
– – fouling resistance and chlorine stability 184–185
– – water permeability and salt rejection 179–184
tin 20
TIP3P water model 284
titanium dioxide nanoparticles
– photoanode fabrication
– – boron-doped diamond (BDD) 38
– – carbon nanotubes 40–41
– – common techniques and substrates 37–38
– – mixed-phase photoanode 39–40
– photocatalysis 33–35
– – gas sensing 56–59
– – sensing application 41–46
– photoelectrocatalysis 36–37
– – analytical signal generation 57–58
– – sensing application 46–56
total organic carbon (TOC) 41
– conversion 17
– photocatalytic determination 42 43
transition-metal colloids 9
transmembrane pressure (TMP) effect 200, 202
transmission electron micrograph (TEM) 252
transport number 126
triacetate (CTA) 163
trichloroethylene (TCE) 21, 25
trimesolylchloride (TMC) 251, 255–260

Triton X-100   176
two-phase model   131
two-step phase inversion technique   146

u
ultrafiltration   3, 85, 249, 279, 286. See also wastewater microfiltration and ultrafiltration, ceramic membrane application in
– and hybrid ozonation   201
ultrasound cleaning   200
UV-induced graft polymerization   99–101
UV-LED, for miniature photoelectrochemical detectors   55

v
vinyl monomers, surface activation with   94–95
volatile organic compounds (VOCs)   41

w
wastewater microfiltration and ultrafiltration, ceramic membrane application in   202–204
– cost   212–213
– membrane microstructure   204–205
– surface charge properties   206–208
– technical process   208–212
– wettability   205–206
water dissociation   129–130
water permeability and salt rejection   179–184
wet burning   41
wet impregnation   10
wettability, of materials   205–206

x
X-ray diffraction (XRD)   171–172, 256
X-ray photoelectron spectrometry (XPS)   251
X-ray photoelectron spectroscopy (XPS)   179

z
zeolite   182–183, 186, 217–219, 267–268
– A-PA TFN RO membranes   165
– force field   270
– membrane preparation   219–220
– – direct *in situ* crystallization   220–221
– – microwave synthesis   223–227
– – postsynthetic treatment   228–229
– – seeded secondary growth   221–223
– membranes, for water treatment
– – for desalination   229–236
– for wastewater treatment   236–239
– – membrane-based reactors   239–240
– modeling   268–270
– nanocrystals   170, 176
– and salt ions   276–278
– and water behavior   270–275
zeolitic imidazolate frameworks (ZIFs)   307–309
zeta (Z) potential   178, 180, 200
zinc   20
zirconia membranes   203, 207